For the students and faculty
of Bates College —
with best regards!

Jim White
4/22/82

The Method of Iterated Tangents
with Applications in Local Riemannian Geometry

The Method of Iterated Tangents with Applications in Local Riemannian Geometry

J. Enrico White
Spelman College

Pitman Advanced Publishing Program
Boston · London · Melbourne

PITMAN BOOKS LIMITED
39 Parker Street, London WC2B 5PB

PITMAN PUBLISHING INC.
1020 Plain Street, Marshfield, Massachusetts

Associated Companies
Pitman Publishing Pty Ltd., Melbourne
Pitman Publishing New Zealand Ltd., Wellington
Copp Clark Pitman, Toronto

First published 1982

AMS Subject Classifications: 53B20, 53A55

Library of Congress Cataloging in Publication Data

White, J. Enrico (James Enrico)
The method of iterated tangents with applications in local Riemannian geometry.
(Monographs and studies in mathematics; 13)
Bibliography: p.
Includes index.
1. Geometry, Riemannian. 2. Tangent Bundles. 3. Manifolds (Mathematics) I. Title. II. Series.
QA649.W48 516.3′73 81-5845
ISBN 0-273-08515-8 AACR2

British Library Cataloguing in Publication Data

White, J. Enrico
The method of iterated tangents with applications in local Riemannian geometry.—(Monographs and studies in mathematics; 13)
1. Geometry, Differential
I. Title II. Series
516.3′73 QA641

ISBN 0-273-08515-8

Filmset and printed in Northern Ireland at The Universities Press (Belfast) Ltd., and bound at the Pitman Press, Bath, Avon.

To Sally, who helped me

Contents

Preface

The aim of this book is to develop and illustrate with applications to Riemannian geometry a method of analysis which takes as its basic geometric objects the elements of iterated tangent bundles of smooth manifolds

$$\begin{array}{c} T_1(T_1(\ldots T_1(M)\ldots)) \\ \downarrow \\ M \end{array}$$

and their duals. The calculus so developed gives a direct generalization of tensor calculus with the advantage that the 'higher-order' geometric objects encountered in calculations and constructions are all *invariants* in this framework and can therefore be treated in the unified and coordinate-independent language appropriate to manifold theory. Of course, the *algebra* of these invariants is somewhat less familiar than the multilinear algebra of tensor invariants, and this often leads to fresh interpretations of classical constructions such as affine connection, fundamental forms, and curvature.

While some knowledge of tensor analysis would be useful to give the reader the support of motivation and examples, that knowledge has not been assumed. The first chapter develops that classical machinery which is to be generalized. It does this in the more modern language of manifold and fiber bundle theory but with an approach which maintains firm contact throughout with the coordinate interpretations in terms of 'transformation laws' which guided the early development of the theory. Knowledge of differential topology at the level of calculus on manifolds and some acquaintance with category theory and vector bundles will be handy.

The first two chapters are basically devoted to establishing the language and the algebra of iterated tangents, and the third chapter develops the properties of their duals. The exercises, illustrations, and constructions which appear in these chapters are *also* meant to give an introduction to local differential geometry and, while they only *survey* the subject, they can be used for that purpose.

The final chapter develops the applications of this method to the particular problems of Riemannian geometry. It develops the various notions of curvature beginning with curves in $\mathbb{R}^2$ (familiar from the calculus) and continuing to study the curvature of surfaces in $\mathbb{R}^3$. The relations between the two notions are discussed, as are the remarkable differences discovered by Gauss (*Theorema Egregium*). The first proof we give of the *Theorema Egregium* essentially follows Gauss, but is formulated in the language of iterated tangents. It leads to the formula expressing curvature in terms of the metric, and to the question why it should be true.

The remainder of the book applies the invariance properties of the duals of iterated tangents to give an answer to that question, along with the construction and interpretation of the Riemann curvature tensor. The somewhat surprising answer is easily formulated in terms of the invariant algebra of iterated tangents, and is presented along with a crisper, simpler form of the formula of Gauss in the second proof of the *Theorema Egregium.*

Some of the ideas which are developed in this book have been discussed from a somewhat more global point of view by Yano and Ishihara [27]; in particular, the metric on the tangent bundle which is here called the *promotion of the metric* A^2 is treated by them (and by others) and has been called *the metric II.* A thought-provoking (but short) discussion of iterated tangents also appears in Abraham and Marsden [1] with their discussion of 'rhombics' in relation to the Legendre diffeomorphism. Without claiming originality in discovering *uses* for iterated tangents (applications made here are essentially classical), the author feels that the *calculus* of these higher-order objects can serve as a unified framework for the presentation of classical local Riemannian geometry, and that it can lead to new methods of analysis in modern differential geometry.

I would like to express my appreciation and my indebtedness to Claude-Paul Bruter of *l'Université de Paris*, a warm and generous *savant* whose clarity of vision and depth of insight offered cool, clean waters in troubled times. Thank you, Claude. Thanks, also, to Alan Levin, a chemist who helped me to learn physics, and to Arthur Seebach, Ken Young, and Dave Strom who sampled and criticized earlier versions of the first chapter in a joint Carleton–St Olaf College seminar on differential geometry and relativity.

Nothing is born of the human imagination unless it begins in dreams. And so my first acknowledgment is to my wife, Sally, who shares her dreams with mine.

James Enrico White

Northfield, Minnesota, June 1980

Symbols and special notation

Introduction

The local differential geometry of smooth manifolds discloses a spectrum of 'invariant objects'. At one end of the spectrum are the tangent and cotangent vectors and their fields; these are materialized and their properties under mappings are interpreted geometrically via the tangent and cotangent bundle functors. The *vectors* are generally referred to the 1-jet of a chart or frame at the point of the manifold. Their *fields* are not entirely specified at a point by *any* order jet: one requires the data of the *germ* of a chart or frame. Higher-order invariant objects such as affine connections or 'variations' of smooth paths may be defined and specified at a point with the data of a 2-jet of a chart or frame at that point. The specification of a geometric object at a point roughly means the determination of certain numbers (components) in the presence of a certain r-jet of a chart or frame together with a rule explaining how these numbers change when one chart is replaced by another; the object has *order* r if for $s \leqslant r-1$ charts with different r-jets and the same s-jet correspond different components to the object.

Classical tensor analysis concerned itself with the classification and the behavior under mappings of first-order objects (tensors) which derived from the imposition of 'additional structure' on the manifold (such as a set of vector fields, a metric, a Lagrangian, and so forth). While the additional structure introduced higher-order objects, the cancellation of contributions in the course of a given construction on them might well yield a first-order object as the result of that construction. Thus, in the *general* case, a pair of smooth vector fields leads immediately to certain second-order objects at a point: the local variation along a solution curve of one field given by pushing the tangent vector determined by the second field at that point along the local flow of the first. Each 'local variation' is, loosely speaking, a second-order object. The Lie bracket construction cancels the higher-order effects and associates at the given point something very much like a tangent vector. The calculation of the coboundary of a differential form is another example of this cancellation of higher-order effects, as is also the Lie derivative of a differential form (or a covariant tensor field in general) with respect to a local vector field. More specifically, in the presence of a metric induced from a codimension-1 immersion in Euclidean space, the 'second fundamental form' arises in this manner. In the presence of a Riemannian metric in general, the associated Levi–Civita connection gives a way of corresponding to a local vector field and a tangent vector at a point, a new tangent vector the covariant derivative. Riemannian curvature is a first-order object with a long history of cancelled higher-order ancestors.

While classical tensor analysis never attained a *coordinate-free calculus of first-order invariants*, it pointed the way to such developments as Cartan's method of moving frames, and the Koszul axiomatization of affine connection which made it possible to dispense with coordinates and to build penetrating and powerful *calculi of first-order invariants*, having absorbed many of the bothersome higher-order effects into their initial data. Of course, in order to adopt the more abstract standpoint, it was necessary to abandon many of the concrete interpretations which tensor analysis was able to give to its constructions: a necessary and desirable exchange for the possibility of eliminating the 'noise' which coordinate-bound calculations must introduce.

Now the *method of iterated tangents* which this book investigates is an intermediate approach. There is a class of higher-order objects, the elements of

$$T_r(M) = \underbrace{T_1(T_1(\ldots T_1}_{r}(M)\ldots)),$$

whose algebraic properties generalize in fairly straightforward ways the linear-algebraic properties of the tangent spaces. Here we denote the tangent bundle of a smooth manifold M as

$$\begin{array}{c} T_1(M). \\ \downarrow \\ M \end{array}$$

If, instead of giving special status to the 'non-tensorial' constructions which arise in the course of coordinate calculations, we can describe them in the language of iterated tangents and can treat them with the *invariant algebra* associated with these higher-order objects, then it will be possible to give coordinate-independent derivations and constructions of useful invariant objects without a need for the sort of special assumption that can tax the intuition. In fact, many of the basic constructions in differential and local Riemannian geometry can be brought easily within the purview of the method, sometimes with novel interpretations.

Iterated tangents, which we call *tangent sectors* (or simply sectors), generalize tangent vectors both formally and conceptually. Formally, the r-sectors on a manifold M, $T_r(M)$, are specified by 'components' which obey a certain transformation law with respect to an action of the group of invertible r-jets of coordinate changes at a point (an extension of the general linear group). If

$$\begin{array}{c} T_r(M) \\ \downarrow \\ M \end{array}$$

denotes the bundle of r-sectors over M, then this is *not* generally a vector bundle over M, and the 'invariant algebra' of the calculus will be defined by the operations in the fiber which are indepedent of coordinatization. The first sector bundle

$$\begin{array}{c} T_1(M) \\ \downarrow \\ M \end{array}$$

is the tangent bundle and in this case *alone* the fiber has an invariant linear structure. Using the tangent bundle as a model, *dual sectors* are defined to be mappings $T_r(M) \to \mathbb{R}$ which, when restricted to fibers, 'preserve' the structure of those fibers, and these will be called *sectorform fields* (in analogy with covector fields). As it happens, all covariant tensor fields are sectorform fields. The r-sectors are certain *equivalence classes* of r-jets of mappings $(\mathbb{R}^r, 0) \to (M, x)$ as x varies in M. This leads to the sense in which sectors give conceptual generalizations of tangent vectors. If tangent vectors are 'infinitesimal paths' to the first order, then r-sectors are *determined by* (but not the same as) infinitesimal r-dimensional surfaces to the rth order. All of the naturality properties of tangent vectors will be extendible to sectors, and in fact there is a natural (invariant) way of differentiating real-valued functions on M by sectors.

The book is divided into four chapters. The first chapter develops most of the invariant algebra and organizes it into a functor which extends the tangent functor, and which we call the *canonical simplicial fiber bundle functor.* It is shown there that to each smooth manifold M there is associated a sequence of fiber bundles

$$\begin{array}{ccccc} T_1(M) & T_2(M) & \ldots & T_r(M) & \ldots \\ \downarrow & \downarrow & & \downarrow & \\ M & M & \ldots & M & \ldots \end{array}$$

and that for *each* s there is a set of $(s+1)$ bundle maps $(D^i(s+1), \mathrm{id})$, $1 \leq i \leq s+1$,

$$\begin{array}{ccc} T_{s+1}(M) & \xrightarrow[D^i(s+1)]{} & T_s(M) \\ \downarrow & & \downarrow \\ M & \xrightarrow[\mathrm{id}]{} & M \end{array}$$

with the additional property that *each*

$$\begin{array}{c} T_{s+1}(M) \\ \downarrow D^i(s+1) \\ T_s(M) \end{array}$$

is itself the projection of a *vector bundle*.

All of these data give a simplicial object in the category of fiber bundles over M, or a simplicial fiber bundle over M (the 'degeneracy operators' will play no role in our development). This association is prolonged to a functor (extending the tangent functor) from the category of smooth manifolds and maps to the category of simplicial fiber bundles and simplicial maps. In particular, the simplicial (graded fiber bundle) maps associated to a smooth map between manifolds commute with the $D^i(s+1)$. Thus the fibers of the $D^i(s+1)$ have intrinsic vector space structure.

Now the 'component' description of an element of $T_r(M)$ reflects this simplicial structure (which makes for a picturesque and suggestive notation). If M is m-dimensional, then an r-sector at a point x_0 in M is specified by the data of an invertible r-jet $(\mathbb{R}^m, 0) \to (M, x_0)$ (an r-jet of a frame) and the association of an m-tuple of real numbers to *each face* of an $(r-1)$ simplex: when r is 1, this reduces to the specification of a tangent vector. The 'transformation law' is linear on the 'vertex' components, and on the components associated with an $(s-1)$ face it requires the data of the s-jet of the coordinate change. In this picture, there is one intrinsic vector space for each vertex of the component simplex. Each r-sector belongs to r such spaces. A fairly obvious action of the symmetric group on r letters is introduced on $T_r(M)$. The chapter ends with an appendix devoted to the proof that the higher-order objects so defined are actually iterated tangents. With this, all but one key element of the invariant algebra is specified.

Chapter 2 completes the discussion of the algebraic properties of tangent sectors and illustrates them with several elementary examples from differential analysis. The picture of tangent sectors as 'infinitesimal simplices' leads to the construction of more general bundles, the K-sector bundles, by taking Whitney products of the various sector bundles in certain ways. The elements of these bundles have components parametrized by *finite simplicial complexes* instead of by simplices alone; if the complex is called K, then the higher-order object is called simply a K-sector. The K-sectors have a natural and important place in the calculus. They are immediately used to give the final algebraic element of structure. The boundary complex of an $(r-1)$ simplex consists of all faces but the $(r-1)$ face. Calling that complex B_r for now, we see that there is a natural transformation of functors

$$\begin{array}{ccc} T_r(M) & \mapsto & B_r(M) \\ \downarrow & & \downarrow \\ M & & M \end{array}$$

where

$$\begin{array}{c} B_r(M) \\ \downarrow \\ M \end{array}$$

denotes the B_r-sector bundle (the components are parametrized by the lower-order faces of the $(r-1)$ simplex). We may safely call this a boundary transformation. Two r-sectors *have the same boundary* if they have the same image under this transformation. For such a pair, there is a *natural* 'bracket', a tangent vector essentially, defined in the following way. The map $T_r(M) \to B_r(M)$ given by 'forgetting' the components of the central face is a fiber bundle projection. Further, there is a *vector bundle* over $B_r(M)$ given by taking the Whitney product of

$$\begin{array}{c} B_r(M) \\ \downarrow \\ M \end{array} \quad \text{with} \quad \begin{array}{c} T_1(M). \\ \downarrow \\ M \end{array}$$

Now each fiber of

$$\begin{array}{c} T_r(M) \\ \downarrow \\ B_r(M) \end{array}$$

is an *affine space* over the corresponding fiber of the vector bundle mentioned above. Thus the boundary transformation is not the projection of a vector bundle in general, nor even the projection of a principal bundle over the additive group of a Euclidean space (an affine bundle), but each fiber is an affine space over the corresponding tangent fiber over the point in M. The bracket of two r-sectors with the same boundary is the tangent vector that 'translates' one to the other. The Lie bracket, the covariant derivative, and the contravariant curvature will all be seen to be examples of this operation. This bracket operation (which is natural) is the final element of algebraic structure for our purposes. The illustrations in this chapter cover such topics as 'Hessians' of smooth maps with vanishing derivative, and generalized Hessians, the equation of variations, Lie brackets and commuting pairs of vector fields, and the Jacobi identity.

Chapter 3 is devoted to the study of *sectorforms*. As we mentioned, these are the objects dual to sectors. They are a great deal less intuitive than sectors but they play a central role in the calculus because they materialize certain invariants which have always been awkward even to formulate in coordinate-dependent language. As we mentioned, all covariant tensors are sectorforms, in particular the metric and all differential forms. We shall be especially interested in sectorforms which 'reduce to' certain K-sector bundles, that is which depend only on the faces which appear in the complex K. When K is the vertex set of an $(r-1)$ simplex, such a sectorform is a covariant tensor of type $\binom{0}{r}$.

Now smooth real-valued functions on M may be thought of as order-0 sectorforms. Given such a function, say $f:M\to\mathbb{R}$, we may form its differential $\mathrm{d}f$ to obtain an order-1 sectorform (covector field) on M. One of the invariant objects which sectorforms bring to the surface is the 'differential' of a covector field. This is a 1-form on the tangent bundle, a second-order object. It is also an order-2 sectorform. In fact, there is a sequence of natural transformations which we will be justified in calling differentials taking order-r sectorform fields to order-$(r+1)$ sectorform fields. For each r we define $(r+1)$ such differentials.

These transformations are natural with respect to the 'pullback' operation which is familiar for covariant tensor fields. If $f:M\to N$ is a smooth map and B^r is a sectorform field of order r on N, the pullback $f^*(B^r)$ is an order-r sectorform field on M. Then for d_i, the ith differential operator taking order-r sectorform fields to order-$(r+1)$ sectorform fields, we have

$$\mathrm{d}_i[f^*(B^r)]=f^*[\mathrm{d}_iB^r].$$

Starting with a differential form A^r, certain signed sums of differentials yield the classical *coboundary* $\mathrm{d}A^r$. On the other hand, starting with a *metric* A^2, the differentials lead to several interesting sectorforms (none of them tensors). For example d_3A^2 yields a K-sectorform field for K a certain subcomplex of the 2-simplex which is easily seen to be a *metric on the tangent bundle.* In fact, the Euler–Lagrange equations giving the geodesic flow on the tangent bundle when the metric is interpreted as 'kinetic energy' are seen to be *gradient* equations—the gradient of the kinetic energy with respect to this (indefinite) metric. The naturality of the association d_3A^2 to M for smooth immersions and A^2 definite then yields the geometrical interpretation of D'Alembert's principle. Another operation obtained from signed sums of differentials of the metric A^2 has an interpretation as a 'bilinear' pairing of 1-sectors with 2-sectors and leads to a concrete (if algebraically novel) interpretation of the Riemann–Christoffel symbols and the Levi-Civita connection associated with A^2.

Most of these topics are taken up in the final chapter on Riemannian geometry. Returning to Chapter 3, once the differentials and certain other operations (like symmetric group action) on sectorform fields are defined, some illustrations from differential geometry proper are taken up. Various classical properties of *differential forms* are derived in this language. In particular, the coboundary operation, and the Lie derivative of a covariant tensor field with respect to a local vector field are constructed. Such applications as the Poincaré lemma, Liouville's theorem and the divergence theorem are given there.

Chapter 4 is devoted to Riemannian geometry. The first part on metric structure and Gaussian curvature derives the two sectorform fields mentioned above via differentials from the metric A^2 which is now given as additional structure on the manifold. The algebraic viewpoint which we have adopted makes it possible to give a new interpretation of affine connection associated with a metric. It is a cross section of the bundle

$$\begin{array}{c} T_2(M) \\ \downarrow \\ B_2(M) \end{array}$$

which as we observed earlier has the property that each of its fibers is an *affine space* over the corresponding tangent fiber. Such a cross section, by giving a choice of 'origin' in each fiber, transforms the afore-mentioned bundle (in a non-natural way) into a *vector* bundle (basically the 'splitting' into horizontal and vertical components of $T_1[T_1(M)]$). From this, an operation on 2-sectors, which we call 'force' and which reduces to the *covariant differentiation* when those sectors are viewed as 'paths of tangent vectors', has an easy definition.

Now the fact that the connection and force constructions are *not* natural with respect to *isometric immersions* leads to a tensor associated with such an immersion via a certain bracket which is very much like the 'second fundamental form'. All of this is assembled to derive classical curvature of curves in $\mathbb{R}^2$, and the Gaussian curvature of surfaces in $\mathbb{R}^3$. The section ends with a proof of the *Theorema Egregium* and the formula for the Gaussian curvature in terms of the metric.

The next section is devoted to some topics in calculus of variations which are necessary to give meaning to the notions of parallel translation and geodesics. The above-mentioned gradient interpretation of the Euler–Lagrange equations for a conservative mechanical system is given there.

In the final section, *Riemannian curvature* in its covariant and contravariant forms is taken up. An order-4 sectorform field derived from the metric as a certain signed sum of *second differentials* which is called the *cross* is studied. In the presence of a pair of vector fields on M, this reduces to the coboundary of a 'rotational 1-form'—a form which measures 'covariant' rotation of one field towards the other—described for example in Cartan's moving-frames method and also in the Ricci and Levi-Civita tensor analysis paper. The 2-form so defined is a form on M, and is equal under certain conditions to the covariant curvature evaluated at the pair of vector fields. An interpretation of the fact that the integral of the Gaussian curvature over a 2-cell gives the total rotation of a parallel translation of a unit vector about its boundary then follows easily.

The 4 sectorform field, the cross, is used to define and to determine the invariance properties of the covariant Riemannian curvature. The usual algebraic properties follow easily from this manner of definition. A second proof of Gauss' *Theorema Egregium* together with the relation between Riemannian and Gaussian curvature are the next topics considered from the viewpoint of the invariance of the cross.

Finally, the differential geometry of the tangent bundle via the metric d_3A^2 is used to generalize certain constructions (such as 'force' and 'affine connection') on M to $T_1(M)$. From these generalized notions, the contravariant curvature tensor emerges as a *bracket of 3-sectors*. In this way, it is given its usual interpretation as a measure of the (non-)commutativity of iterated covariant differentiations. The chapter ends with a characterization of locally flat Riemannian manifolds in terms of the possibility of extending a notion of arclength to higher-dimensional Riemannian manifolds.

1

The canonical simplicial fiber bundle

[All manifolds will be assumed to be *smooth* (C^∞) and finite-dimensional, and each finite-dimensional Euclidean space $\mathbb{R}^m = E$ is equipped *once and for all* with a 'standard' basis (e_i), $i = 1, \ldots, m$, or equivalently with the *dual basis* of linear coordinate functions (x^i), $i = 1, \ldots, m$.]

The canonical simplicial fiber bundle functor 'extends' the tangent functor but shares many of its properties. In order to define it, it will be useful to recall the tangent bundle of a smooth manifold M. This is a vector bundle

$$\begin{array}{c} T_1(M) \\ \downarrow{\scriptstyle q_1} \\ M \end{array}$$

associated with each smooth manifold M and for each smooth $f: M \to N$ there is a bundle map $(T_1(f), f)$ which commutes the rectangle

$$\begin{array}{ccc} T_1(M) & \xrightarrow[T_1(f)]{} & T_1(N) \\ {\scriptstyle q_1}\downarrow & & \downarrow{\scriptstyle q_1} \\ M & \xrightarrow[f]{} & N \end{array}$$

To say that the association

$$M \mapsto \begin{array}{c} T_1(M) \\ \downarrow{\scriptstyle q_1} \\ M \end{array} \qquad f \mapsto (T_1(f), f)$$

gives a functor is to say that $T_1(\mathrm{id}) = \mathrm{id}$ and $T_1(f \circ g) = T_1(f) \circ T_1(g)$. In particular, it gives a *covariant* functor from the category of smooth manifolds and maps to the category of vector bundles and vector bundle maps. We elaborate below.

§1.1. Tangent and cotangent bundles

If f and g are smooth maps between manifolds M and N with $f:(M, x_0)\to(N, y_0)$ [this means $f(x_0)=y_0$] and $g:(M, x_0)\to(N, y_0)$, then f is germ-equivalent to g at x_0 if there is an open neighborhood U of x_0 with the property that $f_{|U}=g_{|U}$. For fixed $x_0\in M$ and $y_0\in N$ an equivalence class with respect to this equivalence relation is called a (smooth) germ with *source* x_0 and *target* y_0. Such germs will be denoted by one of their representative functions. Thus we write the *germ* $f:(M, x_0)\to(N, y_0)$. If $f:(M, x_0)\to(N, y_0)$ and $g:(N, y_0)\to(P, z_0)$ are smooth germs, then the composition of the germs $g\circ f$ is unambiguously represented by the composition of the representative functions. See Poston and Stewart [18].

Now let M and N be as above. Then M is covered with open sets U_i and there are open sets V_i in $\mathbb{R}^m=E$ with homeomorphisms $\phi_i: V_i\to U_i$ satisfying the following 'compatibility conditions': if $U_i\cap U_j$ is not empty, then

$$\phi_j^{-1}\circ\phi_{i\,|_{\phi_i^{-1}(U_i\cap U_j)}}:\phi_i^{-1}(U_i\cap U_j)\to\phi_j^{-1}(U_i\cap U_j)$$

are *smooth*, and hence they are diffeomorphisms. Now call these maps $\phi_i: V_i\to U_i$ *local frames*, and call their inverses $\phi_i^{-1}: U_i\to V_i$ *local charts*. In a similar way, we shall call the germ of a frame $\phi_i:(V_i, z)\to(M, x)$ a *frame-germ* at $x\in M$.

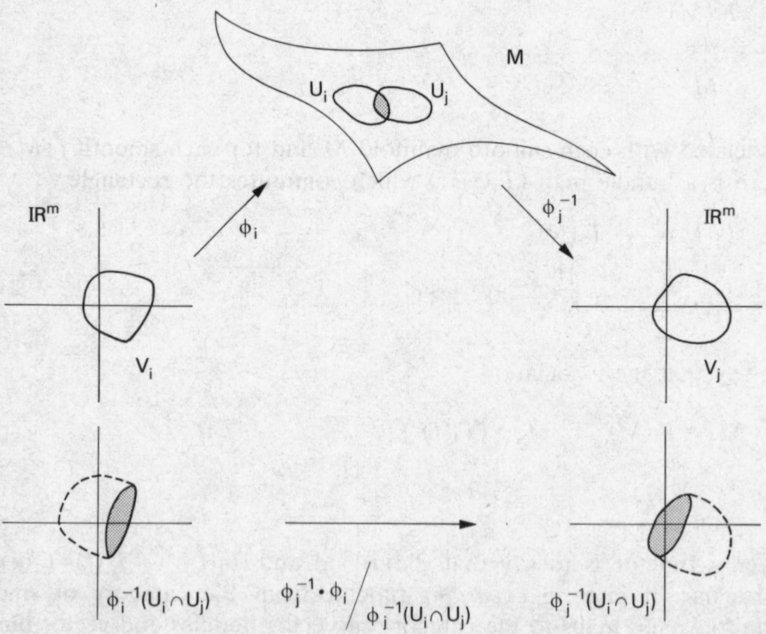

Fig. 1.1

Definition 1.1. Suppose that $f, g:(M, x_0) \to (N, y_0)$ are smooth germs. And suppose that there are frames $\phi_i : V_i \to U_i \subset M$ and $\theta_j : V_j' \to U_j' \subset N$ with $x_0 \in U_i$ and $y_0 \in U_j'$ such that the *germs* $\theta_j^{-1} \circ f \circ \phi_i$ and $\theta_j^{-1} \circ g \circ \phi_i$ taking $(E, \phi_i^{-1}(x_0)) \to (E', \theta_j^{-1}(y_0))$ have the same derivatives.

Under these conditions, say that germ f is 1-equivalent to germ g written $f \underset{1}{\sim} g$. An easy argument with the chain rule guarantees that 1-equivalence is independent of the frames used to test it. Thus 1-equivalence is an equivalence relation. The 1-equivalence class of f will be called the 1-jet of germ $f:(M, x_0) \to (N, y_0)$ and we denote the 1-jet of f by f_1, a subscripted representative germ.

Observe that if $f \underset{1}{\sim} f'$ and $g \underset{1}{\sim} g'$ with the target point of f the source point of g, then it follows from the chain rule again that $g \circ f \underset{1}{\sim} g' \circ f'$. ■

We may thus define a *composition* of the 1-jets of composable germs by the prescription $g_1 \circ f_1 = (g \circ f)_1$. Letting M and N be smooth manifolds, the set of *all* 1-jets of germs of mappings from $M \to N$ we call $J^1(M, N)$. There is a projection $p_1 : J^1(M, N) \to M \times N$ given by $p_1(f_1) = (\text{source } f_1, \text{target } f_1)$. This p_1 is the projection of a *vector bundle:* if $\dim(M) = m$ and $\dim(N) = n$, then the fiber of this bundle has dimension mn and the *total space* $J^1(M, N)$ is a smooth manifold of dimension $(mn + m + n)$. The bundle and manifold structures are given in the following way: for $\phi : V \to U \subset M$ a local frame and $\theta : V' \to U' \subset N$ another local frame, we may harmlessly denote $p_1^{-1}(U \times U')$ by $J^1(U, U')$. Let $L(E, E')$ denote the vector space of linear mappings from $E = \mathbb{R}^m$ to $E' = \mathbb{R}^n$ (coordinatized in the usual way with respect to the 'standard' linear coordinate functions on E and E'). Now using ϕ and θ, $T_{\phi\theta} : J^1(U, U') \to L(E, E') \times U \times U'$ by the rule

$$T_{\phi\theta}(f_1) = ([D_{|\phi^{-1}(x_0)}(\theta^{-1} \circ f \circ \phi)], x_0, y_0)$$

for f any representative of f_1, x_0, and y_0 the respective source and target, and D the derivative, a linear map from E to E'. We use the symbol $[D]$ to represent the matrix in standard coordinates of linear map D. The maps $T_{\phi\theta}$ so defined are then used to give $J^1(M, N)$ the smallest topology which makes all of them continuous. With respect to this topology, they become local charts for the vector bundle structure for projection p_1.

Consider the category whose objects consist of *fiber bundles* $p : X \to B$ and whose morphisms are fiber bundle mappings $(F, f) : p \to p'$—that is, commuting the diagram

$$\begin{array}{ccc} X & \xrightarrow{F} & X' \\ {\scriptstyle p}\downarrow & & \downarrow{\scriptstyle p'} \\ B & \xrightarrow{f} & B' \end{array}$$

Also consider the category whose objects are pairs of smooth manifolds (M, N) and whose morphisms are pairs of smooth mappings

$(h, k):(M, N) \to (M', N')$ with $h: M \to M'$ a *diffeomorphism* and $k: N \to N'$ simply smooth. Then the association $(M, N) \mapsto p_1: J^1(M, N) \to M \times N$ with $J^1(h, k) f_1 \to (k \circ f \circ h^{-1})_1$ defines a *covariant functor* from the category of pairs to the category of fiber bundles. Denote the functor J^1. Since we intend to generalize the *tangent* bundle and *cotangent* bundle constructions we shall examine their derivation from J^1 closely.

Example **(*The tangent bundle of a smooth manifold*).** Suppose that M is a manifold. If $x_0 \in M$, let $\alpha: (\mathbb{R}, 0) \to (M, x_0)$ be a smooth germ, and let α_1 be its 1-jet. Now if $\phi: (\mathbb{R}^m, 0) \to (M, x_0)$ is the *germ of a frame* at x_0, and $\phi^{-1}: (M, x_0) \to (\mathbb{R}^m, 0)$ is the germ of the associated chart, then $(\phi^{-1} \circ \alpha)_1$ is the 1-jet of a germ $(\mathbb{R}, 0) \to (\mathbb{R}^m, 0)$. Let (t) and $(x^i)_{i=1,\ldots,m}$ be the (standard) coordinate functions for $\mathbb{R}$ and $\mathbb{R}^m$ respectively which we specified initially. Then the matrix representing $D_{|0}(\phi^{-1} \circ \alpha)$ in terms of the standard bases has the form

$$\begin{bmatrix} X^1 \\ X^2 \\ \vdots \\ X^m \end{bmatrix}.$$

We represent this matrix simply as X^i (it being understood that i takes values from 1 to m). The data of the pair (ϕ_1, X^i) *entirely* specifies the 1-jet α_1, where ϕ_1 is the 1-jet of the frame-germ ϕ.

Now if $f: (M, x_0) \to (N, y_0)$ is the germ of a smooth map, and if $\theta: (\mathbb{R}^n, 0) \to (N, y_0)$ is the germ of a frame at y_0 in N, then the 'infinitesimal' path α transforms to germ $f \circ \alpha: (\mathbb{R}, 0) \to (N, y_0)$, and $(f \circ \alpha)_1$ is specified by the pair (θ_1, Y^j) $(j = 1, \ldots, n)$ in a similar fashion. Now the relation between the matrix X^i and the matrix Y^j is given by the chain rule (matrix product):

$$Y^j = [D_{|0}(\theta^{-1} \circ f \circ \phi)] \cdot X^i.$$

Recall that we represent the matrix of a linear transformation A between Euclidean spaces in terms of standard bases as $[A]$.

In particular, if f is the germ of the identity map $(M, x_0) \to (M, x_0)$, this formula gives us the relation between X^i and $\bar{X}^j$ $(i, j = 1, \ldots, m)$ where (ϕ_1, X^i) and $(\bar{\theta}_1, \bar{X}^j)$ specify the 1-jet α_1 in terms of two frame-germs at x_0 in M. It then becomes the familiar 'transformation law' for a change of frame at x_0.

Definition 1.2. If M is a smooth manifold, and $x_0 \in M$, then a *tangent vector* at x_0 is a 1-jet α_1 for a smooth germ $\alpha: (\mathbb{R}, 0) \to (M, x_0)$. If $\phi: (\mathbb{R}^m, 0) \to (M, x_0)$ is the germ of a frame at x_0, then the tangent vector is specified by the pair (ϕ_1, X^i) where X^i is an $m \times 1$ matrix. Further, it is clear that for ϕ_1 fixed, there is a bijective correspondence between $m\times 1$ matrices and tangent vectors at x_0 which associates to each X^i the 1-jet $(\phi\circ X)_1$ where $X:\mathbb{R}\to\mathbb{R}^m$ is the linear map such that $[X]=X^i$. ■

We now make a systematic construction which organizes the tangent vectors at various points into a vector bundle associated with M. Consider the vector bundle

$$\begin{array}{c} J^1(\mathbb{R}, M) \\ {\scriptstyle p_1}\downarrow \\ \mathbb{R}\times M \end{array}$$

associated to M. Restrict p_1 to $p_1^{-1}(O\times M)=T_1(M)$ and call the restriction p_1 also. Then it is clear that by identifying $O\times M$ with M in the obvious way, the map

$$\begin{array}{c} T_1(M) \\ {\scriptstyle p_1}\downarrow \\ M \end{array}$$

is the projection of a vector bundle when we give $T_1(M)$ the relative topology. [Check that it inherits a smooth manifold structure for which p_1 is a smooth map.] The fiber of $T_1(M)$ is $L(\mathbb{R}, \mathbb{R}^m)\cong\mathbb{R}^m$. Thus, for $x_0\in M$, $p_1^{-1}(x_0)$ is the set of 1-jets from $(\mathbb{R}, 0)\to(M, x_0)$.

Now suppose that $f:M\to N$ is a smooth map. Recalling that J^1 is a functor, we have a commutative diagram:

$$\begin{array}{ccc} T_1(M)=p_1^{-1}(O\times M) & \xrightarrow{J^1(\mathrm{id},f)} & T_1(N)=p_1^{-1}(O\times N) \\ {\scriptstyle p_1}\downarrow & & {\scriptstyle p_1}\downarrow \\ M\cong O\times M & \xrightarrow{\mathrm{id}\times f} & N\cong O\times N \end{array}$$

From now on, the map $J^1(\mathrm{id}, f)_{|p_1^{-1}(O\times M)}$ will be called $T_1(f)$.

Observation 1.3. The association

$$\begin{array}{c} M\to T_1(M) \\ \downarrow{\scriptstyle p_1} \\ M \end{array} \qquad \text{and} \qquad f\to(T_1(f), f)$$

gives a functor from the category of smooth manifolds to the category of vector bundles. Thus, $T_1(\mathrm{id})=\mathrm{id}$, and $T_1(g\circ f)=T_1(g)\circ T_1(f)$. If $x_0\in M$, then $T_{x_0}(M)$ will denote $p_1^{-1}(x_0)$. This carries, as we mentioned, the structure of $\mathbb{R}^m$. ■

In the case that M and N are Euclidean spaces $\mathbb{R}^m$ and $\mathbb{R}^n$, we may choose for frames $\mathrm{id}:\mathbb{R}^m\to\mathbb{R}^m$ and $\mathrm{id}:\mathbb{R}^n\to\mathbb{R}^n$. We then have

$T_{\mathbb{R},\mathbb{R}^m}: J^1(\mathbb{R},\mathbb{R}^m) \to L(\mathbb{R},\mathbb{R}^m)\times\mathbb{R}\times\mathbb{R}^m$ and similarly $T_{\mathbb{R},\mathbb{R}^n}: J^1(\mathbb{R},\mathbb{R}^n)\to L(\mathbb{R},\mathbb{R}^n)\times\mathbb{R}\times\mathbb{R}^n$. These 'global' frames give $J^1(\mathbb{R},\mathbb{R}^m)$ and $J^1(\mathbb{R},\mathbb{R}^n)$ their bundle (and manifold) structure. In particular, if $\alpha_1 \in J^1(\mathbb{R},\mathbb{R}^m)$ with source and target x_0 and y_0, then $T_{\mathbb{R},\mathbb{R}^m}(\alpha_1) = (D_{|x_0}(\alpha_1), x_0, y_0)$. In these frames, the map $J^1(\text{id}, f): J^1(\mathbb{R},\mathbb{R}^m)\to J^1(\mathbb{R},\mathbb{R}^n)$ for a given $f:\mathbb{R}^m\to\mathbb{R}^n$ can be represented as the map taking $L(\mathbb{R},\mathbb{R}^m)\times\mathbb{R}\times\mathbb{R}^m \to L(\mathbb{R},\mathbb{R}^n)\times\mathbb{R}\times\mathbb{R}^n$ for $(A, x_0, y_0)\to(D_{|y_0}(f)\circ A, x_0, f(y_0))$.

Now specializing to the tangent bundle $T_1(\mathbb{R}^m)$, we see that

$$T_{\mathbb{R},\mathbb{R}^m}{}_{|T_1(\mathbb{R}^m)}: T_1(\mathbb{R}^m)\to L(\mathbb{R},\mathbb{R}^m)\times O\times\mathbb{R}^m \cong L(\mathbb{R},\mathbb{R}^m)\times\mathbb{R}^m$$

gives the bundle and manifold structure for $T_1(\mathbb{R}^m)$. And for $f:\mathbb{R}^m\to\mathbb{R}^n$, $T_1(f)$ can be represented as a map from $L(\mathbb{R},\mathbb{R}^m)\times\mathbb{R}^m\to L(\mathbb{R},\mathbb{R}^n)\times\mathbb{R}^n$ taking $(A, y_0)\mapsto(D_{|y_0}(f)\circ A, f(y_0))$ for $A\in L(\mathbb{R},\mathbb{R}^m)$ arbitrary. Now if $f:(\mathbb{R}^m, y_0)\to(\mathbb{R}^n, w_0)$ and $g:(\mathbb{R}^n, w_0)\to(\mathbb{R}^p, z_0)$ are smooth germs, then the statement $T_{y_0}(g\circ f) = T_{w_0}(g)\circ T_{y_0}(f)$ which expresses the functoriality of T_1 is simply the *chain rule.*

We thus see that the functor T_1 allows a *generalized interpretation of the chain rule* to the case of smooth mappings between manifolds (instead of Euclidean spaces). It does this by giving an invariant meaning to the *derivative* of the smooth map $f: M\to N$ at any point $y_0\in M$. The latter is simply the *linear* map $T_{y_0}(f)$ taking $T_{y_0}(M)$ to $T_{f(y_0)}(N)$ by $\alpha_1\to(f\circ\alpha)_1$ for α the germ of a map $(\mathbb{R}, 0)\to(M, y_0)$.

We turn now to the *dual* of this construction for the last example of this section.

Example (The cotangent bundle of a smooth manifold). For a smooth manifold M and point $x_0\in M$, let $\beta:(M, x_0)\to(\mathbb{R}, 0)$ be a smooth germ, and let β_1 be its 1-jet. Now if $\phi:(\mathbb{R}^m, 0)\to(M, x_0)$ is the germ of a frame at x_0, then $(\beta\circ\phi)_1$ is the 1-jet of a map $(\mathbb{R}^m, 0)\to(\mathbb{R}, 0)$. The matrix representing $D_{|0}(\beta\circ\phi)$ in standard bases has the form $(X_1, X_2, \ldots, X_m)$. [Note the use of *subscripts* instead of *superscripts* here. This notational device will allow us to distinguish row from column matrices (covectors from vectors according to a fairly standard convention). We shall thus represent this matrix simply by X_i.]

As in the case with tangent vectors, the data of the pair (ϕ_1, X_i) entirely specifies the 1-jet β_1 (where ϕ_1 is again the 1-jet of the frame-germ). The 1-jet β_1 will then be a *cotangent vector* (a covector, or a 1-form) at $x_0\in M$.

If $f:(M, x_0)\to(N, y_0)$ is the germ of a smooth map and $\theta:(\mathbb{R}^n, 0)\to(N, y_0)$ a frame-germ, suppose that $\beta_1:(N, y_0)\to(\mathbb{R}, 0)$ is a cotangent vector at y_0 specified by the pair (θ_1, Y_j). Then we may associate to β_1 the cotangent vector $(\beta\circ f)_1$ at $x_0\in M$. Notice that this association goes the opposite way to the association for tangent vectors; it takes cotangent vectors on N to cotangent vectors on M for $f: M \to N$. Now suppose that $(\beta \circ f)_1$ is specified by (ϕ_1, X_i). What is the relation between the Y_j and the X_i? The chain rule tells us that (matrix product)

$$Y_j \cdot [D_{|0}(\theta^{-1} \circ f \circ \phi)] = X_i.$$

Letting f be the germ of the identity map, we have a relation between Y_j and $\bar{Y}_i$ for specifications by different frames at y_0: (θ_1, Y_j) and $(\bar{\theta}_1, \bar{Y}_i)$. It should be kept in mind that matrices are computed with respect to our standard bases in all cases.

Definition 1.4. If M is a smooth manifold, $x_0 \in M$, then a *cotangent vector* at x_0 is a 1-jet β_1 of a smooth germ $\beta: (M, x_0) \to (\mathbb{R}, 0)$. If $\phi: (\mathbb{R}^m, 0) \to (M, x_0)$ is a frame-germ at x_0, the cotangent vector is specified by the data of the pair (ϕ_1, X_i) where X_i is a $1 \times m$ matrix. There is a bijective correspondence (for fixed x_0) between cotangent vectors at x_0 and $1 \times m$ matrices X_i associating to each X_i the 1-jet $(X \circ \phi^{-1})_1$ for X the linear map $\mathbb{R}^m \to \mathbb{R}$ whose matrix is X_i. ■

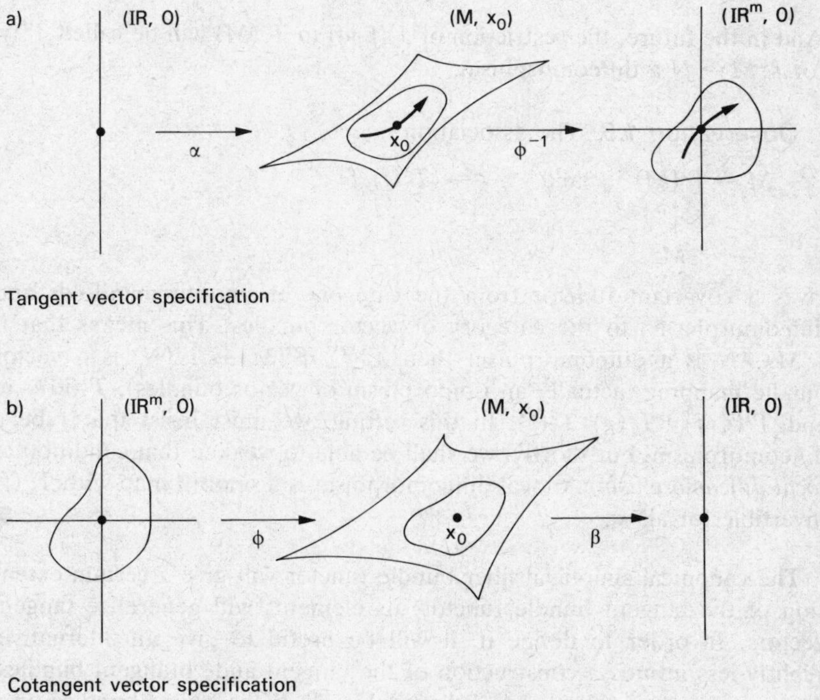

Fig. 1.2. (a) Tangent vector specification. (b) Cotangent vector specification

Now consider the vector bundle

$$\begin{array}{c} J^1(M,\mathbb{R}) \\ {\scriptstyle p_1}\downarrow \\ M\times\mathbb{R} \end{array}$$

which can be associated to a smooth manifold M. Restrict p_1 to $p_1^{-1}(M\times O)$, and call the restriction p^1; $p_1^{-1}(M\times O)$ will be called $T^1(M)$. Then, as in the previous case, it will be a straightforward verification that

$$\begin{array}{c} T^1(M) \\ {\scriptstyle p^1}\downarrow \\ M \end{array}$$

is the projection of a vector bundle when we give $T^1(M)$ the relative topology. The fiber of this bundle is $L(\mathbb{R}^m,\mathbb{R})\cong\mathbb{R}^m$.

Now suppose that $f:M\to N$ is a (smooth) *diffeomorphism.* We have a commutative diagram:

$$\begin{array}{ccc} T^1(M) & \xrightarrow{J^1(f,\mathrm{id})} & T^1(N) \\ {\scriptstyle p^1}\downarrow & & \downarrow{\scriptstyle p^1} \\ M\cong M\times O & \xrightarrow{\quad f \quad} & N\cong N\times O \end{array}$$

And in the future, the restriction of $J^1(f,\mathrm{id})$ to $T^1(M)$ will be called $T^1(f)$ for $f:M\to N$ a diffeomorphism.

Observation 1.5. The association

$$\begin{array}{c} M\to T^1(M) \\ \downarrow{\scriptstyle p^1} \\ M \end{array} \qquad \text{and} \qquad f\to(T^1(f),f)$$

gives a covariant functor from the category of smooth manifolds and diffeomorphisms to the category of vector bundles. This means that if $f:M\to N$ is a diffeomorphism then $T^1(f):T^1(M)\to T^1(N)$ is a vector bundle mapping (actually an isomorphism of vector bundles), $T^1(\mathrm{id})=\mathrm{id}$ and $T^1(g\circ f)=T^1(g)\circ T^1(f)$. In this setting, we must insist that f be a diffeomorphism, but shortly, we shall be able to weaken that condition to *local diffeomorphism.* A local diffeomorphism is a smooth map with $T_{x_0}(f)$ invertible for all x_0. ■

The canonical simplicial fiber bundle functor will give a certain extension of the tangent bundle functor; its elements will generalize tangent vectors. In order to define it, it will be useful to give an alternative, slightly less intuitive, construction of the tangent and cotangent bundles. This construction will also make available all of the *tensor* bundles on a smooth manifold. It would be possible to proceed with the definition that we already have, but there are certain advantages to be gained by viewing the later constructions from the viewpoint to follow: it almost *materializes* the notion of the *invariance* of a geometric object.

Exercises

1. Show that $J^1(M, N)$ has a linear structure on its fibers by giving a frame-independent definition of the addition and scalar multiplication of 1-jets.

2. Let $T_1(T_1(M))$ be denoted $T_2(M)$. What is its dimension as a smooth manifold in terms of the dimension of M? Define a projection $T_2(M) \to M$ for a fiber bundle by composing $T_2(M) \to T_1(M) \to M$. Is this the projection of a vector bundle? (Do the fibers have a frame-independent linear structure for which the bundle maps induced by smooth $f: M \to N$ are linear?)

§1.2. 1-Frame bundles and tensor bundles

The aim of this section is to give a construction which can serve as the starting point for a precise definition of a large and important class of geometric objects, namely *tensors*. For a smooth manifold, we shall construct its bundle of 1-frames. [For the entire discussion which follows in this section, M will be assumed to be an m-dimensional manifold, and $\mathbb{R}^m$ (with its standard linear coordinates (x^i), $i = 1, \ldots, m$) will be denoted E.]

Definition 1.6. A smooth germ $f: (M, x_0) \to (N, y_0)$ is *invertible* if there is a smooth germ $g: (N, y_0) \to (M, x_0)$ such that $g \circ f =$ identity germ $(M, x_0) \to (M, x_0)$, and $f \circ g$ is the identity germ $(N, y_0) \to (N, y_0)$.

A 1-jet $f_1: (M, x_0) \to (N, y_0)$ is *invertible* if there is a 1-jet $g_1: (N, y_0) \to (M, x_0)$ such that $g_1 \circ f_1 (= (g \circ f)_1) = \mathrm{id}_1 : (M, x_0) \to (M, x_0)$ and $f_1 \circ g_1 = \mathrm{id}_1 : (N, y_0) \to (N, y_0)$. ■

According to the inverse function theorem, a germ is invertible if and only if its 1-jet is invertible (Lang [12] or Spivak [20]).

For $x \in M$, let $F^x(M)$ be the set of *invertible germs* $\alpha: (E, 0) \to (M, x)$. By choosing linear representatives, we may identify the *group* of invertible 1-jets $(E, 0) \to (E, 0)$ (with respect to composition) with the group $Gl(m)$ of linear *isomorphisms* from E to E. We shall denote that group $Gl(m)$ and let $F_1^x(M)$ be the set of 1-jets of germs in $F^x(M)$.

Then $Gl(m)$ acts on $F_1^x(M)$ *on the right* in the following way: if $\alpha_1 \in F_1^x(M)$ and $g \in Gl(m)$, say $\alpha_1 \cdot g = \alpha_1 \circ g_1 = (\alpha \circ g)_1$. To say that this is a *right action* (as distinct from an action on the *left*, or a *left action*) is to insist that for all g and h in $Gl(m)$ and $\alpha_1 \in F_1^x(M)$, $\alpha_1 \cdot (gh) = (\alpha_1 \cdot g) \cdot h$, and $\alpha_1 \cdot 1 = \alpha_1$. For typographical convenience, we shall delete the ubiquitous '·' when describing the action.

Now let $F_1(M)$ be the union of all sets $F_1^x(M)$. There is an obvious mapping $\pi_1 : F_1(M) \to M$ (associate to each 1-jet its target), and there is then a global action of $Gl(m)$ on $F_1(M)$ *on the right* which carries the π_1 fibers to themselves. The action of $Gl(m)$ on each π_1 fiber satisfies this property. If α_1 and β_1 belong to $\pi_1^{-1}(x)$, the fiber over x, there is a *unique* g in $Gl(m)$ such that $\alpha_1 g = \beta_1$.

Now $F_1(M)$ is identified with the subset of $J^1(E, M)$: the invertible elements of $p_1^{-1}(O \times M)$. We could argue that the latter is a smooth submanifold of $J^1(E, M)$ for which the restriction of p_1 is the projection of a fiber bundle if it is given the relative topology, but it is better to topologize $F_1(M)$ *directly* to facilitate later calculations.

Thus suppose that $\phi : V \to U \subset M$ is a *frame*, for V and U open. Then define the mapping $\tilde{\phi} : Gl(m) \times U \to \pi_1^{-1}(U)$ by the following prescription: $\tilde{\phi}(g, x) = (\phi \circ t_z \circ g)_1$ where the latter is a 1-jet $(E, 0) \to (M, x)$ and $\phi(z) = x$, t_z is the germ of the *translation* $(E, 0) \to (E, z)$, $y \mapsto y + z$. $\tilde{\phi}$ is then easily seen to be bijective and $\pi_1 \tilde{\phi}(g, x) = x$ for all $x \in U$. Now if $\theta : V' \to U' \subset M$ is another frame and if $\tilde{\theta}(g, x) = \tilde{\phi}(h, x)$, then if $\phi(z_1) = \theta(z_2) = x$ we will have $g = (t_{z_2}^{-1} \circ \theta^{-1} \circ \phi \circ t_{z_1})_1 h$ since $(\phi \circ t_{z_1} \circ h)_1 = (\theta \circ t_{z_2} \circ g)_1$.

From this latter observation, we see that g is a *left* multiple of h by an element of $Gl(m)$ which varies smoothly with x in $U \cap V$. Now topologize $F_1(M)$ with the *smallest* topology making all $\tilde{\phi}^{-1} : \pi_1^{-1}(U) \to Gl(m) \times U$ continuous, and all $\pi_1^{-1}(U)$ open, and π_1 will be the projection of a fiber bundle.

Now

$$\begin{array}{c} F_1(M) \\ \downarrow{\scriptstyle \pi_1} \\ M \end{array}$$

is *not* a vector bundle over $\mathbb{R}$. It is, however, a special type of fiber bundle called a *principal bundle* with structure group $Gl(m)$. In general, if $p : X \to M$ is the projection of a fiber bundle, we shall say that p is the projection of a *principal bundle with structure group and fiber* G (for G a Lie group) if there is a smooth action of G on X on the right such that each fiber is an orbit under that action and such that, for any pair α and β in the same fiber, there is a *unique* element $g \in G$ such that $\alpha g = \beta$. Finally, the 'frames' like $\tilde{\phi}$ above are restricted to satisfy $\tilde{\phi}(hg, b) = \tilde{\phi}(h, b)g$ for h, g in G, and b in V.

We shall call π_1 the projection of the (principal) *bundle of 1-frames* over M. The action of $Gl(m)$ on fibers materializes the effect of all possible *linear* changes of frame at each point.

Observation 1.7. It would be possible to carry this procedure further. Let $F(M)$ be the union of all $F^x(M)$. There is an obvious projection $\pi : F(M) \to M$. What is not so obvious is how to topologize $F(M)$ or how to topologize the group $\mathcal{G}$ of invertible germs from $(E, 0) \to (E, 0)$. To take advantage of the theory of fiber bundles, it seems to be necessary to define certain equivalence relations on $\mathcal{G}$ (like 1-equivalence) and to pass these equivalences on to $F(M)$ via the obvious right action of $\mathcal{G}$ on $F(M)$. This will often be done by factorization by an invariant subgroup, giving a principal bundle, with fiber being the quotient group suitably topologized.

In $F(M)$, however, is contained *all* information about all possible smooth local changes of frame. This makes $F(M)$ *universal* in a certain sense, and we shall see that it can be reduced by stronger and stronger equivalence relations to give principal bundles capable of serving as the basis for a description of more and more subtle geometric objects. In this section, however, we shall restrict ourselves to the very weak equivalence relation (1-equivalence) in $\mathcal{G}$ to give the definition of the most elementary type of geometric objects, namely tensors. ■

Returning to the bundle

$$\begin{array}{c} F_1(M) \\ \downarrow{\scriptstyle \pi_1} \\ M \end{array}$$

we now suppose that V is a (finite-dimensional) vector space and that $Gl(m)$ acts *on the left* on V. This means that if $w \in V$, g and $h \in Gl(m)$, then $(gh)w \in V$ and $(gh)w = g(hw)$. Also $1w = w$ for 1 the identity of $Gl(m)$. We shall say that the action of $Gl(m)$ on V is *linear* in the case that $w \to gw$ is a linear isomorphism for each g in $Gl(m)$. [In any case, we shall always insist that $w \to gw$ be a smooth diffeomorphism of V for all g, and that $Gl(m) \times V \to V$, $(g, w) \to gw$ be smooth.]

Form the product $F_1(M) \times V$, and define the following smooth action of $Gl(m)$ *on the right* in $F_1(M) \times V$:

$$(\alpha_1, w)g = (\alpha_1 g, g^{-1}w).$$

Say that two elements of $F_1(M) \times V$ are *equivalent* if they belong to the same orbit under this $Gl(m)$ action; that is, if some element of $Gl(m)$ multiplies one to give the other. In general, we shall have

$$(\alpha_1 g, w) \cong (\alpha_1, gw),$$

a very useful form for the equivalence relation induced by the $Gl(m)$ action.

Definition 1.8. For M a smooth manifold, V a finite-dimensional vector space on which $Gl(m)$ acts on the left, $x \in M$, then a V-*tensor* at x

is an *equivalence class* of pairs (α_1, w) such that $\pi_1(\alpha_1) = x$. It is first of all obvious that if $(\alpha_1, w) \cong (\beta_1, u)$ then $\pi_1(\alpha_1) = \pi_1(\beta_1)$. Next, if g is the *unique* element of $Gl(m)$ such that $\alpha_1 g = \beta_1$, then $w = gu$. In particular, if the equivalence classes of $F_1(M) \times V$ are denoted $F_1(M) \times_G V$, then there is a projection, which we shall call π_V,

$$\begin{array}{c} F_1(M) \times_G V \\ \downarrow \pi_V \\ M \end{array}$$

taking each *class* to the target point of the first coordinate of any representative pair. Further, if $x \in M$, and if $\pi_1(\alpha_1) = x$ for *fixed* α_1, then if $[\alpha_1, w]$ denotes the class of (α_1, w), the map $w \to [\alpha_1, w]$ is a *bijection* from $V \to \pi_V^{-1}(x)$.

Now if the action of $Gl(m)$ on V is *linear*, and if α_1 and β_1 are two elements of $\pi_1^{-1}(x)$ and if $\alpha_1 g = \beta_1$, then the induced maps $V \to \pi_V^{-1}(x)$ described above are related in the following way: if $[\alpha_1, w] = [\beta_1, u]$, then $w = gu$. Therefore $\pi_V^{-1}(x)$ inherits a linear structure (the linear structure of V) *independently* of the particular jet α_1 which is used to define it! Thus in the case of a linear action we may say that each $\pi_V^{-1}(x)$ has the structure of a linear space. ■

All that remains is to topologize $F_1(M) \times_G V$. While this construction will appear slightly 'abstract' at this point, we shall in a moment show how it provides a direct and simple interpretation of the 'transformation laws' which are classically used to define tangent and cotangent vectors and other tensors. For a more detailed discussion of this construction (but from a 'dual' point of view) see Sternberg [24].

Exercises

3. Show that the action of $Gl(m)$ on the fibers of $F_1(M)$ is *simply transitive;* that is, for two frame-jets in a given fiber, there is one and only one element of $Gl(m)$ carrying one to the other.

4. Suppose that $\phi : W \to U \subset M$ is a local frame and let V be a finite-dimensional vector space with $Gl(m)$ acting *linearly* on the left. For $z \in V$ let $t_z : (E, 0) \to (E, z)$ be the translation germ: $y \to y + z$. If we write $\phi(z) = x$, consider the 1-jet $(\phi \circ t_z)_1 : (E, 0) \to (M, x)$. Now define $\tilde{\phi} : V \times U \to \pi_V^{-1}(U)$ by the rule (for $w \in V$, $x \in U$) $\tilde{\phi}(w, x) = [(\phi \circ t_z)_1, w] \in F_1(M) \times_G V$. Show that if we give $F_1(M) \times_G V$ the smallest topology with $\pi_V^{-1}(U)$ open and all $\tilde{\phi}^{-1}$ continuous, then the $\tilde{\phi}$ are the local trivializations for a vector bundle structure.

The proof of the result of the exercise 4 above will show that if the action of $Gl(m)$ on V is not linear, then one has a fiber bundle with fiber diffeomorphic to V but *no* natural linear structure on that fiber. This situation has been encountered already in exercise 2. We shall examine it more closely eventually, but turn our attention now to the important cases of a linear action.

Definition 1.9. For M a smooth manifold, $F_1(M)$ the bundle of 1-frames, and V a finite-dimensional vector space on which $Gl(m)$ acts *linearly* on the left, the vector bundle

$$\begin{array}{c} F_1(M)\times_G V \\ \downarrow{\scriptstyle \pi_V} \\ M \end{array}$$

is the *bundle of V-tensors over M.* For $x\in M$, a V-tensor at x will be a *class* $[\alpha_1, w]$ for $\alpha_1:(E,0)\to(M,x)$, that is, an *orbit* under the action of $Gl(m)$ on $F_1(M)\times V$ defined above. This representation of the class is not unique, but *depends on the choice of 1-frames.* The following relation connects the different representations of the same V-tensor, and it expresses the 'transformation laws' of these tensors:

$$[\alpha_1 g, w]=[\alpha_1, gw] \tag{1.1}$$

for g in $Gl(m)$. ■

Equation (1.1) will be of central importance for the remainder of the section and will be generalized in various ways in the sequel. It expresses the 'invariance' of the geometric object with respect to changes of 1-frame. The object is to be thought of as the *class* [,] which survives the variation of representations due to 1-frame changes. Classically, the dependence of the tensor on the 1-frame is suppressed, and one is presented merely with the element w of V with the understanding that, when coordinates are changed, the element w must transform according to the law of the group action on V.

Definition 1.10. A *V-tensor field* is a smooth section of the bundle

$$\begin{array}{c} F_1(M)\times_G V \\ {\scriptstyle \pi_V}\downarrow \\ M \end{array}$$

That is, it is a map $s:M\to F_1(M)\times_G V$ which is smooth such that $\pi_V\circ s=\mathrm{id}_{|M}$. Thus it associates to each $x\in M$ a V-tensor at x, and it does this in a smooth way. It can be considered an *assignment* to each frame on M $\phi:O\to U\subset M$ of a smooth function $S:O\to V$ such that if $\theta:O'\to U'$ is another frame, and $S':O'\to V$ is its assignment, and if

$\phi(z_1)=\theta(z_2)=x$, then

$$[(\phi\circ t_{z_1})_1, S(z_1)]=[(\theta\circ t_{z_2})_1, S'(z_2)].$$ ■

Observation 1.11. Suppose that V and V' are finite-dimensional Euclidean spaces with linear $Gl(m)$ actions defined for them. Suppose further that $\tau: V\rightarrow V'$ is a *$Gl(m)$-equivariant linear map* (that is, $\tau(gw)=g\tau(w)$ for w in V, g in $Gl(m)$), then the map $\mathrm{id}\times\tau: F_1(M)\times V\rightarrow F_1(M)\times V'$ 'passes to quotients' giving a well-defined map $\mathrm{id}\times_G\tau: F_1(M)\times_G V\rightarrow F_1(M)\times_G V'$. It is clear that

$$\begin{array}{ccc} F_1(M)\times_G V & \xrightarrow[\mathrm{id}\times_G\tau]{} & F_1(M)\times_G V' \\ {\scriptstyle\pi_V}\downarrow & & \downarrow{\scriptstyle\pi_{V'}} \\ M & \xrightarrow[\mathrm{id}]{} & M \end{array}$$

will then be a vector bundle mapping. This observation will be useful in defining such operations as *contraction, tensor multiplication,* and *index raising and lowering.* ■

We shall show shortly that the tangent and cotangent bundles are examples of V-tensor bundles. Now we have established certain 'functorial' properties for these cases, that is we have shown how certain types of smooth maps from M to N can be used to give maps between these bundles. While the following fact is the most general that can be stated in this setting, it will be necessary to extend it in special cases where more structure is available.

Observation 1.12. Suppose that M and N are m-dimensional manifolds, V and V' m-dimensional Euclidean spaces, and $\tau: V\rightarrow V'$ a linear map. Suppose also that V and V' have linear $Gl(m)$ actions for which τ is $Gl(m)$-equivariant. Then if $f: M\rightarrow N$ is a *local diffeomorphism,* there is a smooth map $F: F_1(M)\times_G V\rightarrow F_1(N)\times_G V'$ with the property that

$$\begin{array}{ccc} F_1(M)\times_G V & \xrightarrow{F} & F_1(N)\times_G V' \\ {\scriptstyle\pi_V}\downarrow & & \downarrow{\scriptstyle\pi_{V'}} \\ M & \xrightarrow[f]{} & N \end{array}$$

commutes and is a vector bundle mapping.

Proof. First observe that there is a mapping $H: F_1(M)\rightarrow F_1(N)$ defined by $H(\alpha_1)=(f\circ\alpha)_1$ (here is where we need that f be a local diffeomorphism). It is easy to see that H commutes with the right action of $Gl(m)$, that is $H(\alpha_1 g) = (H\alpha_1)g$, and that

$$\begin{array}{ccc} F_1(M) & \xrightarrow[H]{} & F_1(N) \\ \downarrow & & \downarrow \\ M & \xrightarrow[f]{} & N \end{array}$$

is a fiber bundle mapping which is bijective on fibers.

Now consider $H \times \tau : F_1(M) \times V \to F_1(N) \times V'$. $Gl(m)$-equivariance of τ guarantees that it passes to quotients giving

$$H \times_G \tau : F_1(M) \times_G V \to F_1(N) \times_G V'.$$

Let F be $H \times_G \tau$. ∎

For various choices for V this will allow us to construct functors from the category of manifolds and *smooth maps which are local diffeomorphisms* to the category of vector bundles (the V-tensor bundles). Given the fact that the cotangent bundle is a V-tensor bundle, we see that the statement in §1.1 that the covariant form of the cotangent functor is defined for diffeomorphisms can now be strengthened to local diffeomorphisms.

We now give examples of some of the most important V-tensor bundles.

Example (The tangent bundle $T_1(M)$ revisited). Here V is a linear space isomorphic with E. In particular, let V be the space of 1-jets of maps $(\mathbb{R}, 0) \to (E, 0)$. (Recall that $\mathbb{R}$ and E have 'standard' coordinates). Suppose that $\gamma_1 : (\mathbb{R}, 0) \to (E, 0)$ is such a 1-jet, $\gamma_1 \in V$. And suppose that the derivative of γ at 0 is represented in standard coordinates by the $m \times 1$ column matrix X^i where for $i = 1, \ldots, m$,

$$X^i = \frac{d\gamma^i}{dt}_{|0}.$$

As we observed earlier, each 1-jet has *unique* representation as such a column matrix.

Now $Gl(m)$ is the group of linear automorphisms of E (invertible linear maps from E to E). We identify such an automorphism likewise with its (invertible) 1-jet as a map from $(E, 0) \to (E, 0)$. If $g \in Gl(m)$ then the derivative of g at 0 has representation in terms of standard coordinates as a $m \times m$ matrix:

$$\frac{\partial g^j}{\partial x^i}_{|0} \quad \text{where } i, j = 1, \ldots, m.$$

[In the entire sequel, all derivatives not assigned a point at which they are to be evaluated will be assumed to be evaluated at 0.]

The action of $Gl(m)$ on V is given by the rule

$$g\gamma_1 = (g \circ \gamma)_1, \tag{1.2}$$

and this action can be represented with matrices

$$Y^j = \frac{\partial g^j}{\partial x^i} X^i,$$

where Y^j $(j = 1, \ldots, m)$ is the $m \times 1$ column matrix representing the transform $g\gamma_1$. [The *Einstein summation convention* will be systematically invoked here and in the sequel: summation will be implied over any index which appears once as an upper index and once as a lower index.]

The equivalence relation

$$[\phi_1 g, X^i] = \left[\phi_1, \frac{\partial g^j}{\partial x^i} X^i\right]$$

then merely expresses the commutativity of the diagram:

$$\begin{array}{ccccc} & & (E, 0) & \xrightarrow{\phi_1} & (M, x) \\ & {\scriptstyle g\gamma_1}\nearrow & \uparrow{\scriptstyle g} & \nearrow{\scriptstyle \phi_1 g} & \\ (\mathbb{R}, 0) & \xrightarrow{\gamma_1} & (E, 0) & & \end{array} \tag{1.3}$$

Each equivalence class $[\phi_1, \gamma^i]$ in $\pi_V^{-1}(x)$ is unambiguously associated with the 1-jet $(\phi \circ \gamma)_1 : (\mathbb{R}, 0) \to (M, x)$. Thus it can be identified with a tangent vector at x.

In this scheme, g is a '*change of 1-frame*'. It has a dual role. On the one hand, it transforms the 'components' X^i of the tangent vector in 1-frame $\phi_1 g$ to the components of the same tangent vector in 1-frame ϕ_1 at x. On the other hand, it 'transforms' 1-frame ϕ_1 to 1-frame $\phi_1 g$ at x.

Now suppose that $\phi : O \to U \subset M$ and $\theta : O' \to U' \subset M$ are frames on M with $x \in U \cap U'$. A *vector field over* ϕ (see Definition 1.10) will be a smooth map $S : O \to V$, and a *vector field over* θ will be a smooth map $S' : O' \to V$, and if S and S' are the restrictions to U and U' of the same smooth vector field on M, then we must have the following consistency condition satisfied for points such as x (where $\phi(z_1) = \theta(z_2) = x$):

$$[(\phi \circ t_{z_1})_1, S(z_1)] = [(\theta \circ t_{z_2})_1, S'(z_2)]$$

and this means

$$S'(z_2) = (t_{z_1}^{-1} \circ \phi^{-1} \circ \theta \circ t_{z_2})_1 S(z_1). \tag{1.4}$$

In particular, the 1-jet $(t_{z_1}^{-1} \circ \phi^{-1} \circ \theta \circ t_{z_2})_1$ plays the role of the change of 1-frame $g \in Gl(m)$ above. It varies smoothly with x and gives a smooth set of component changes for $z_1 \in \phi^{-1}(U \cap U')$.

Finally, it should be observed here that Observation 1.12 can be strengthened for *this* choice of V. We have shown how to define $T_1(f)$ for

$f: M \to N$ an *arbitrary* smooth map (Observation 1.3). [Henceforth, tangent vectors will be denoted $[\phi_1, X^i]$, a 1-frame at x, and an $m \times 1$ matrix representing a 1-jet $\gamma_1: (\mathbb{R}, 0) \to (E, 0)$. This will represent $(\phi \circ \gamma)_1: (\mathbb{R}, 0) \to (M, x)$.]

***Example** (The cotangent bundle $T^1(M)$ revisited).* Here V is chosen to be the linear space isomorphic to E^* (the dual of E). V is the space of 1-jets of maps $(E, 0) \to (\mathbb{R}, 0)$.

Suppose that $\alpha_1: (E, 0) \to (\mathbb{R}, 0)$ is such a 1-jet, $\alpha_1 \in V$, and its *derivative* is represented in standard coordinates by the $1 \times m$ matrix A_i, $i = 1, \ldots, m$. Thus

$$A_i = \frac{\partial \alpha}{\partial x^i}_{|0}.$$

Each such 1-jet has a unique representation as such a row matrix.

Now let $Gl(m)$ act *on the left* on V by the rule

$$g\alpha_1 = (\alpha \circ g^{-1})_1. \tag{1.5}$$

This *is* a left action $(gh)\alpha_1 = g(h\alpha_1)$ as is easily seen.

In terms of matrices, this action is given by

$$B_j = A_i \frac{\partial \hat{g}^i}{\partial y^j}$$

where we represent g^{-1} by $\hat{g}$ for notational convenience. Now the equivalence relation

$$[\phi_1 g, A_j] = \left[\phi_1, A_i \frac{\partial \hat{g}^i}{\partial y^j}\right]$$

simply expresses the commutativity of the diagram:

$$\begin{array}{ccccc} & & (E, 0) & \xleftarrow{\phi_1^{-1}} & (M, x) \\ & \swarrow{\scriptstyle g\alpha_1} & \downarrow{\scriptstyle g^{-1}} & \swarrow{\scriptstyle (\phi g)_1^{-1}} & \\ (\mathbb{R}, 0) & \xleftarrow{\alpha_1} & (E, 0) & & \end{array} \tag{1.6}$$

Thus the class $[\phi_1, A_i]$ is unambiguously associated with the 1-jet $(\alpha \circ \phi^{-1})_1: (M, x) \to (\mathbb{R}, 0)$, a *cotangent vector* according to our earlier definition at $x \in M$.

Now according to Observation 1.12, the associations $M \to T^1(M)$ and $f \to T^1(f)$ for $f: M \to N$ a local diffeomorphism, where $T^1(f)$ is defined by $T^1(f)[\phi_1, A_i] = [(f \circ \phi)_1, A_i]$, gives a *covariant* functor from the category of manifolds and local diffeomorphisms to the category of vector bundles. [Henceforth, cotangent vectors will be denoted $[\phi_1, A_i]$, ϕ_1 a 1-frame at x and A_i the $1 \times m$ matrix representing α_1.] This pair gives $(\alpha \circ \phi^{-1})_1: (M, x) \to (\mathbb{R}, 0)$.

Classically, the tangent vectors and cotangent vectors are called, respectively, *contravariant* and *covariant* tensors of rank 1. These names derive from the nature of the action of $Gl(m)$ on components of elements of V. This use of the words 'contravariant' and 'covariant' is difficult to align with the more modern denotation of those words for describing functors in category theory; in fact, from a certain point of view, their meaning is precisely opposite to the category-theoretic one (see Spivak [21]). We shall take a contravariant tensor to be one for which the transformation law makes use of the *Jacobian matrix* of g and a covariant tensor to be one for which the *inverse* of the Jacobian matrix is used in the transformation law. This is consistent with the classical terminology. The meaning of the rank of a tensor will be clearer with further examples.

Example (Covariant tensors of higher rank). The definition of V in these cases will be slightly more elaborate than it was in the previous examples. These tensors will actually generalize the cotangent vectors. They will not, however, be given an immediate interpretation in terms of the jet bundles. While a cotangent vector at a point is a 1-jet of a mapping of a certain type, the higher-rank covariant tensors will not have an interpretation as the 1-jet of a mapping.

Let U and W be vector spaces of arbitrary finite dimension. And let $L(U; W)$ denote the vector space of linear maps from U to W. When bases are chosen, these maps can be represented in the usual way as matrices with addition and scalar multiplication being defined componentwise. The dimension of $L(U; W)$ is the product of the dimension of U with the dimension of W. In particular, $L(U;\mathbb{R})$ has dimension equal to the dimension of U and is sometimes denoted U^*, the *dual space of* U.

There is an obvious mapping $U^*\times U\to\mathbb{R}$ which takes $(f, u)\to f(u)$ and this mapping is *linear in each variable separately*. Such a mapping is called *bilinear* or 2-*linear.*

Definition 1.13. If $U_1, U_2, \ldots, U_k$ are finite-dimensional vector spaces, we denote by the symbol $L(U_1, U_2, U_3, \ldots, U_k;\mathbb{R})$ the vector space $L(U_1; L(U_2; \ldots; L(U_k;\mathbb{R}))\ldots)$. This space is isomorphic with the vector space of *functions* $f: U_1\times U_2\times U_3\times\ldots\times U_k\to\mathbb{R}$ satisfying

$$\text{a) } f(x_1, \ldots, x_i+x_i', \ldots, x_k)=f(x_1, \ldots, x_i, \ldots, x_k) + f(x_1, \ldots, x_i', \ldots, x_k), \qquad 1\leq i\leq k$$

$$\text{b) } f(x_1, \ldots, ax_i, \ldots, x_k)=af(x_1, \ldots, x_i, \ldots, x_k), \qquad \text{for scalar } a,\ 1\leq i\leq k.$$

$L(U_1, U_2, \ldots, U_k;\mathbb{R})$ is the vector space of k-*linear maps from* $U_1\times\ldots\times U_k$ to $\mathbb{R}$. ■

These functions, while they are not linear for $k > 1$, are 'linear in each variable separately'. The dimension of L is given by $\dim L(U_1, \ldots, U_k; \mathbb{R}) = \prod_{i=1}^k \dim(U_i)$.

Now return to our standard m-dimensional space E. Let $E_i = E$ for $i = 1, \ldots, k$. Then an element of $L(E_1, \ldots, E_k; \mathbb{R})$ has a simple representation in terms of the standard basis. Call that basis $(e_1, e_2, \ldots, e_m)$. Then any f in $L(E_1, \ldots, E_k; \mathbb{R})$ is entirely determined by its values on the m^k k-tuples $(e_{i_1}, e_{i_2}, \ldots, e_{i_k})$ for i_j in $\{1, 2, \ldots, m\}$, as follows easily from k-linearity.

Now suppose that we are given an assignment of m^k *numbers*, one to each k-tuple $(e_{i_1}, e_{i_2}, \ldots, e_{i_k})$. Let us agree to represent this assignment as a *matrix array* in the form of a k-dimensional cube (m entries per side) with the symbol $A_{i_1 i_2 \ldots i_k}$ simultaneously representing the *matrix itself* and the *generic entry* in the $(i_1, i_2, \ldots, i_k)$ position. This defines a unique element $f \in L(E_1, E_2, \ldots, E_k; \mathbb{R})$ by the prescription $f(e_{i_1}, e_{i_2}, \ldots, e_{i_k}) = A_{i_1 i_2 \ldots i_k}$. Addition and scalar multiplication in $L(E_1, E_2, \ldots, E_k; \mathbb{R})$ can be done *componentwise* with these matrix representatives.

In particular, if $A_{i_1 i_2 \ldots i_k}$ represents f and if (using Einstein summation convention)

$$\begin{aligned} x_1 &= \gamma_1^i e_i \\ x_2 &= \gamma_2^i e_i \\ &\ \vdots \\ x_k &= \gamma_k^i e_i \end{aligned}$$

then

$$f(x_1, x_2, \ldots, x_k) = A_{i_1 i_2 \ldots i_k} \gamma_1^{i_1} \gamma_2^{i_2} \ldots \gamma_k^{i_k} \tag{1.7}$$

(where this sum is over *all* repeated indices, and hence involves m^k summands indexed by $(i_1, i_2, \ldots, i_k)$). This follows at once from k-linearity.

Thus V will be the vector space $L(E_1, E_2, \ldots, E_k; \mathbb{R})$, and we shall have to define the left action of $Gl(m)$ on V. Suppose that $g \in Gl(m)$ is an invertible linear map $g: E \to E$, and we think of g as the 1-jet of a coordinate change at the origin as usual. We *motivate* the definition of the action of g on V with the following diagram which is in a sense a generalization of diagram (1.6).

$$\begin{array}{ccccccc} & & (E^k, 0) & \ldots & (E, 0) & \xleftarrow{\phi_1^{-1}} & (M, x) \\ & \swarrow^{gf} & \downarrow{\scriptstyle (g^{-1})^k} & & \downarrow{\scriptstyle g^{-1}} & \swarrow_{(\phi g)_1^{-1}} & \\ (\mathbb{R}, 0) & \xleftarrow{f} & (E^k, 0) & \ldots & (E, 0) & & \end{array} \tag{1.8}$$

For g the 1-jet of a coordinate change, the right triangle should commute (this simply *defines* the new frame-germ ϕg). Now $E^k = E_1 \times \ldots \times E_k$ and $(g^{-1})^k$ is the Cartesian product of g^{-1} with itself k

times. Since g^{-1} is linear in each factor, it is clear that $f\circ(g^{-1})^k$ is an element of $L(E_1, E_2, \ldots, E_k; \mathbb{R})$. Insisting that the left triangle commutes then gives the definition of $gf = f\circ(g^{-1})^k$.

We shall represent the elements of V *as matrices* as we did in earlier cases, and so we shall require the matrix representing gf when the matrix for g and the matrix for f, *i.e.* $A_{i_1 i_2 \ldots i_k}$, are given. A straightforward application of formula (1.7) letting

$$x_j = \frac{\partial \hat{g}^i}{\partial y^j} e_i$$

(where again $\hat{g} = g^{-1}$) shows that the matrix for

$$gf = B_{j_1 j_2 \ldots j_k} = A_{i_1 i_2 \ldots i_k} \frac{\partial \hat{g}^{i_1}}{\partial y^{j_1}} \frac{\partial \hat{g}^{i_2}}{\partial y^{j_2}} \cdots \frac{\partial \hat{g}^{i_k}}{\partial y^{j_k}}. \tag{1.9}$$

Of course, this action of $Gl(m)$ on $L(E_1, E_2, \ldots, E_k; \mathbb{R})$ is only *motivated* by diagram (1.8). There is no question here of interpreting f or gf as 1-jets which compose with the 1-jets of charts.

In any case, with these choices for V and the left $Gl(m)$ action, the tensors $[\phi_1, A_{i_1 i_2 \ldots i_k}]$ in $F_1(M) \times_G V$ are the *covariant tensors of rank k*, also called *tensors of type* $\binom{0}{k}$. Letting $k = 1$, we see that the covariant tensors of type $\binom{0}{1}$ are simply the cotangent vectors defined earlier.

Example (Contravariant tensors of higher rank). Consider the bilinear map $E \times E^* \to \mathbb{R}$ taking $(w, f) \to f(w)$. This element of $L(E, E^*; \mathbb{R})$ is in $L(E; L(E^*; \mathbb{R}))$; in fact, it is an *isomorphism* $E \to L(E^*; \mathbb{R})$. In the case of the tangent bundle, the choice for V was E. Generalizing to the higher-rank contravariant tensors leads us to consider for the vector space V, $L(E_1^*, E_2^*, \ldots, E_k^*; \mathbb{R})$, where $E_i^* = E^*$ for $i = 1, \ldots, k$ (Lang [13]).

Suppose that we represent the standard basis of E^* as $(e^1, e^2, \ldots, e^m)$. Then an element of $L(E_1^*, E_2^*, \ldots, E_k^*; \mathbb{R})$ is entirely determined by its effect on the m^k k-tuples $(e^{i_1}, e^{i_2}, \ldots, e^{i_k})$. Notice the use of superscripts to label linear forms (dual vectors). Recall also that the e^i are defined to be the elements of $L(E; \mathbb{R})$ satisfying $e^i(e_j) = \delta^i_j$ (the Kronecker delta), which is 1 if $i = j$ and is 0 if $i \neq j$.

Thus if $f \in L(E_1^*, E_2^*, \ldots, E_k^*; \mathbb{R})$ then it may be represented as a k-dimensional matrix array with m entries per side giving a total of m^k entries. We shall represent the matrix itself by its generic entry in the $(i_1, i_2, \ldots, i_k)$ position, namely $X^{i_1 i_2 \ldots i_k}$.

Thus if

$$x^1 = \gamma_i^1 e^i$$
$$x^2 = \gamma_i^2 e^i$$
$$\vdots$$
$$x^k = \gamma_i^k e^i$$

then k-linearity gives

$$f(x^1, x^2, \ldots, x^k) = X^{i_1 i_2 \cdots i_k} \gamma_{i_1}^1 \gamma_{i_2}^2 \ldots \gamma_{i_k}^k. \tag{1.10}$$

Now in order to define the appropriate action of $Gl(m)$ on V in this case, we must introduce the notion of the 'adjoint' of a linear map. Suppose U and V are finite-dimensional vector spaces and $H : U \to V$ a linear map. Then the *adjoint of* H is the linear map denoted $H^* : V^* \to U^*$ defined by the prescription $H^*(f) = f \circ H$ for $f \in V^*$. The association $H \to H^*$ is itself a linear isomorphism from $L(U; V)$ to $L(V^*; U^*)$.

Now for *motivation* for the definition of the left action of $Gl(m)$ on $V = L(E_1^*, E_2^*, \ldots, E_k^*; \mathbb{R})$ we inspect the diagram:

$$\begin{array}{ccccc} & ((E^*)^k, 0) \ldots (E, 0) & \xrightarrow{\phi_1} & (M, x) \\ {}^{gf}\swarrow & \downarrow{\scriptstyle (g^*)^k} \qquad \uparrow{\scriptstyle g} & \nearrow{\scriptstyle \phi_1 g} & \\ (\mathbb{R}, 0) \xleftarrow[f]{} & ((E^*)^k, 0) \ldots (E, 0) & & \end{array} \tag{1.11}$$

where g^* is the adjoint of g, and ϕ_1 is the 1-jet of a frame at x. Commutativity of the right triangle gives, as always, the definition of the *right action* of $Gl(m)$ on 1-jets of frames. Then commutativity of the left triangle requires that

$$gf = f \circ (g^*)^k. \tag{1.12}$$

Now the operation of taking the adjoint satisfies $(F \circ H)^* = H^* \circ F^*$ for composable linear maps F and H. It follows from this that

$$h(gf) = f \circ (g^*)^k \circ (h^*)^k = f \circ ((hg)^*)^k = (hg)f.$$

Thus, the action we have defined is a left action on V.

Now suppose that the matrix representing f is $X^{i_1 i_2 \cdots i_k}$ and that the matrix representing coordinate change g is $\partial g^j / \partial x^i$, then a straightforward application of formula (1.10) letting

$$x^j = \frac{\partial g^j}{\partial x^i} e^i$$

gives the matrix representing gf. This matrix is

$$Y^{j_1 j_2 \cdots j_k} = X^{i_1 i_2 \cdots i_k} \frac{\partial g^{j_1}}{\partial x^{i_1}} \frac{\partial g^{j_2}}{\partial x^{i_2}} \cdots \frac{\partial g^{j_k}}{\partial x^{i_k}}. \tag{1.13}$$

We shall thus represent *contravariant tensors of rank k* or *tensors of type* $\binom{k}{0}$ with the symbol $[\phi_1, X^{i_1 i_2 \cdots i_k}]$ representing classes in $F_1(M) \times_G V$.

Exercise

5. Letting $V = L(E_1, E_2, \ldots, E_n; E_1^*, E_2^*, \ldots, E_k^*; \mathbb{R})$ we have the tensors of 'mixed type': covariant of rank n and contravariant of rank k. These are described simply as tensors of type $\binom{k}{n}$. Imitate the previous examples and give the 'transformation law' for such tensors.

We end this section with some comments on notation. In practice, we shall be interested in *tensor fields* (particularly vector fields and covariant tensor fields of higher rank). Given a local frame $\phi : O \to U \subset M$ a V-tensor field over ϕ is described by a smooth map $S : O \to V$; that is, for such a map we have a cross section of the bundle

$$\begin{array}{c} F_1(M) \times_G V \\ \downarrow \pi_V \\ M \end{array}$$

over U as $z \to [(\phi \circ t_z)_1, S(z)]$. Now when it is safe to omit mention of the local frame ϕ, such a field will be colloquially denoted $\tilde{S}(z)$, where the 'tilde' serves as a reminder that there is a frame around. The $\tilde{S}(z)$ will be referred to as a *local V-tensor field on U* (even though z lives in O). Thus local vector fields on U such as $X^i(z)\partial/\partial x^i$ will be denoted simply $\tilde{X}^i(z)$ and local covector fields such as $Y_i(z)dx^i$ will be denoted simply $\tilde{Y}_i(z)$. Here $\partial/\partial x^i$ is simply the classical name for the 'standard' basis vector for E, that is e_i, and dx^j is the classical name for the 'standard' linear coordinate e^j for E, this being simultaneously a 'standard' basis vector for E^*.

§1.3. Higher-order jet bundles and frame bundles

We have given some indication in the previous sections of the way in which 1-jets and the 1-jet bundles $J^1(M, N)$ can be used to define and develop the classical tensors. Tensors capture properties of germs of mappings which are determined by their 1-jet and, therefore, in order to define them, we must sacrifice any information which those germs might convey which is too subtle to be detected by 1-jets.

In this section, we shall construct the higher-order jet bundles in order to lay the groundwork for certain generalizations of tensors to be developed in this chapter. The spirit of this construction (which is due to

Ehresmann) is the following: 1-jets correspond, essentially, to the *linear part* of the Taylor expansion of a germ; they are defined by declaring 'equivalent' two germs with the same linear part. The higher-order jets will be defined by introducing *stronger* equivalence relations, ones capable of distinguishing germs with the same 1-jet. The classes of these equivalence relations materialize the idea of the kth-*order part* of the Taylor expansion of a germ. Our discussion must be somewhat sketchy and the interested reader is referred to Thom and Levine [26], Yano and Ishihara [27] or Golibutsky and Guillemin [9] for further details.

We begin this discussion with the case of a germ $f:(\mathbb{R}^m, 0) \to (\mathbb{R}^n, 0)$ where the Euclidean spaces are referred to linear coordinates $(x^1, \ldots, x^m)$ and $(y^1, \ldots, y^n)$ respectively. The germ f is uniquely represented as $(f^1, f^2, \ldots, f^n)$ where $f^i = y^i \circ f$.

Now we introduce some special notation. Let $\alpha = (\alpha_1, \alpha_2, \ldots, \alpha_m)$ denote a sequence of *non-negative integers*, and let $|\alpha|$ denote $\sum_{i=1}^m \alpha_i$. Then the symbol $\partial^{|\alpha|} f^i/\partial x^\alpha$ will denote the partial derivative $\partial^{|\alpha|} f^i/\partial(x^1)^{\alpha_1} \ldots \partial(x^m)^{\alpha_m}$ where these partial derivatives are evaluated at 0.

The symbol $\alpha!$ represents the integer $(\alpha_1!) \ldots (\alpha_m!)$; $x^\alpha = (x^1)^{\alpha_1} \ldots (x^m)^{\alpha_m}$.

Definition 1.14. Two *germs* $f, g:(\mathbb{R}^m, 0) \to (\mathbb{R}^n, 0)$ are *k-equivalent*, denoted $f \underset{k}{\sim} g$, if for *each* r, $1 \leq r \leq k$, *each* j, $1 \leq j \leq n$,

$$\frac{\partial^{|\alpha|} f^j}{\partial x^\alpha} = \frac{\partial^{|\alpha|} g^j}{\partial x^\alpha}$$

for *all* α such that $|\alpha| = r$. ■

It is obvious that k-equivalence is an equivalence relation, and that it generalizes what earlier we called 1-equivalence.

The first thing to show is that this definition respects composition.

Proposition 1.15. Suppose that f and g are as above, $f \underset{k}{\sim} g$, and suppose that $h:(\mathbb{R}^q, 0) \to (\mathbb{R}^m, 0)$ and $e:(\mathbb{R}^n, 0) \to (\mathbb{R}^p, 0)$ are germs. Let $(\bar{x}^1, \ldots, \bar{x}^q)$ and $(\bar{y}^1, \ldots, \bar{y}^p)$ be linear coordinates, respectively, for $\mathbb{R}^q$ and $\mathbb{R}^p$. Then $(e \circ f \circ h) \underset{k}{\sim} (e \circ g \circ h)$.

Proof. We argue by induction on k. The case $k = 1$ has already been shown to be a consequence of the chain rule. For larger k, what is needed is a higher-order version of the chain rule. Assume that the proposition is true for $1 \leq k' \leq k-1$. Then suppose that $|\alpha| = k$, $\alpha = (\alpha_1, \alpha_2, \ldots, \alpha_q)$. Then show that

$$\frac{\partial^k (e^j \circ f \circ h)}{\partial \bar{x}^\alpha} = \frac{\partial^k (e^j \circ g \circ h)}{\partial \bar{x}^\alpha} \qquad \text{for } all\ j = 1, 2, \ldots, p.$$

Suppose for some s, $\alpha_s \geqslant 1$, and let $\alpha' = (\alpha_1, \ldots, \alpha_s - 1, \ldots, \alpha_q)$. Then consider the *germ*

$$\frac{\partial(e^i \circ f \circ h)}{\partial \bar{x}^s} : (\mathbb{R}^q, 0) \to (\mathbb{R}, t_0).$$

This can be written as a matrix product with variable entries according to the chain rule in the following form:

$$\left[\frac{\partial e^i}{\partial y^1}_{|f[h(\bar{x})]}, \ldots, \frac{\partial e^i}{\partial y^n}_{|f[h(\bar{x})]}\right] \begin{bmatrix} \frac{\partial f^1}{\partial x^1}_{|h(\bar{x})}, \ldots, \frac{\partial f^1}{\partial x^m}_{|h(\bar{x})} \\ \vdots \qquad\qquad \vdots \\ \frac{\partial f^n}{\partial x^1}_{|h(\bar{x})}, \ldots, \frac{\partial f^n}{\partial x^m}_{|h(\bar{x})} \end{bmatrix} \begin{bmatrix} \frac{\partial h^1}{\partial \bar{x}^s}_{|\bar{x}} \\ \vdots \\ \frac{\partial h^m}{\partial \bar{x}^s}_{|\bar{x}} \end{bmatrix}.$$

This germ is then a *sum* of germs (summing over i and r) of the form

$$\frac{\partial e^j}{\partial y^i}_{|f[h(\bar{x})]} \cdot \frac{\partial f^i}{\partial x^r}_{|h(\bar{x})} \cdot \frac{\partial h^r}{\partial \bar{x}^s}_{|\bar{x}} : (\mathbb{R}^q, 0) \to \mathbb{R}$$

where the source of *each* summand is 0, and the target of *each* summand is unspecified in $\mathbb{R}$. Addition of such germs is defined by taking the germ of the sum of representative functions.

Now the germ

$$\frac{\partial(e^j \circ g \circ h)}{\partial \bar{x}^s} : (\mathbb{R}^q, 0) \to (\mathbb{R}, t_0)$$

has a similar expression as a sum of germs:

$$\frac{\partial e^j}{\partial y^i}_{|g[h(\bar{x})]} \cdot \frac{\partial g^i}{\partial x^r}_{|h(\bar{x})} \cdot \frac{\partial h^r}{\partial \bar{x}^s}_{|\bar{x}} : (\mathbb{R}^q, 0) \to \mathbb{R}.$$

First, it is obvious that corresponding summands have the same target ($h(0) = 0$ and $f(0) = g(0)$; f and g have the same Jacobian matrices).

Suppose that F and G are corresponding summands with respect to (i, r). It will be enough to show that

$$\frac{\partial F}{\partial \bar{x}^{\alpha'}} = \frac{\partial G}{\partial \bar{x}^{\alpha'}}$$

for each such pair. Now

$$\frac{\partial f^i}{\partial x^r} \text{ and } \frac{\partial g^i}{\partial x^r} : (\mathbb{R}^m, 0) \to \mathbb{R}$$

are $(k-1)$ equivalent. Thus applying the inductive hypothesis, the germs

$$\frac{\partial f^i}{\partial x^r}_{|h(\bar{x})} \text{ and } \frac{\partial g^i}{\partial x^r}_{|h(\bar{x})} : (\mathbb{R}^q, 0) \to \mathbb{R}$$

are also $(k-1)$ equivalent. It is already clear that $f[h(\bar{x})]$ is $(k-1)$ equivalent to $g[h(\bar{x})]$ as germs $(\mathbb{R}^q, 0) \to (\mathbb{R}^n, 0)$ by applying the inductive hypothesis directly to f and g. Finally, we may apply the inductive hypothesis yet again to see that

$$\frac{\partial e^j}{\partial y^i}_{|f[h(\bar{x})]}$$

is $(k-1)$ equivalent to

$$\frac{\partial e^j}{\partial y^i}_{|g[h(\bar{x})]}$$

in light of this last observation.

Thus, the corresponding *factors* in F and G are $(k-1)$ equivalent germs each with source 0 in $\mathbb{R}^q$, and each with the same target as its correspondent. Now the *product formula* for partial derivatives can be used to express $(\partial/\partial \bar{x}^{\alpha'})(uvw)$ for *functions* $u, v, w : (\mathbb{R}^q, 0) \to \mathbb{R}$ as a canonical polynomial in u, v, w and various partial derivatives of u, v, w of degree less than or equal to $(k-1)$. Since these polynomials *are* canonical, it will follow that

$$\frac{\partial F}{\partial \bar{x}^{\alpha'}} = \frac{\partial G}{\partial \bar{x}^{\alpha'}}.$$

Applying this analysis to the sum of corresponding Fs and Gs gives the inductive step. ∎

An immediate consequence of this proposition is that we can now speak of k-equivalence of smooth germs of maps of manifolds.

Definition 1.16. Let M and N be smooth manifolds and suppose that f and g are germs $(M, x_0) \to (N, y_0)$ and that $\phi : (\mathbb{R}^m, 0) \to (M, x_0)$ and $\theta : (\mathbb{R}^n, 0) \to (N, y_0)$ are invertible germs (frame-germs in $F(M)$ and $F(N)$, respectively, as discussed in §1.2). Then $f \underset{k}{\sim} g$, f is k-equivalent to g, if the germs $\theta^{-1} \circ f \circ \phi$ and $\theta^{-1} \circ g \circ \phi$ are k-equivalent in the sense of Definition 1.14. According to the previous proposition, k-equivalence with respect to one pair of frame-germs guarantees k-equivalence with respect to *all* pairs of frame-germs. This also guarantees that k-equivalence is an equivalence relation.

The k-equivalence class of a germ f will be called the *k-jet* of that germ, and will be denoted f_k. ∎

Another important consequence of Proposition 1.15 is that k-jets can be *composed* as germs can. That is, we may define $f_k \circ g_k$ to be $(f \circ g)_k$ for composable germs f and g. This fact is especially useful as we shall see.

Now suppose that $f:(\mathbb{R}^m, 0) \to (\mathbb{R}^n, 0)$ is a germ. We choose now a *representative element* of f_k which plays a role analogous to the linear representative of a 1-jet and which generalizes it. Consider the degree-k polynomial in (x^i) (the linear coordinates for $\mathbb{R}^m$) with values in $\mathbb{R}^n$ whose jth component is

$$\left[\sum_{i=1}^{m} \left(\frac{\partial f^j}{\partial x^i}\right) x^i + \sum_{|\alpha|=2} \frac{1}{\alpha!}\left(\frac{\partial^2 f^j}{\partial x^\alpha}\right) x^\alpha + \ldots + \sum_{|\alpha|=k} \frac{1}{\alpha!}\left(\frac{\partial^k f^j}{\partial x^\alpha}\right) x^\alpha\right]. \tag{1.14}$$

A direct calculation shows that this function $(\mathbb{R}^m, 0) \to (\mathbb{R}^n, 0)$ is k-equivalent to f, and hence can be taken as a representative of f_k. We shall call it the *polynomial representative* of $f_k:(\mathbb{R}^m, 0) \to (\mathbb{R}^n, 0)$.

Just as composition of Jacobian matrices can be thought of as the composition of linear functions, the composition of k-jets can be thought of as the composition of the degree-k polynomial representatives of those jets. In fact, this gives the most effective general procedure for calculating the composition of jets, and we shall use it frequently in the sequel.

It is clear that if $h \leq k$, then $f \underset{k}{\sim} g \to f \underset{h}{\sim} g$; therefore k-equivalence gives stronger relations for increasing k.

With these preliminaries now put aside, we construct the k-jet bundles and the corresponding covariant functors in a manner directly generalizing the construction for 1-jet functors. Surprisingly, the k-jet bundles will *not* have a natural linear structure for their fibers for $k > 1$.

Let M and N be smooth manifolds of respective dimensions m and n. Denote by $J^k(M, N)$ the set of all k-jets of germs of mappings from M to N.

As in the case of 1-jets studied earlier, there is a projection which we shall call p_k where $p_k: J^k(M, N) \to M \times N$ given by $p_k(f_k) = (x_0, y_0)$ and x_0 is the source of f_k and y_0 is the target of f_k.

We shall provide $J^k(M, N)$ with the structure of a smooth manifold for which p_k is the projection of a *fiber* bundle. The dimension of $J^k(M, N)$ will be

$$m+n+n\left[\sum_{i=1}^{k} \binom{m+i-1}{i}\right] = m+n\binom{m+k}{k}$$

where $\binom{a}{b}$ represents the binomial coefficient $a!/(a-b)!\,b!$.

The fiber of this bundle is the set of all k-jets of germs $(\mathbb{R}^m, 0) \to (\mathbb{R}^n, 0)$ and will be denoted $J^k(m, n)$ for convenience. This fiber space can be given the structure of a vector space, but if $k > 1$ this structure cannot be globalized in any natural way (for reasons to be discussed later).

Now suppose that $\phi: O \to U \subset M$ is a frame on M, and $\theta: O' \to V \subset N$ is a frame on N. Denote $p_k^{-1}(U \times V)$ by $J^k(U, V)$. We shall define a family of 'charts' on the sets $J^k(U, V)$ in the following way. Let $T_{\phi\theta}: J^k(U, V) \to J^k(m, n) \times U \times V$ be associated with the pair (ϕ, θ) by

$$T_{\phi\theta}(f_k) = ((t_{z'}^{-1} \circ \theta^{-1} \circ f \circ \phi \circ t_z)_k, (x_0, y_0))$$

where $(x_0, y_0) = p_k(f_k)$ and $\phi(z) = x_0$, $\theta(z') = y_0$.

The next step is to give $J^k(m, n)$ its structure as a topological vector space in terms of the standard linear coordinates (x^i) for $\mathbb{R}^m$ and (y^j) for $\mathbb{R}^n$. According to Proposition 1.15, each k-jet in $J^k(m, n)$ has a unique polynomial representative. Since germs $(\mathbb{R}^m, 0) \to (\mathbb{R}^n, 0)$ can be added and multiplied by scalars, $J^k(m, n)$ has a natural vector space structure. In the case that $k = 1$, this was simply the structure of $L(\mathbb{R}^m, \mathbb{R}^n)$. We may take for basis the k-jets of the following polynomials:

1) polynomials with jth component x^i, all other components being 0, for $i = 1, \ldots, m$ and $j = 1, \ldots, n$;

2) polynomials with jth component

$$\frac{x^\alpha}{\alpha!} = \frac{(x^1)^{\alpha_1} \ldots (x^m)^{\alpha_m}}{\alpha!}, \qquad |\alpha| = 2,$$

all other components being 0, for $j = 1, \ldots, n$;

⋮

k) polynomials with jth component

$$\frac{x^\alpha}{\alpha!}, \qquad |\alpha| = k,$$

all other components being 0, for $j = 1, \ldots, n$.

This follows fairly easily from the general form of polynomial representatives given in equation (1.14). The dimension of $J^k(m, n)$ is

$$\sum_{i=1}^{k} n \binom{m+i-1}{i}$$

and there is a *unique* topology for $J^k(m, n)$ determined by any Banach space structure. Choose, for convenience, the Euclidean norm for $J^k(m, n)$ with respect to the basis above.

Thus, for the frames ϕ and θ assigned above, we may write

$$T_{\phi\theta}(f_k) = \left(\left(\sum_{1 \le |\alpha| \le k} \frac{\partial^{|\alpha|}(\theta^{-1} \circ f \circ \phi)}{\partial x^\alpha}\bigg|_{z_0} \cdot \frac{x^\alpha}{\alpha!} \right)_k, (\phi(z_0), f(\phi(z_0))) \right)$$

$$\text{for } z_0 \in O. \quad (1.15)$$

It is clear that $T_{\phi\theta}$ defined in this way gives a *bijection* from $J^k(U, V)$ to $J^k(m, n) \times U \times V$. Topologize $J^k(M, N)$ with the *smallest* topology which makes the $J^k(U, V)$ open and *all* maps $T_{\phi\theta}$ continuous. In order to show that the $T_{\phi\theta}$ give an atlas of charts for smooth manifold structure on $J^k(M, N)$, we must consider *another* pair of frames $\phi': W \to U' \subset M$ and $\theta': W' \to V' \subset N$. Suppose also that $U \cap U' = U''$ and $V \cap V' = V''$. Then $T_{\phi'\theta'} \circ T_{\phi\theta}^{-1}$ maps $J^k(m, n) \times U'' \times V'' \to J^k(m, n) \times U'' \times V''$ and has the form

$$(A_k, (x_0, y_0)) \to (B(x_0, y_0)_k \circ A_k \circ C(x_0, y_0)_k, (x_0, y_0))$$

where $A_k \in J^k(m, n)$, $B(x_0, y_0)_k \in J^k(n, n)$ and $C(x_0, y_0)_k \in J^k(m, m)$, respectively, and they depend *smoothly* on (x_0, y_0). Since $T_{\phi'\theta'} \circ T_{\phi\theta}^{-1}$ is smooth, we see that $J^k(M, N)$ has the structure of a smooth manifold. Also, the maps $T_{\phi\theta}^{-1}$ are easily seen to provide appropriate local trivialization so that p_k is the projection of a fiber bundle with fiber $J^k(m, n)$.

If $J^k(m, n)$ is thought of as the vector space of polynomials truncated at degree k, then it is easy to see that the maps $T_{\phi'\theta'} \circ T_{\phi\theta}^{-1}$ do *not* induce *linear* maps of the fiber but give instead *polynomial* mappings. Therefore, we shall not be able to endow $J^k(M, N)$ with vector bundle structure for $k > 1$. We shall call $T_{\phi\theta}$ the *Taylor map* with respect to frames ϕ and θ.

Suppose we call $M \times N$ by the label $J^0(M, N)$. There is a sequence of smooth surjections

$$\begin{array}{l} \vdots \\ \downarrow \\ J^k(M, N) \\ \downarrow p_{k,k-1} \\ J^{k-1}(M, N) \\ \downarrow p_{k-1,k-2} \\ \vdots \\ \downarrow p_{3,2} \\ J^2(M, N) \\ \downarrow p_{2,1} \\ J^1(M, N) \\ \downarrow p_{1,0} \\ J^0(M, N) = M \times N \end{array}$$

where $p_k = p_{1,0} \circ p_{2,1} \circ \cdots \circ p_{k,k-1}$ and where *each* $p_{k,k-1}$ is the projection of a fiber bundle. The fiber of $p_{k,k-1}$ will easily be seen to be a vector space of dimension $\binom{m+k-1}{k} n$ (the homogeneous polynomials of degree k). This tower of jet bundles represents a way to materialize finer and finer structure in smooth map germs.

One useful construction is that of the k-*jet of a smooth map* $f: M \to N$. To each $x_0 \in M$, associate the k-jet of the germ of f at x_0 (that is, of the germ $f:(M, x_0) \to (N, f(x_0))$. This defines a map $j^k(f): M \to J^k(M, N)$ such that $p_k \circ j^k(f) = \mathrm{id} \times f$.

We shall conclude with a brief discussion of the covariant functor J^k. Recall that the category of pairs of smooth manifolds has objects (M, N) and morphisms (h, g) where $h: M \to M'$ is a smooth diffeomorphism and $g: N \to N'$ is an arbitrary smooth map. Then (h, g) is a morphism from (M, N) to (M', N'). Now J^k is the following functor from the category of pairs to the category of fiber bundles and bundle mappings. $J^k(M, N)$ is the k-jet bundle of the same name and, if (h, g) is a morphism from (M, N) to (M', N'), then $J^k(h, g): J^k(M, N) \to J^k(M', N')$ by the rule $J^k(h, g): f_k \to (g \circ f \circ h^{-1})_k$. This is easily seen to be a bundle map over $h \times g: M \times N \to M' \times N'$.

The last topic will be the construction of the higher-order frame bundles. These will be the generalization of the 1-frame bundles which were used to define all of the tensors. The principal bundles which we shall construct will then be shown to be the support of geometric objects which are interesting generalizations of tensors.

Exercises

6. Give the explicit law of composition in $J^2(1, 1)$. This will be a 'second-order' version of the 'chain rule'.
7. A cross section of the 1-frame bundle $\pi_1: F_1(M) \to M$ is called a 'moving frame' on M (H. Cartan [6] or E. Cartan [5]). Describe such a section in components *locally* by defining a local coordinate frame for $F_1(M)$ in terms of a frame on M.

Let M be a smooth m-dimensional manifold, and let E denote $\mathbb{R}^m$ with its standard linear coordinates (x^i), $i = 1, 2, \ldots, m$. In §1.2 we defined the set of invertible germs $\alpha: (E, 0) \to (M, x)$ and denoted that set $F^x(M)$. The *group* of invertible 1-jets of germs $(E, 0) \to (E, 0)$ was denoted $Gl(m)$, and the group of invertible *germs* $(E, 0) \to (E, 0)$ was denoted $\mathcal{G}$.

In this section, we turn attention to the group of k-jets of invertible germs in $\mathcal{G}$. Again, the inverse function theorem guarantees that a germ in $\mathcal{G}$ is invertible if and only if its k-jet is invertible. This follows from the following consideration. Whenever $h \geq k$ there is a homomorphism of *algebras* $J^h(m, m) \to J^k(m, m)$ which is easily seen to be surjective, obtained essentially by truncation. $J^1(m, m)$ is the $m \times m$ matrix algebra and, in general, the (non-commutative) multiplication is given by jet composition. Now denote the *group* of invertible elements of $J^k(m, m)$ by $G_k(m)$. This homomorphism gives by restriction a surjective group homomorphism $G_h(m) \to G_k(m)$. Therefore an invertible element of $G_k(m)$ has an invertible Jacobian, and the result follows.

Thus $G_k(m)$ is a generalization of $Gl(m)$. The elements of $G_k(m)$ may be thought of as truncated polynomials $(E, 0) \to (E, 0)$ whose degree-1 part is an invertible linear map. With its inherited topology, $G_k(m)$ is a Lie group.

Now let $F_k^x(M)$ denote the set of k-jets of invertible germs $\alpha_k : (E, 0) \to (M, x)$ and let $F_k(M)$ denote the union of the $F_k^x(M)$ over all x in M. There is a surjection $\pi_k : F_k(M) \to M$ which associates with each jet its target.

$G_k(m)$ acts on $F_k^x(M)$ *on the right* in the following way. If $\alpha_k \in F_k^x(M)$, and $g_k \in G_k(m)$ say $\alpha_k g_k = (\alpha \circ g)_k$. This definition is entirely analogous to the one given in §1.2 where k was unity.

The action of $G_k(m)$ on $F_k^x(M)$ satisfies the following crucial condition. If α_k and β_k belong to $F_k^x(M)$, then there is a *unique* $g_k \in G_k(m)$ such that $\alpha_k g_k = \beta_k$. The action is simply transitive.

The right action of $G_k(m)$ on the individual $F_k^x(M)$ can be 'globalized' to give a fiber-preserving right action of $G_k(m)$ over $F_k(M)$. We shall topologize $F_k(M)$ in such a way as to make it a smooth manifold, in such a way that the right action of $G_k(m)$ is smooth, and in such a way that the projection $\pi_k : F_k(M) \to M$ is the projection of a *principal bundle* with group $G_k(m)$. $F_k(M)$ is the total space of what we shall call the *bundle of k-frames* on M.

The topology for $F_k(M)$ is constructed in the following way. Suppose that $\phi : O \to U \subset M$ is a frame. Then define the map $\tilde{\phi} : G_k(m) \times U \to \pi_k^{-1}(U)$ by $\tilde{\phi}(g_k, x) = (\phi \circ t_z \circ g)_k$ where $\phi(z) = x$, g is the germ of a frame at x, that is g belongs to $F^x(M)$, and t_z is the germ of the translation $(E, 0) \to (E, z)$. It is clear that $\tilde{\phi}$ is bijective, and that $\pi_k \tilde{\phi}(g_k, x) = x$ for all x in U.

Now suppose that $\theta : W \to V$ is another frame on M, and suppose that $\tilde{\theta}(g_k, x) = \tilde{\phi}(h_k, x)$. Then if $\phi(z_1) = \theta(z_2) = x$, we have $g_k = (t_{z_2}^{-1} \circ \theta^{-1} \circ \phi \circ t_{z_1})_k h_k$ since $(\phi \circ t_{z_1} \circ h)_k = (\theta \circ t_{z_2} \circ g)_k$.

Thus, as in the case studied in §1.2, g_k is a left multiple of h_k by an element of $G_k(m)$, and this element varies smoothly with x in $U \cap V$. Thus if $F_k(M)$ is given the *smallest* topology which makes all $\pi_k^{-1}(U)$ open and for which all $\tilde{\phi}^{-1} : \pi_k^{-1}(U) \to G_k(m) \times U$ are continuous, then it has the structure of a smooth manifold and π_k is the projection of a fiber bundle. According to the criterion stated in §1.2, this bundle is a principal $G_k(m)$ bundle via the right action of this Lie group on $F_k(M)$.

We observed earlier that for a smooth manifold M, the collection of all frame-germs $F(M)$ is universal in the sense that, via the right action of $\mathcal{G}$, it materializes all local changes of frame at each point of M. Unfortunately it is too big, and there is no obvious way to topologize either it *or* $\mathcal{G}$. We use the following device to 'approximate' $F(M)$ with a tower of fiber bundles.

Proposition 1.17. Let M be a smooth manifold. The *tower of frame bundles* is the sequence of smooth manifolds and surjective maps:

$$\begin{array}{c} \vdots \\ \downarrow \\ F_k(M) \\ \downarrow \pi_{k,k-1} \\ F_{k-1}(M) \\ \downarrow \pi_{k-1,k-2} \\ \vdots \\ \downarrow \\ F_2(M) \\ \downarrow \pi_{2,1} \\ F_1(M) \\ \downarrow \pi_1 \\ M \end{array}$$

The map $\pi_{k,k-1}$ associates to each k-jet of a frame the $(k-1)$ jet to which it belongs, and $\pi_k = \pi_1 \circ \pi_{2,1} \circ \ldots \circ \pi_{k,k-1}$. *Each* map $\pi_{k,k-1}: F_k(M) \to F_{k-1}(M)$ is a *principal bundle projection* with fiber isomorphic to the *additive group of* homogeneous polynomials of degree k taking $(\mathbb{R}^m, 0) \to (\mathbb{R}^m, 0)$.

Proof. All that needs proving is the last sentence. We begin by observing that $\pi_{k,k-1}$ is the *quotient map* obtained by associating to each k-jet in $F_k(M)$ its orbit under the action of the following invariant subgroup of $G_k(m)$. Let $\gamma_k(m)$ be the *kernel* of the projection $G_k(m)$ to $G_{k-1}(m)$ given by truncation. Then $\pi_{k,k-1}(\alpha_k) = \pi_{k,k-1}(\beta_k)$ if and only if $\alpha_k = \beta_k g_k$ for some g_k in $\gamma_k(m)$.

We argue that if α_k and β_k have the same $(k-1)$ jet, then they are in the same fiber for π_k and so $\alpha_k = \beta_k g_k$ for some g_k in $G_k(m)$. Then $\alpha_{k-1} = \beta_{k-1} g_{k-1}$ in $F_{k-1}(M)$. And from the *uniqueness* of g_{k-1} as solution to this equation, it follows that $g_{k-1} = \mathrm{id} \in G_{k-1}(m)$. The other direction is trivial.

Now

$$\begin{array}{c} F_k(M) \\ \downarrow \pi_k \\ M \end{array}$$

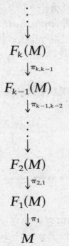

is the projection of a principal bundle with group $G_k(m)$. This implies that M can be covered with open sets U, where $\phi: O \to U$ is a frame, and there is a diffeomorphism $\tilde{\phi}: G_k(m)\times U \to \pi_k^{-1}(U)$ with the property that for *any* α in $G_k(m)$, $\tilde{\phi}(x\alpha, b) = \tilde{\phi}(x, b)\alpha$.

Since $\gamma_k(m) \in G_k(m)$ is the kernel of the projection $G_k(m) \to G_{k-1}(m)$ given by truncation, it is clear that $\gamma_k(m)$ is a *closed, invariant* Lie subgroup of $G_k(m)$. Standard arguments (Steenrod [23], Sternberg [24]) give a general proof of the conclusion that we shall draw in this special case.

$\pi_{k-1}^{-1}(U)$ is open in $F_{k-1}(M)$ for U as above. Also, $\pi_{k,k-1}^{-1}(\pi_{k-1}^{-1}(U))$ is equal to $\pi_k^{-1}(U)$. What we need, then, is a *diffeomorphism*

$$T: \gamma_k(m)\times \pi_{k-1}^{-1}(U) \to \pi_k^{-1}(U)$$

with the properties:

a) $\pi_{k,k-1}(T(g_k, \bar{x})) = \bar{x}$ for $\bar{x} \in \pi_{k-1}^{-1}(U)$, $g_k \in \gamma_k(m)$,

b) $T(g_k\gamma, \bar{x}) = T(g_k, \bar{x})\gamma$ for $\gamma \in \gamma_k(m)$.

In order to define T, we refer to the map $\tilde{\phi}$ above, where $\tilde{\phi}: G_k(m)\times U \to \pi_k^{-1}(U)$, and to the map, similarly defined at the $(k-1)$ level, $\tilde{\theta}: G_{k-1}(m)\times U \to \pi_{k-1}^{-1}(U)$, and remember that the elements of $G_k(m)$ may be thought of simply as degree-k truncated polynomials in standard coordinates, their structure being independent of any particular manifold M that they are used to k-frame.

Now define the following smooth mapping:

$$\tau: \gamma_k(m)\times G_{k-1}(m) \to G_k(m)$$

by the rule

$$\tau(g_k, p_{k-1}) = p_{k-1}\circ g_k$$

where $p_{k-1} \in G_{k-1}(m)$ is a polynomial of degree $(k-1)$ and $g_k \in \gamma_k(m) \subseteq G_k(m)$ is a degree-k polynomial which may be represented in the form

$$g_k = \delta_i^j x^i + \sum_{|\alpha|=k} A_\alpha^j \frac{x^\alpha}{\alpha!} \tag{1.16}$$

where the superscript j indicates the jth component of the 'vector-valued' polynomial, δ_i^j is as usual the Kronecker delta, and $A_\alpha^j = A_{\alpha_1\alpha_2\ldots\alpha_m}^j$, a scalar coefficient for each choice of α. This mapping is *not* a homomorphism, but it *is* bijective, since the map gives, essentially, a right coset decomposition of $G_k(m)$.

We use this mapping together with $\tilde{\phi}$ and $\tilde{\theta}$ to define T as the composition

$$\gamma_k(m)\times \pi_{k-1}^{-1}(U) \xrightarrow{1\times\tilde{\theta}^{-1}} \gamma_k(m)\times(G_{k-1}(m)\times U) \xrightarrow{\tau\times 1} G_k(m)\times U \xrightarrow{\tilde{\phi}} \pi_k^{-1}(U). \tag{1.17}$$

It is a straightforward verification to see that conditions a) and b) are satisfied for this definition of T. And since this is a composition of diffeomorphisms, T is a diffeomorphism.

Now to finish the proof, observe that if g_k presented in the form of equation (1.16) is associated to the homogeneous degree-k polynomial

$$\sum_{|\alpha|=k} A^j_\alpha \frac{x^\alpha}{\alpha!},$$

then *composition* of k-jets corresponds to *addition* of the associated polynomials. ■

Although the fiber of $\pi_{k,k-1}$ is group isomorphic to a Euclidean space of dimension $m\binom{m+k-1}{k}$, the bundle is *not* a vector bundle. It has the structure of what we shall call an *affine bundle.* It will be seen later that the bundle $\pi_{2,1}: F_2(M) \to F_1(M)$ plays a central role in the definition of *affine connection.*

Definition 1.18. An m-dimensional *affine space* is a set X together with a left or right action of the additive group of $E = \mathbb{R}^m$ (owing to the commutativity of the latter group, we make no distinction between its left and right actions on sets) where that action satisfies the following condition:

for any x_1 and x_2 in X there is a *unique* $\alpha \in E$ such that $x_1\alpha = x_2$.

Fixing any point in X as origin, x_0, there is then a bijective correspondence between points in X and points in E. Associate to each $x \in X$ the unique element of E that carries x_0 to x. If there is need to topologize X, then topologize it in such a way that this correspondence is a homeomorphism; for smooth structure, insist that it be a diffeomorphism.

In fact, the standard affine space of dimension m is simply E itself with the group action being given by addition in E. In general, however, there is no natural choice for 'origin' in X, so the affine structure is weaker than the group structure which X inherits through choice of origin. ■

The 'affine' structure on an affine space is easy to picture as a group of translations (the group being isomorphic to E), and an affine map is then a pair (A, f) where $A: X \to Y$, and f is a linear map $E \to E$ such that $A(x\alpha) = A(x)f(\alpha)$ for all $x \in X$ and $\alpha \in E$.

Definition 1.19. Let M be a smooth manifold, and $p: W \to M$ the projection of a fiber bundle with fiber X. Then p is the projection of an *m-dimensional affine bundle* if (1) X is an m-dimensional affine space, and (2) if $\tilde{\phi}: X \times O \to p^{-1}(O)$ and $\tilde{\theta}: X \times O' \to p^{-1}(O')$ are a pair of local trivializations for open sets O, O' in M, then if $b \in O \cap O'$, the diffeomorphism $\tilde{\theta}^{-1} \circ \tilde{\phi} : X \times b \to X \times b$ is a *translation* by an element of E via the obvious identification of $X \times b$ with X.

Under these conditions, *each* fiber $p^{-1}(b)$ inherits the structure of m-dimensional affine space. [Note that if t_α is the translation of X given by $\alpha \in E$ then (t_α, id) is an affine map.] Now the affine structure on the fiber is given in this way. Let $y \in p^{-1}(b)$ and suppose $y = \tilde{\phi}(x, b)$. *Define* $y\alpha = \tilde{\phi}(x\alpha, b)$ for $\alpha \in E$. This clearly gives $p^{-1}(b)$ an affine structure. We must, of course, show that it is well-defined. Thus, suppose that for another frame θ, $y = \tilde{\theta}(x', b)$. The relation between x and x' for this b is, by our assumption, $x' = x\beta$ for some fixed β in E. Thus, with respect to this frame, $y\alpha = \tilde{\theta}(x'\alpha, b) = \tilde{\theta}(x\beta\alpha, b) = \tilde{\theta}(x\alpha\beta, b)$. Now $\tilde{\theta}(x\alpha\beta, b) = \tilde{\phi}(x\alpha, b)$. ■

Finally, we show that the projections $\pi_{k,k-1}$ are affine bundle projections. Accepting that they are the projections of principal bundles with group $\gamma_k(m)$, then we must show that each fiber inherits the structure of $m\binom{m+k-1}{k}$-dimensional affine space. We argue, in general, that if $p : W \to M$ is the projection of a principal bundle with group G, then if $\tilde{\phi} : G \times O \to p^{-1}(O)$ and $\tilde{\theta} : G \times O' \to p^{-1}(O')$ are the corresponding local trivializations for open sets O, O' in M, then for $b \in O \cap O'$, $\tilde{\theta}^{-1} \circ \tilde{\phi} : G \times b \to G \times b$ is via the identification of $G \times b$ with G, *left* translation by an element of G: $\tilde{\theta}^{-1} \circ \tilde{\phi}(g, b) = (\gamma g, b)$ for some γ in G which depends only on b.

Suppose, then, that $\tilde{\phi}(x, b) = \tilde{\theta}(x', b)$ in $p^{-1}(b)$. Then for a unique γ in G, we have $x' = \gamma x$. Now, according to the additional criterion which $\tilde{\phi}$ and $\tilde{\theta}$ must satisfy for a principal bundle (commutativity with the action of G), for *any* $\alpha \in G$, $\tilde{\phi}(x\alpha, b) = \tilde{\phi}(x, b)\alpha = \tilde{\theta}(x'\alpha, b)$, and so $\gamma(x\alpha) = x'\alpha$. Thus, γ does not depend on x. It follows that $\tilde{\theta}^{-1} \circ \tilde{\phi}(g, b) = (\gamma g, b)$ for all $g \in G$.

Exercises

8. Let $M = \mathbb{R}$ and let $I : \mathbb{R} \to \mathbb{R}$ be the *identity* global frame on $\mathbb{R}$ ($I(x) = x$). Denote $G_1(1)$: $\mathbb{R}^*$ (non-zero reals) via the identification of a linear map with its representation as scalar multiplication. Let $\tilde{I}_1 : \mathbb{R}^* \times M \to F_1(M)$ by the rule $\tilde{I}_1[g_1, x] = (t_x \circ g)_1$ which is the 1-jet at 0 of the *affine map* $t \to x + gt$ (where g is the non-zero scalar identified with the 1-jet g_1 in $G_1(1)$).

   Now consider $\pi_2 : F_2(\mathbb{R}) \to F_1(\mathbb{R})$. Let an element of $G_2(1)$, say g_2, be represented by its polynomial representative

$$g_2(t) = gt + ht^2/2 \qquad \text{for } g \in \mathbb{R}^*,\ h \in \mathbb{R}.$$

Thus an element of $\gamma_2(1)$ will have the representation

$$\gamma_2(t) = t + ht^2/2, \qquad h \in \mathbb{R}.$$

Now represent the element $gt + ht^2/2$ of $G_2(1)$ as an ordered pair (g, h) in $\mathbb{R}^* \times \mathbb{R}$ for simplicity.

a) In these components, describe the group multiplication in $G_2(1)$.
b) The elements of the form $(1, h)$ form an *invariant subgroup*; describe the effect of inner automorphism on $\gamma_2(1)$ by an arbitrary element of $G_2(1)$.

9. In the setup for exercise 8 construct the trivialization $\tilde{I}_2 : G_2(1) \times M \to F_2(M)$ following Proposition 1.17 by observing that each element g_2 of $G_2(1)$ has unique representation as a product $(g, 0)(1, h)$, and defining $\tilde{I}_2[(g, 0)(1, h), x]$ to be the polynomial mapping $t \to x + gt + ght^2/2$. Now with respect to these identifications, let $\pi_{2,1}$ be presented as the map $(x + gt + ght^2/2) \to (x + gt)$ on the fiber over $x \in \mathbb{R}$. Show that there is a well-defined way to associate a simply transitive action of $\mathbb{R}$ on the fiber of $\pi_{2,1}$. Give a coordinate-independent interpretation for that action (that is, independent of the initial choice of global frame I).

§1.4. Tangent sectors and sector bundles

We shall use the term *tangent sector* or *sector* to denote the geometric object which generalizes the tangent vector. A sector is an element of an *iterated* tangent bundle $T_1[T_1[\ldots T_1(M)\ldots]]$. Following the procedure used in the first two sections to define the tangent bundle, we construct a *graded set* of sector bundles associated with each smooth manifold M. This association will give a functor from the category of smooth manifolds to the category of 'simplicial fiber bundles' which extends the tangent functor.

In order to motivate the discussion which follows, we give an intuitive description of the 1-*sectors* and the 2-*sectors* on a smooth manifold M. First of all, if $x_0 \in M$, then a 1-sector at x_0 is simply a *tangent vector* at $x_0 : [\phi_1, X^i]$ for frame-germ ϕ at x_0. Thus a 1-sector at x_0 is simply the 1-jet of a germ $\alpha : (\mathbb{R}, 0) \to (M, x_0)$ and α_1 may be thought of as an element of $T_{x_0}(M)$.

Now a 2-sector at $x_0 \in M$ will be an element of $T_{\alpha_1}[T_1(M)]$ (where $\alpha_1 \in T_{x_0}(M)$). Since $T_1(M)$ is a smooth manifold, it makes sense to speak of its tangent bundle. In particular, if M is m-dimensional, the projection

$$\begin{array}{c} T_1[T_1(M)] \\ \big\downarrow q_{2,1} \\ T_1(M) \end{array}$$

is the projection of a vector bundle with $2m$-dimensional fiber. $T_1[T_1(M)]$ is then a $4m$-dimensional manifold. The *composed* projection

$$\begin{array}{c} T_1[T_1(M)] \\ \downarrow q_{2,1} \\ T_1(M) \\ \downarrow q_1 \\ M \end{array} \qquad q_1 \circ q_{2,1} : T_1[T_1(M)] \to M$$

is *not* a vector bundle projection (exercise 2). It is the projection of a fiber bundle with fiber a $3m$-dimensional manifold. The definition of 2-sectors will make this clear. We can, at this stage, give a *heuristic* account of the structure of the fiber of the projection $q_1 \circ q_{2,1}$.

Suppose that α_1 belongs to $T_{x_0}(M)$ and suppose that $\phi : O \to U$ is a frame such that $\phi(z_0) = x_0$. Then α_1 may be represented as $[(\phi \circ t_{z_0})_1, X^i(z_0)]$. Now we want to define an element of $T_{\alpha_1}(T_1(M))$, a tangent vector *on* $T_1(M)$ with *source* α_1. Thus, we require a *germ* $(\mathbb{R}, 0) \to (T_1(M), \alpha_1)$.

To obtain such a germ, consider a germ $A : (\mathbb{R}^2, 0) \to (M, x_0)$. The standard linear coordinates for $\mathbb{R}^2$ we call (t^1, t^2). Then $\phi^{-1} \circ A$ is the germ of a smooth map $(\mathbb{R}^2, 0) \to (\mathbb{R}^m, z_0)$. Now choose A so that

$$\frac{\partial(\phi^{-1} \circ A)^i}{\partial t^1} = X^i.$$

[Recall that unsubscripted partial derivatives are evaluated at '0'.] Then A defines a *path* of tangent vectors parametrized by t^2 for t^2 close to 0:

$$t^2 \to \alpha(t^2)_1$$

where $\alpha(t^2)_1$ is the tangent vector on M,

$$\left[(\phi \circ t_{\phi^{-1} \circ A(0, t^2)})_1, \frac{\partial(\phi^{-1} \circ A)^i}{\partial t^1}_{|(0, t^2)} \right].$$

In particular, $\alpha_1 = \alpha(0)_1$ in this notation. This device simply makes use of the natural isomorphism (where A^B is the set of maps from B to A) $A^{B \times C} \cong (A^C)^B$ interpreted at the level of smooth germs, and then at the level of 1-jets. This is why the discussion remains heuristic at this point. Even if A, B, and C are smooth manifolds, A^C has no obvious manifold structure.

Thus the choice of smooth germ $A : (\mathbb{R}^2, 0) \to (M, x_0)$ such that $\alpha_1 = T_0(A)(\partial/\partial t^1)$ gives, intuitively, the germ of a *path of tangent vectors on M:*

$$T_{(0, t^2)}(A)\left(\frac{\partial}{\partial t^1}\right) = \alpha(t^2)_1,$$

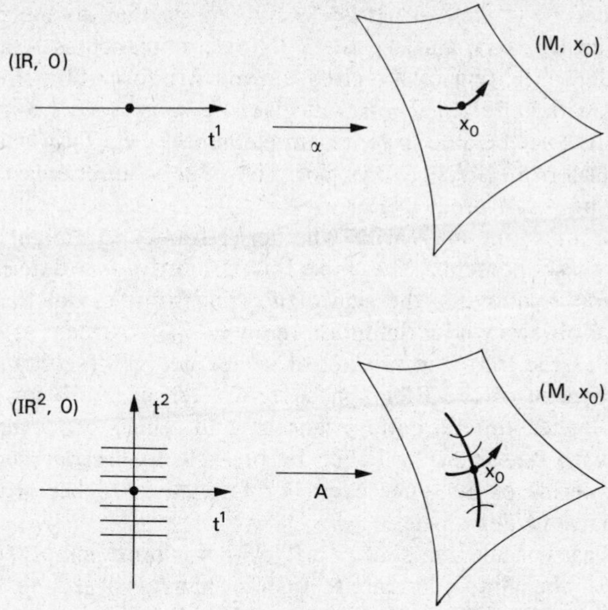

Fig. 1.3. A as 'path-germ' (parametrized by t^2) of path-germs

a path with source the tangent vector α_1. It is clear that we need *much less* than the data of a germ A to specify an element of $T_{\alpha_1}(T_1(M))$: different germs will specify the same element. In fact, referring to the frame ϕ, the tangent vector $[(\phi \circ t_z)_1, X^i(z)]$ is entirely specified by the element (z, X^i) of $O \times \mathbb{R}^m$.

If α_1 is then identified with the point $(z_0, X^i(z_0))$, then an element of $T_{\alpha_1}(T_1(M))$ will be determined by the $2m$ numbers $\partial z^j/\partial t^2$, $\partial X^i/\partial t^2$ where $\alpha(t^2)_1$ is identified with $(z^j(t^2), X^i(z^j(t^2)))$ for small t^2. Now with respect to frame ϕ, $\partial z^j/\partial t^2$ is just $\partial(\phi^{-1} \circ A)/\partial t^2$ and $\partial X^i/\partial t^2$ is just $\partial^2(\phi^{-1} \circ A)/\partial t^1 \delta t^2$.

We therefore see that *given frame-germ* $\phi:(O, z_0) \to (M, x_0)$, then a smooth germ $A:(\mathbb{R}^2, 0) \to (M, x_0)$ determines an element of $(q_1 \circ q_{2,1})^{-1}(x_0)$ of $T_1(T_1(M))$ by the data of the $3m$ numbers:

$$\frac{\partial(\phi^{-1} \circ A)}{\partial t^1} \qquad \text{(specifying } \alpha_1\text{)},$$

$$\frac{\partial(\phi^{-1} \circ A)}{\partial t^2},$$

$$\frac{\partial^2(\phi^{-1} \circ A)}{\partial t^1 \partial t^2}$$

(all evaluated at $(t^1, t^2) = 0$). It is clearly enough then to specify ϕ_2, the 2-jet of a frame at x_0, and A_2, the 2-jet of a map such as A above. But *even* the determination of A_2 gives too much information since, for ϕ_2 fixed, A's with different 2-jets will give the same set of $3m$ numbers. Thus, we shall *not* be able to represent elements of $T_1(T_1(M))$ as 2-jets of mappings (say from $(\mathbb{R}^2, 0) \to (M, x_0)$). They will be 'quotients' of 2-jets in a sense to be made precise later.

Now the question may arise whether 1-frames ϕ_1 might suffice to determine these elements. The 'mixed partial derivative' determining the last m numbers above is the sign of possible trouble, but there will be examples of objects whose definition requires higher-order partial derivatives (such as the local Lie bracket of vector fields, the coboundary of a differential form, or the Riemann curvature tensor itself) for which the effects of higher differentiation cancels, and which can eventually be identified with tensors. It will then be possible to *describe* such objects entirely in terms of 1-frames even if reference to higher-order frames must be made to *define* them.

In the case of elements of $T_1(T_1(M))$ we must simply check the behavior of the $3m$ components under changes of frame to decide whether or not those elements should be classified as tensors over M. We do that now.

Let $\phi^{-1} \circ A$ be as above, and suppose that $\theta : (\mathbb{R}^m, 0) \to (M, x_0)$ is the germ of another frame at x_0. Then $\theta^{-1} \circ A = (\theta^{-1} \circ \phi) \circ (\phi^{-1} \circ A)$. Computing this gives

$$\frac{\partial(\theta^{-1} \circ A)}{\partial t^i} = \frac{\partial(\theta^{-1} \circ \phi)}{\partial x^j}\bigg|_{z_0} \frac{\partial(\phi^{-1} \circ A)^j}{\partial t^i}, \qquad i = 1, 2 \tag{1.18}$$

and

$$\frac{\partial^2(\theta^{-1} \circ A)}{\partial t^1 \partial t^2} = \frac{\partial^2(\theta^{-1} \circ \phi)}{\partial x^j \partial x^k}\bigg|_{z_0} \frac{\partial(\phi^{-1} \circ A)^j}{\partial t^1} \frac{\partial(\phi^{-1} \circ A)^k}{\partial t^2} + \frac{\partial(\theta^{-1} \circ \phi)}{\partial x^j}\bigg|_{z_0} \frac{\partial^2(\phi^{-1} \circ A)^j}{\partial t^1 \partial t^2}. \tag{1.19}$$

Equations (1.18) look promising. The first two m-tuples transform individually like contravariant tensors. But equation (1.19) fails to conform. In particular, the *first* summand on the right requires the 2-jet of the coordinate change $\theta^{-1} \circ \phi$. Thus, coordinate changes with different 2-jets but with the same 1-jet will yield different values for the components of an element of $T_1(T_1(M))$ in general. Intuitively, then, the elements of $T_1(T_1(M))$ require the data of the 2-jet of a frame at a point x_0 of M, *and* they require the data of $3m$ components which can be interpreted as the left-hand sides of equations (1.18) and (1.19) above for some germ $A : (\mathbb{R}^2, 0) \to (M, x_0)$. It is clear that the projection $q_1 \circ q_{2,1}$ is not the projection of a vector bundle since the coordinate changes given by (1.18) and (1.19) will in general be polynomial mappings of degree 2 on fibers, and not linear mappings. Finally, it is clear from this intuitive treatment that notation will very quickly become strained as this construction is generalized (for example to $T_1(T_1(T_1(M)))$) because of the asymmetry implicit in the stepwise construction of these bundles as sequences of tangent bundles. This asymmetry is illusory and disappears from the viewpoint which we shall develop shortly. (For a nice discussion of the structure of the bundle $T_1(T_1(M))$, see Abraham and Marsden [1].)

Exercise

10. Let 0_x denote the zero tangent vector at $x \in M$. Calculate the coordinate change for elements of $T_{0_x}[T_1(M)]$. In a similar fashion, consider the projection $q_1 : T_1(M) \to M$. Say that an element of $T_{\alpha_1}[T_1(M)]$ is *vertical* (Yano and Ishihara [27], or Abraham and Marsden [1]) if its image under $T_1(q_1)$ is zero. Calculate the coordinate change formula for a vertical element of $T_{\alpha_1}[T_1(M)]$.

The 1-sectors at $x_0 \in M$ are simply the tangent vectors at x_0. On the other hand, the 2-sectors generalize the tangent vector construction (§1.2) in several directions. It will be convenient to defer the general definition of k-sectors until the nature of the generalization required for the case $k=2$ is understood. Thus, we precede the most general treatment with a close study of the case of 2-sectors, the elements of $T_1(T_1(M))$.

Let M be a smooth manifold and let

$$\begin{array}{c} F_2(M) \\ \downarrow \pi_2 \\ M \end{array}$$

be the 2-frame bundle over M. This is, as we observed, a principal bundle with fiber $G_2(m)$, for m the dimension of M, and where $G_2(m)$ is the group of invertible 2-jets of maps $(E,0) \to (E,0)$ for $E = \mathbb{R}^m$ with standard linear coordinates (x^i). Using equation (1.14), an element $g_2 \in G_2(m)$ has the representation

$$g_2 = \left(A_t^u x^t + \sum_{|\alpha|=2} B_\alpha^u \frac{x^\alpha}{\alpha!}\right) \tag{1.20}$$

for $u, t = 1, 2, \ldots, m$, and A_t^u an invertible $m \times m$ matrix.

For ease of calculation, we present this generic element in a different form. For $A \in L(E, E^*; \mathbb{R})$ and $B \in L(E, E, E^*; \mathbb{R})$, B being *symmetric in the first two variables*, then in the notation of §1.3, $B_{rs}^u = B_{sr}^u$, A_t^u is invertible, and

$$g_2 = (A_t^u x^t + \tfrac{1}{2} B_{rs}^u x^r x^s). \tag{1.21}$$

The right-hand sides of equations (1.20) and (1.21) give the same polynomial representative.

Now $G_2(m)$ acts on the right on $F_2(M)$ by composition of jets. What we wish to do is to define a *left action* of $G_2(m)$ on a $3m$-dimensional real vector space and then to form the 'tensor' bundle in the sense of §1.2.

Definition 1.20. A (standard) k-simplex is the collection of *non-empty* subsets of the set $\{1, 2, \ldots, k, k+1\}$. An i-face of the simplex, $i \leq k$, is a subset with $(i+1)$ elements.

Thus, a k-simplex has $2^{k+1}-1$ faces. It has $\binom{k+1}{i+1}$ i-faces in general. We shall designate an i-face of the k-simplex as an ordered $(k+1)$-tuple $(a_1, a_2, \ldots, a_{k+1})$ with $a_j \in \{0, 1\}$ and $(i+1)$ of the entries equal to unity. This corresponds to the subset containing each j such that $a_j = 1$. The tuple $(0, 0, 0, \ldots, 0)$ is *excluded.* ∎

In particular, the 0-simplex has face (1); the 1-simplex has faces (1, 0), (0, 1), (1, 1); the 2-simplex has faces (1, 0, 0), (0, 1, 0), (0, 0, 1), (1, 1, 0), (1, 0, 1), (0, 1, 1), (1, 1, 1); and so forth.

Now in order to define the 2-sectors on M (of dimension m), let E_2 be the vector space of functions from the 1-simplex to E. Using the standard basis for E, we could represent such an element of E_2, namely $(X^i_{10}, X^k_{11}, X^j_{01})$, as a triple of $m \times 1$ matrices with subscripts for the face to which they correspond. For higher-order sectors, we shall have to adopt a notation of this sort, but for the 2-sectors we leave the subscripts off and allow the *position* of the matrix to indicate its face (X^i——Z^k——Y^j); the 10 face corresponds to the first entry, the 11 face corresponds to the middle entry, and the 01 face corresponds to the third entry. Denote a general element of E_2 by $X(2)$.

Now we define a *left action* of the group $G_2(m)$ on E_2. Suppose then that $X(2) = (X^i$——Z^k——$Y^j) \in E_2$ and let $g_2 = A^u_t x^t + \frac{1}{2}B^u_{rs}x^r x^s \in G_2(m)$. Then define

$$g_2 X(2) = (A^u_i X^i \text{——} A^u_k Z^k + B^u_{ij} X^i Y^j \text{——} A^u_j Y^j). \tag{1.22}$$

Observation 1.21. The rule stated in equation (1.22) gives a left action of $G_2(m)$ on E_2.

Proof. The identity element of $G_2(m)$ is $\delta^u_t x^t$, and this clearly induces the identity map on E_2.

Now suppose that $X(2)$ is as above and that $h_2 = (\bar{A}^u_t x^t + \frac{1}{2}\bar{B}^u_{rs} x^r x^s)$ is another element of $G_2(m)$. Then $h_2 g_2$ is computed by jet composition to be

$$\begin{aligned} &\bar{A}^u_t(A^t_i x^i + \tfrac{1}{2}B^t_{rs}x^r x^s) + \tfrac{1}{2}\bar{B}^u_{vw}(A^v_i x^i + \tfrac{1}{2}B^v_{jk}x^j x^k)(A^w_i x^i + \tfrac{1}{2}B^w_{jk}x^i x^j) \\ &\quad = \bar{A}^u_t A^t_i x^i + \tfrac{1}{2}[\bar{A}^u_t B^t_{rs} + \bar{B}^u_{vw} A^v_r A^w_s] x^r x^s. \end{aligned}$$

Therefore

$$(h_2g_2)X(2)=(\bar{A}^u_tA^t_iX^i\text{——}\bar{A}^u_tA^t_kZ^k \\ +[\bar{A}^u_tB^t_{ij}+\bar{B}^u_{vw}A^v_iA^w_j]X^iY^j\text{——}\bar{A}^u_tA^t_jY^j)$$

and computing $h_2(g_2X(2))$ by equation (1.22) we have

$$h_2(g_2X(2))=(\bar{A}^u_tA^t_iX^i\text{——}\bar{A}^u_tA^t_kZ^k+\bar{A}^u_tB^t_{ij}X^iY^j \\ +\bar{B}^u_{rs}A^r_iA^s_jX^iY^j\text{——}\bar{A}^u_tA^t_jY^j).$$

These are obviously equal. ■

Now let $G_2(m)$ act on the product $F_2(M)\times E_2$ *on the left* by the rule:

$$g_2(\phi_2, X(2))=(\phi_2g_2^{-1}, g_2X(2)).$$

Then if we represent an orbit under this action by a representative element with square brackets $[\phi_2, X(2)]$, we have the relation, just as we did in §1.2, equation (1.1),

$$[\phi_2g_2, X(2)]=[\phi_2, g_2X(2)]. \tag{1.23}$$

Call the collection of orbits with the quotient topology $F_2(M)\times_{G_2}E_2$. Then we shall have a fiber bundle projection

$$\begin{array}{c} F_2(M)\times_{G_2}E_2 \\ \downarrow q_2 \\ M \end{array}$$

and this bundle is finally the bundle of 2-sectors on M. Its fiber is a $3m$-dimensional manifold diffeomorphic to $\mathbb{R}^{3m}=E_2$ but it does not carry linear structure. If ϕ_2 is the 2-jet of a frame at x_0 in M, a 2-sector at x_0 is a point in the fiber over x_0: $[\phi_2, (X^i\text{——}Z^k\text{——}Y^j)]$. The appendix to this chapter (§1.5) will justify our interpretation of 2-sectors as elements of $T_1(T_1(M))$.

This gives the formal definition of a 2-sector. But, as in the case of tangent vectors (see §1.1) there is an informal definition which both motivates and elucidates this construction. We want to think of a 2-sector as an *equivalence class* of 2-jets of maps $(\mathbb{R}^2, 0)\to(M, x_0)$. Thus, $X(2)$ should determine a *class* of 2-jets $(\mathbb{R}^2, 0)\to(\mathbb{R}^m, 0)$, and the composition of elements in this class with $\phi_2:(\mathbb{R}^m, 0)\to(M, x_0)$ should give the desired class.

We turn now to a description of the equivalence relation.

Definition 1.22. Suppose that F is a k-dimensional Euclidean space equipped with linear coordinates (t^i), $i=1,\ldots,k$, $F=\mathbb{R}^k$ with these coordinates. Let M be an m-dimensional manifold.

Suppose that γ_r and $\bar{\gamma}_r$ are r-jets $(\mathbb{R}^k, 0)\to(M, x_0)$. Say that γ_r ana $\bar{\gamma}_r$

are *F-related* if for some frame-jet $\phi_r: (\mathbb{R}^m, 0) \to (M, x_0)$ the *representative polynomials* (1.14) for $(\phi^{-1} \circ \gamma)_r$ and $(\phi^{-1} \circ \bar{\gamma})_r$ have the *same* coefficients for *all* terms $x^\alpha/\alpha!$ where $\alpha = (\alpha_1, \alpha_2, \ldots, \alpha_k)$ and all $\alpha_i \in \{0, 1\}$. (In other words, they agree on all terms for which $\alpha! = 1$.)

It is a straightforward argument in polynomial algebra that if this condition is met for a single frame-jet ϕ_r, then it is met for all frame-jets at x_0. Essentially the same argument will show that if $f: (M, x_0) \to (N, y_0)$ is a smooth germ, and if γ_r and $\bar{\gamma}_r$ are F-related, then $(f \circ \gamma)_r$ and $(f \circ \bar{\gamma})_r$ are also F-related.

Once it is clear that F-relatedness is well-defined, it follows directly that it is an equivalence relation. The important point here is that coordinate changes act on Taylor polynomials *on the left*. F-relatedness does not survive frame changes on $(\mathbb{R}^k, 0)$. ■

Observation 1.23. If $r \geq k$ then γ_r is F-related to $\bar{\gamma}_r$ if and only if γ_k is F-related to $\bar{\gamma}_k$. Also γ_1 is F-related to $\bar{\gamma}_1$ if and only if $\gamma_1 = \bar{\gamma}_1$. ■

If now we set $F = \mathbb{R}^2$ with coordinates (t^1, t^2), then the 2-sectors at a point $x_0 \in M$ are just the classes of F-related 2-jets $(\mathbb{R}^2, 0) \to (M, x_0)$. In particular, for each such class we may choose a 'polynomial' representative as follows.

Associate to $(X^i \text{---} Z^k \text{---} Y^j)$ the polynomial map $(\mathbb{R}^2, 0) \to (\mathbb{R}^m, 0)$ with jth component

$$X^j t^1 + Y^j t^2 + Z^j t^1 t^2. \tag{1.24}$$

Given $g_2: (\mathbb{R}^m, 0) \to (\mathbb{R}^m, 0)$ the 2-jet of a frame change with form

$$g_2 = (A^u_t x^t + \tfrac{1}{2} B^u_{rs} x^r x^s),$$

then

$$g_2(X^j t^1 + Y^j t^2 + Z^j t^1 t^2) = (A^u_j X^j t^1 + A^u_j Y^j t^2 + (A^u_j Z^j + B^u_{rs} X^r Y^s) t^1 t^2).$$

Comparing with equation (1.22) we see that the action of $G_2(m)$ on E_2 is given by composition with F-related classes of maps $(\mathbb{R}^2, 0) \to (\mathbb{R}^m, 0)$. In particular, the expression $[\phi_2 g_2, X(2)] = [\phi_2, g_2 X(2)]$ simply states the commutativity of the diagram of *F-related classes of maps:*

$$\begin{array}{ccccc} & & (E, 0) & \xrightarrow{\phi_2} & (M, x_0) \\ & \overset{g_2\bar{f}}{\nearrow} & \uparrow g_2 & \overset{\phi_2 g_2}{\nearrow} & \\ (F, 0) & \xrightarrow{\bar{f}} & (E, 0) & & \end{array}$$

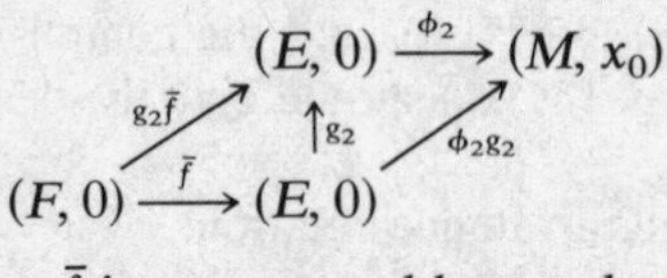

where $\bar{f}$ is represented by a polynomial such as (1.24) and is an F-related *class* of 2-jets, and the same for $g_2\bar{f}$. The 2-sector $[\phi_2 g_2, (X^i \text{---} Z^k \text{---} Y^j)]$ is then the class of maps $\phi_2 \circ g_2 \circ \bar{f}$. This completes the analogy with tangent vectors. We shall refer to these *F-related classes* as *sectors* in the future.

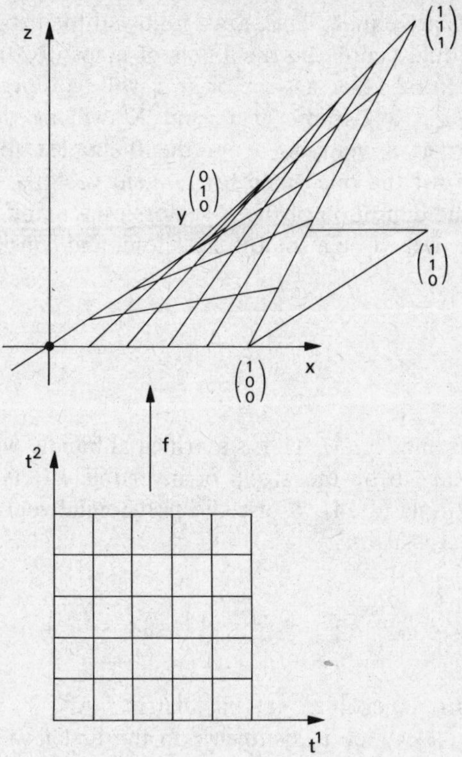

Fig. 1.4. The polynomial $f(t^1, t^2) = (t^1, t^2, t^1t^2)$

We shall generally represent a 2-sector at a point $x_0 \in M$ as a *pair* $[\phi_2, (X^i \text{——} Z^k \text{——} Y^j)]$ where ϕ_2 is a 2-frame at x_0 and the triple $(X^i \text{——} Z^k \text{——} Y^j)$ represents the *F-related class* to which belongs the polynomial $(\mathbb{R}^2, 0) \to (\mathbb{R}^m, 0)$:

$$(t^1, t^2) \xrightarrow{f} (X^i t^1 + Y^i t^2 + Z^i t^1 t^2).$$

Then the 2-sector represented by this pair is the *class* of $\phi_2 \circ f$.

The triples $(X^i \text{——} Z^k \text{——} Y^j)$ belong to the vector space E_2 of maps from the 1-simplex to $E = \mathbb{R}^m$. This vector space is, of course, isomorphic to $\mathbb{R}^{3m}$.

From this, we see the way in which the 2-sectors generalize tangent vectors. E itself is isomorphic to the vector space of maps from the 0-simplex to E (since the 0-simplex has a single face). According to Observation 1.23 if $F = \mathbb{R}$, two jets $(F, 0) \to (M, x_0)$ are F-related if and only if their 1-jets are equal. Thus, if we followed the program above, the 1-sectors at x_0 would simply be the 1-jets of maps $(\mathbb{R}, 0) \to (M, x_0)$, that is, the tangent vectors. Thus, a 1-sector at x_0 will be represented as a pair $[\phi_1, X^i]$ where ϕ_1 is a 1-frame at x_0 and X^i will be thought of as an element of E, or as a mapping from the 0-simplex to E. The transformation law is just the one given for tangent vectors.

We now give the definition of the k-sectors on a manifold for arbitrary positive integer k. Let M be a smooth manifold, and consider the bundle

$$\begin{array}{c} F_k(M) \\ \downarrow{\scriptstyle \pi_k} \\ M \end{array}$$

the bundle of k-frames on M. This is a principal bundle with fiber $G_k(m)$. We consider $G_k(m)$ to be the group of invertible k-jets $(E, 0) \to (E, 0)$, and following formula (1.14) we use the polynomial representation for a general element $g_k \in G_k(m)$:

$$g_k = \left(A^u_t x^t + \tfrac{1}{2} A^u_{t_1 t_2} x^{t_1} x^{t_2} + \ldots + \frac{1}{k!} A^u_{t_1 t_2 \ldots t_k} x^{t_1} x^{t_2 \ldots t_k}\right), \tag{1.25}$$

where A^u_t is an invertible $m \times m$ matrix, $A^u_{t_1 t_2 \ldots t_i}$ an element of $L(E_1, \ldots, E_i, E^*; \mathbb{R})$ which is *symmetric* in the first i variables.

We have, as usual, the *right* action of $G_k(m)$ on $F_k(M)$ by composition of k-jets. In order to define the *bundle of k-sectors on M*, we mimic the treatment given in the cases $k = 1, 2$. First, we define E_k to be the $m(2^k - 1)$-dimensional Euclidean space of mappings from the $(k-1)$ *simplex* to E (recall that the $(k-1)$ simplex has $2^k - 1$ 'elements' or faces).

Now let $\Gamma(k-1)$ be the set of (non-empty) faces of the $(k-1)$ simplex. Then $\gamma \in \Gamma(k-1)$ is a sequence $(\gamma_1, \gamma_2, \ldots, \gamma_k)$ where $\gamma_i \in \{0, 1\}$, not all γ_i zero. We represent an element of E_k by (X^i_γ) where $X^i_\gamma \in E$ is the image of γ for each $\gamma \in \Gamma(k-1)$.

We shall associate to *each* (X^i_γ) in E_k a polynomial representing a k-jet from $(F, 0) \to (E, 0)$ where F is $\mathbb{R}^k$ with standard linear coordinates $(t^1, t^2, \ldots, t^k)$:

$$(t^1, \ldots, t^k) \to \sum_{\gamma \in \Gamma(k-1)} X^i_\gamma t^\gamma \qquad \text{where } t^\gamma = (t^1)^{\gamma_1} \ldots (t^k)^{\gamma_k}$$

as usual. We shall adopt the Einstein summation convention to this to represent the polynomial simply as $X^i_\gamma t^\gamma$.

We are now in a position to give the definition of the bundle of k-sectors on M.

Definition 1.24. Let M be a smooth manifold. Let

$$\begin{array}{c} F_k(M) \\ \downarrow{\scriptstyle \pi_k} \\ M \end{array}$$

be the bundle of k-frames. Let E_k be the $m(2^k-1)$-dimensional vector space of mappings from the $(k-1)$ simplex to E, *i.e.* (X^i_γ), $\gamma \in \Gamma(k-1)$, where we identify (X^i_γ) with the *F-related class* of k-jets $(F, 0) \to (E, 0)$ (for $F = \mathbb{R}^k$ with linear coordinates (t^i)), *i.e.* $X^i_\gamma \equiv X^i_\gamma t^\gamma$.

Now there is the standard right action of $G_k(m)$ on $F_k(M)$ by k-jet composition, which maps fibers of π_k to themselves. Define the following *left* action of $G_k(m)$ on E_k: for g_k in $G_k(m)$ in the form of formula (1.25) and X^i_γ an F-related class of k-jets $(F, 0) \to (E, 0)$ define $g_k \circ X^i_\gamma$ to be the F-related class of the *composition* of k-jets $g_k \circ (X^i_\gamma t^\gamma)$. The *class* of this composition is well-defined as we observed earlier.

There is then a left action of $G_k(m)$ on $F_k(M) \times E_k$ given by $g_k(\phi_k, X^i_\gamma) = (\phi_k g_k^{-1}, g_k X^i_\gamma)$. The set of orbits with respect to this Lie group action will be denoted $T_k(M) = F_k(M) \times_{G_k(m)} E_k$. With the quotient topology, this has a smooth manifold structure and there is a 'factored' projection

$$\begin{array}{c} T_k(M) = F_k(M) \times_{G_k(m)} E_k \\ \downarrow{\scriptstyle q_k} \\ M \end{array}$$

the projection of a fiber bundle with fiber an $m(2^k-1)$-dimensional manifold diffeomorphic to Euclidean $m(2^k-1)$ space.

An element of $T_k(M)$ will be represented as a class $[\phi_k, (X^i_\gamma)]$ where ϕ_k is a k-frame at some $x_0 \in M$, and $X^i_\gamma \in E_k$, and equivalence is defined by the usual relation:

$$[\phi_k g_k, (X^i_\gamma)] = [\phi_k, (g_k X^i_\gamma)]$$

and this equation expresses the commutativity of the diagram of F-related classes of k-jets:

$$\begin{array}{ccccc} & & (E, 0) & \xrightarrow{\phi_k} & (M, x_0) \\ & {\scriptstyle g_k \bar{f}} \nearrow & \uparrow{\scriptstyle g_k} & \nearrow {\scriptstyle \phi_k g_k} & \\ (F, 0) & \xrightarrow{\bar{f}} & (E, 0) & & \end{array}$$

where $\bar{f}$ is the F-related class of k-jet $X^i_\gamma t^\gamma$.

Finally, we observe that if $f: M \to N$ is a smooth map of manifolds, there is induced a map $T_k(f): T_k(M) \to T_k(N)$ such that

$$\begin{array}{ccc} T_k(M) & \xrightarrow[T_k(f)]{} & T_k(N) \\ {\scriptstyle q_k}\downarrow & & \downarrow{\scriptstyle q_k} \\ M & \xrightarrow[f]{} & N \end{array}$$

commutes. The pair $(T_k(f), f)$ is a fiber bundle mapping. To show this, we should follow the procedure used in the case $k=1$ (tangent bundle) and give a *direct* local trivialization of $T_k(M)$ from frames on M. The result will yield that if $\bar{\alpha}:(F,0)\to(M,x_0)$ is the F-related class of a k-jet, then $T_k(f)(\bar{\alpha})=\overline{(f\circ\alpha)}\in T_k(N)_{f(x_0)}$. From this it follows that

$$M\mapsto \begin{array}{c} T_k(M) \\ \downarrow{\scriptstyle q_k} \\ M \end{array} \qquad \text{and} \qquad f\mapsto (T_k(f), f)$$

is a covariant functor from the category of smooth manifolds to the category of fiber bundles: $T_k(\mathrm{id})=\mathrm{id}$ and $T_k(g)\circ T_k(f)=T_k(g\circ f)$. ■

Exercise

11. Show explicitly that 'F-relatedness' is an equivalence relation for $F=\mathbb{R}^k$ for r-jets $(F,0)\to(M,x_0)$.

Now in order to define the canonical simplicial fiber bundle functor, it will be useful to study the properties of a certain (simplicial) vector space. For each $k\geq 1$, E_k is a Euclidean space of dimension $m(2^k-1)$; and for fixed m these data provide us with a certain *graded vector space* $(E_1, E_2, \ldots, E_k, \ldots)$.

For each k, the elements of E_k may be thought of in two ways:

1) They are the *mappings* from $\Gamma(k-1)$, the $(k-1)$ simplex, to the vector space $E=\mathbb{R}^m$.
2) They are the F-related classes of the k-*jets* $(F,0)\to(E,0): X^i_\gamma t^\gamma$, $\gamma\in\Gamma(k-1)$ (in this case, the extended Einstein summation convention is used).

Addition and scalar multiplication are defined in the usual way in both cases. We shall employ *either* interpretation at various times.

The group $G_k(m)$ acts *simultaneously* on the spaces $E_1, E_2, \ldots, E_k$. That action can be interpreted as the 'transformation law' on the *tangent sectors*. We have observed that this action is *not* in general linear; the algebraic properties of sectors will be determined by those properties of E_k which are 'equivariant' in some sense with respect to this action. In particular, the action of all $G_k(m)$ on E_1 *is* linear (they reduce to the action of $G_1(m)=Gl(m)$ on $E_1=E$) and this is reflected in the fact that $T_1(M)$ has a *natural* linear structure on its fibers.

Recall that $J^k(m,n)$ is the vector space of k-jets of germs $(\mathbb{R}^m,0)\to(\mathbb{R}^n,0)$. If we denote the graded vector spaces corresponding, respectively, to $\mathbb{R}^m$ and $\mathbb{R}^n$ as $(E_1(m), E_2(m), \ldots, E_k(m), \ldots)$ and $(E_1(n), E_2(n), \ldots, E_k(n), \ldots)$, then it is clear from previous discussion that for $f_k\in J^k(m,n)$ and $\bar{g}_r\in E_r(m)$, $r\leq k$, jet composition gives an action

$f_k \cdot \bar{g}_r = \overline{(f \circ g)}_r \in E_r(n)$. This action generalizes the action of $G_k(m)$ and it will be useful to study it *explicitly* for the following reason.

Suppose that M, N are respectively m- and n-dimensional manifolds and $f:(M, x_0) \to (N, y_0)$ is the germ of a smooth map. Suppose also that frame-germs $\phi:(\mathbb{R}^m, 0) \to (M, x_0)$ and $\theta:(\mathbb{R}^n, 0) \to (N, y_0)$ are given. Let $[\phi_k, X^i_\gamma]$ be a k-sector at $x_0 \in M$. Then according to Definition 1.24 $T_k(f)[\phi_k, X^i_\gamma] = [\theta_k, Y^i_{\gamma'}]$ where $Y^i_{\gamma'} t^{\gamma'}$ belongs to the F-related class of $(\theta^{-1} \circ f \circ \phi)_k(X^i_\gamma t^\gamma)$. For ease of calculation, the germ $\theta^{-1} \circ f \circ \phi$ may be called $\bar{f}$. Then in order to calculate the effect of $T_k(f)$ on the k-sector we must determine the action of $\bar{f}_k \in J^k(m, n)$ on elements of $E_k(m)$: this amounts to determining the $Y^i_{\gamma'} \in E_1(n)$ for each $\gamma' \in \Gamma(k-1)$.

In principle, that calculation has already been made since it is given by the composition of representative jets. But since it will be necessary to examine the action of $G_k(m)$ on $E_k(m)$ closely in order to determine the equivariant algebraic properties of the simplicial vector space, we give now the *computational scheme* for the calculation of the coefficients $Y^i_{\gamma'}$ above; that is, the determination of the effect of $\bar{f}_k \in J^k(m, n)$ on elements of the vector space $E_k(m)$. Since $G_k(m)$ is the group of invertible elements of $J^k(m, m)$. this scheme will also give the coordinate change laws.

Definition 1.25. Let $\gamma \in \Gamma(k-1)$ where $\gamma = (\gamma_1, \gamma_2, \ldots, \gamma_k)$. Suppose that γ' and γ'' belong to $\Gamma(k-1)$ where $\gamma' = (\gamma'_1, \ldots, \gamma'_k)$ and $\gamma'' = (\gamma''_1, \ldots, \gamma''_k)$ and for *each* $1 \leq i \leq k$ $\gamma'_i \cdot \gamma''_i = 0$ and $\gamma'_i + \gamma''_i = \gamma_i$ (the subset represented by γ is the *disjoint union* of the subsets represented by γ' and by γ''), then we say that γ is the *join* of γ' and γ''. Represent this relationship as $\gamma = \gamma' \cup \gamma''$. ■

Then for $\bar{f}$ as in the previous discussion and $T_k(f):[\phi_k, X^i_\gamma] \to [\theta_k, Y^i_{\gamma'}]$, we have the following generalization of the coordinate change law for tangent vectors. For each $\gamma' \in \Gamma(k-1)$

$$Y^i_{\gamma'} = \sum_{\gamma_{i_1} \cup \gamma_{i_2} \cup \ldots \cup \gamma_{i_r} = \gamma'} \frac{\partial^r \bar{f}^i}{\partial x^{k_1} \ldots \partial x^{k_r}} X^{k_1}_{\gamma_{i_1}} X^{k_2}_{\gamma_{i_2}} \ldots X^{k_r}_{\gamma_{i_r}}. \tag{1.26}$$

This sum is taken over *all* decompositions of γ' as join of faces (but considering two decompositions where one is obtained by permuting the order in which the faces appear in the other as *identical*).

In the cases of 1-, 2-, and 3-sectors, this formula is easy to apply, as follows.

For a 1-*sector*

$$[\phi_1, X^i] \to \left[\theta_1, \frac{\partial \bar{f}^i}{\partial x^i} X^i\right]. \tag{1.27}$$

For a 2-sector

$$[\phi_2, X^i \text{---} Z^r \text{---} Y^j] \to \left[\theta_2, \frac{\partial \bar{f}^i}{\partial x^u} X^u \text{---} \frac{\partial \bar{f}^r}{\partial x^v} Z^v + \frac{\partial^2 \bar{f}^r}{\partial x^u \partial x^v} X^u Y^v \text{---} \frac{\partial \bar{f}^j}{\partial x^w} Y^w\right]. \tag{1.28}$$

For a 3-sector

It is necessary to adopt a convention concerning the way in which elements of E_3 are to be denoted; that is, we must specify for a given 2-simplex which position will correspond to which face. This has already been done for the elements of E_1 and E_2 (see the remarks following Definition 1.20). We shall in the entire sequel make the following choice:

$$\begin{array}{ccccc} 100 & \text{———} & 110 & \text{———} & 010 \\ & \diagdown & & \diagup & \\ & 101 & 111 & 011 & \\ & & \diagdown \; \diagup & & \\ & & 001 & & \end{array}$$

(*e.g.*, 111 is the 2-face, etc.). Thus we have

$$\begin{array}{ccccc} [\phi_3, X^i & \text{———} & A^r & \text{———} & Y^j] \to \\ & \diagdown & & \diagup & \\ & B^s & E^u & C^t & \\ & & \diagdown \; \diagup & & \\ & & Z^k & & \end{array}$$

$$\begin{array}{ccccc} \left[\theta_3, \frac{\partial \bar{f}^i}{\partial x^a} X^a\right. & \text{———} & \frac{\partial \bar{f}^r}{\partial x^b} A^b + \frac{\partial^2 \bar{f}^r}{\partial x^a \delta x^c} X^a Y^c & \text{———} & \left.\frac{\partial \bar{f}^j}{\partial x^c} Y^c\right] \\ & \diagdown & & \diagup & \\ \left[\frac{\partial \bar{f}^s}{\partial x^d} B^d + \frac{\partial^2 \bar{f}^s}{\partial x^a \partial x^v} X^a Z^v\right] & & (*) & & \left[\frac{\partial \bar{f}^t C^e}{\partial x^e} C^e + \frac{\partial^2 \bar{f}^t}{\partial x^c \delta x^v} Y^c Z^v\right] \\ & \diagdown & & \diagup & \\ & & \frac{\delta \bar{f}^k}{\delta x^v} Z^v & & \end{array} \tag{1.29}$$

where $(*) =$

$$\left[\frac{\partial^3 \bar{f}^u}{\partial x^a \partial x^c \partial x^v} X^a Y^c Z^v + \frac{\partial \bar{f}^u}{\partial x^w} E^w + \frac{\partial^2 \bar{f}^u}{\partial x^a \partial x^e} X^a C^e \right.$$

$$\left. + \frac{\partial^2 \bar{f}^u}{\partial x^c \partial x^d} Y^c B^d + \frac{\partial^2 \bar{f}^u}{\partial x^v \partial x^b} Z^v A^b \right].$$

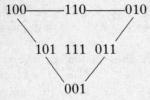

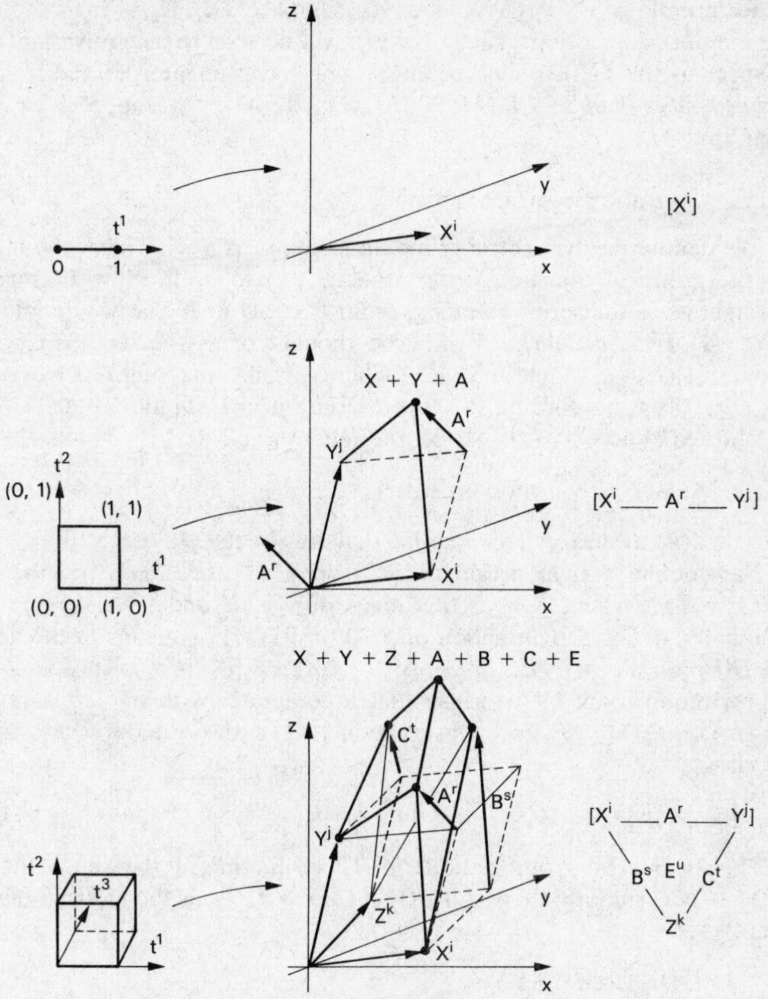

Fig. 1.5

Despite their first appearance, these formulas exhibit a very simple regularity. Still, to represent the transformation law for 4-sectors according to these pictures, we would need a 3-dimensional notation! For higher-order sectors, the formula (1.26) will suffice.

Exercises

12. Give an inductive proof of formula (1.26).

13. Interpret the results of exercise 10 in terms of formula (1.28).

Returning now to the graded vector space $(E_1, E_2, \ldots, E_k, \ldots)$, we define *three* 'operations' each of which will be seen to be equivariant with respect to the $G_k(m)$ actions and so will have an interpretation on the *graded fiber bundle* $(T_1(M), T_2(M), \ldots, T_k(M), \ldots)$ for each smooth manifold M.

A) The symmetric group action

We denote the symmetric group on k letters as $S(k)$. This is thought of as the group of bijections from $\{1, 2, \ldots, k\}$ to itself. Now E_k may be thought of as the set of mappings from $\Gamma(k-1)$ to E and we denote this $[\Gamma(k-1), E]$. Also, $\Gamma(k-1)$ may be thought of as the set of mappings from $\{1, 2, \ldots, k\}$ to $\{0, 1\}$, where each mapping is denoted $(\gamma_1, \gamma_2, \ldots, \gamma_k)$ according to the prescription in Definition 1.20.

Thus $S(k)$ acts on $\Gamma(k-1)$ on the *left* by the rule

$$s(\gamma_1, \gamma_2, \ldots, \gamma_k) = (\gamma_{s^{-1}(1)}, \gamma_{s^{-1}(2)}, \ldots, \gamma_{s^{-1}(k)})$$

for $s \in S(k)$ arbitrary. This implies that $(ts)\gamma = t(s(\gamma))$.

Now define a *right* action of $S(k)$ on $E_k = [\Gamma(k-1), E]$ by the rule $(X^i_\gamma)s = X^i_{s(\gamma)}$ where $X^i_{s(\gamma)}$ is the image of γ in E under $(X^i_\gamma)s$.

Finally, define a *right* action of $S(k)$ on $T_k(M)$ by means of the action of $S(k)$ on $F_k(M) \times E_k$ given by $(\phi_k, X^i_\gamma)s = (\phi_k, X^i_\gamma s)$. This action is $G_k(m)$ equivariant in the sense that it commutes with the left action of $G_k(m)$ on $F_k(M) \times E_k$ (see Observation 1.11) and so it induces a mapping of classes

$$[\phi_k, X^i_\gamma]s = [\phi_k, X^i_\gamma s] \qquad \text{for } s \in S(k). \tag{1.30}$$

The symmetric group action on T_k is *natural* in the sense that, if $f: M \to N$ is smooth, then, if $T_k(f): T_k(M) \to T_k(N)$ is the induced bundle map,

$$s \circ T_k(f) = T_k(f) \circ s. \tag{1.31}$$

Exercises

14. Show that the action of $S(k)$ on E_k commutes with the action of $G_k(m)$ on E_k, thus justifying our assertion of equivariance.

15. Interpret the action of $S(k)$ on $T_k(M)$ as being induced by permutation of the variables in $F = \mathbb{R}^k$ and argue naturality (formula (1.31)) from this.

B) The side operators

Returning to the sequence of vector spaces $(E_1, E_2, \ldots, E_k, \ldots)$, define for each $k \geq 1$ the *inclusion* $\lambda_{k+1}: \Gamma(k-1) \to \Gamma(k)$ by

$\lambda_{k+1}(\gamma_1, \gamma_2, \ldots, \gamma_k) = (\gamma_1, \gamma_2, \ldots, \gamma_k, 0)$. This gives for each $k \geq 1$ a *vector space projection* $\lambda^*_{k+1}: E_{k+1} \to E_k$ by associating to each $f \in E_{k+1} = [\Gamma(k), E]$ the map $f \circ \lambda_{k+1} \in E_k = [\Gamma(k-1), E]$.

Now let $C(k+1) \subset S(k+1)$ be the subgroup generated by the 'shift' σ where $\sigma(i) = i+1$ for $1 \leq i \leq k$ and $\sigma(k+1) = 1$. When we speak of the *shift* in $S(k)$ we shall always mean the permutation analogously defined. Using the various powers of σ we define $(k+1)$ linear projections from E_{k+1} to E_k by letting $d^i = (\sigma^i \circ \lambda_{k+1})^*$.

From now on, we consider the graded vector space $(E_1, E_2, \ldots, E_k, \ldots)$ to be equipped with this collection of linear projections: for each $k \geq 1$ there are $(k+1)$ projections $\{d^1, d^2, \ldots, d^k, d^{k+1}\}$ taking E_{k+1} to E_k. Here d^i is called the *i-side operator* and it is easy to see that d^{k+1} is just λ^*_{k+1}.

These data make the sequence a *simplicial object in the category of vector spaces* or a *simplicial vector space* (see May [17]).

Now it is easy to see that the maps $d^i: E_{k+1} \to E_k$ $(1 \leq i \leq k+1)$ satisfy the following equivariance property: for $g_{k+1} \in G_{k+1}(m)$ and g_k its k-jet in $G_k(m)$,

$$g_k \circ d^i = d^i \circ g_{k+1} \qquad \text{on } E_{k+1}. \tag{1.32}$$

This allows us to define for each $k \geq 1$, for each i, $1 \leq i \leq k+1$, a fiber bundle mapping $(D^i(k+1), \mathrm{id})$

$$\begin{array}{ccc} T_{k+1}(M) & \xrightarrow[D^i(k+1)]{} & T_k(M) \\ {\scriptstyle q_{k+1}}\downarrow & & \downarrow{\scriptstyle q_k} \\ M & \xrightarrow[\mathrm{id}]{} & M \end{array}$$

from $d^i: E_{k+1} \to E_k$ by 'passing to quotients' in the usual way. Further, it is easy enough to see that each fiber bundle mapping $(D^i(k+1), \mathrm{id})$ is a *natural transformation of functors* in the sense that for smooth $f: M \to N$ we shall have

$$T_k(f) \circ D^i(k+1) = D^i(k+1) \circ T_{k+1}(f). \tag{1.33}$$

The maps $\{D^1(k+1), D^2(k+1), \ldots, D^{k+1}(k+1)\}$ taking $T_{k+1}(M) \to T_k(M)$ will be called the *side operators* ($D^i(k+1)$ the i-side operator). Finally, the association to each smooth manifold M of the sequence of fiber bundles

$$\begin{array}{ccccccc} T_1(M) & T_2(M) & & T_k(M) & \\ \downarrow{\scriptstyle q_1} & \downarrow{\scriptstyle q_2} & \cdots & \downarrow{\scriptstyle q_k} & \cdots \\ M & M & & M & \end{array}$$

together with the $(k+1)$ side operators $\{D^i(k+1)\}$ taking $T_{k+1}(M) \to T_k(M)$ gives, for each such manifold, a *simplicial fiber bundle* (May [17]) and a *functor* from the category of smooth manifolds to the category of simplicial fiber bundles. Morphisms in the latter category are graded fiber bundle maps which commute with side operators.

This latter functor obviously generalizes the tangent bundle functor and it is what we shall call the *canonical simplicial fiber bundle functor.*

Exercise

16. Prove formula (1.33) for the case that $i = k+1$.

The last general element of structure which we wish to discuss here will allow us to interpret each side operator $D^i(k+1)$ as the projection of a *vector bundle*

$$\begin{array}{c} T_{k+1}(M) \\ \downarrow D^i(k+1) \\ T_k(M) \end{array}$$

C) The intrinsic spaces $E_{k+1,i} \subset E_{k+1}$

Returning to the simplicial vector space $(E_1, E_2, \ldots, E_k, \ldots)$, we recall that the action of the groups $G_k(m)$ is not linear in general. Suppose $d^i : E_{k+1} \to E_k$ is the i-side operator. For $X \in E_{k+1}$, define $E_{k+1,i}(X) \subset E_{k+1}$ to be $(d^i)^{-1}[d^i(X)]$. Then for $g \in G_{k+1}(m)$ we have for each X, and for each i, $1 \leq i \leq k+1$,

$$g : E_{k+1,i}(X) \to E_{k+1,i}(gX). \tag{1.34}$$

This follows from formula (1.32). We may think of $E_{k+1,i}(X)$ as the 'fiber' of the projection $d^i : E_{k+1} \to E_k$.

We intend to give, for each $X \in E_{k+1}$, *vector space* structure to the $(k+1)$ fibers $E_{k+1,i}(X)$ in such a way that the action of $G_{k+1}(m)$ induces *linear isomorphisms* connecting these spaces. When this is done, we shall be entitled to speak of an 'intrinsic' vector space structure on each fiber $E_{k+1,i}(X)$ for each X, and this intrinsic structure will pass to the tangent sector bundle fibers.

The vector space structure on $E_{k+1,i}(X)$ is defined in the following way. Let $\gamma' = (\gamma'_1, \gamma'_2, \ldots, \gamma'_i, \ldots, \gamma'_{k+1}) \in \Gamma(k)$ with $\gamma'_i = 0$ be arbitrary. And suppose that f_1 and $f_2 \in [\Gamma(k), E]$ both belong to $E_{k+1,i}(X)$. Then, it is clear that $f_1(\gamma') = f_2(\gamma')$. Call a face such as γ' a *type*-1 face. A face $\gamma = (\gamma_1, \ldots, \gamma_i, \ldots, \gamma_{k+1})$ with $\gamma_i = 1$ will be called a *type*-2 face. Then for f_1 and $f_2 \in E_{k+1,i}(X)$, define $f_1 + f_2$ by the rule

$$(f_1 + f_2)(\gamma) = f_1(\gamma) + f_2(\gamma) \qquad \text{for } \gamma \text{ of type 2,}$$

$$(f_1 + f_2)(\gamma) = f_1(\gamma) \qquad \text{for } \gamma \text{ of type 1,}$$

and define scalar multiplication by the rule

$$(af_1)(\gamma) = a[f_1(\gamma)] \qquad \text{for } \gamma \text{ of type 2,}$$
$$(af_1)(\gamma) = f_1(\gamma) \qquad \text{for } \gamma \text{ of type 1.}$$

Now suppose that $g \in G_{k+1}(m)$ so that $g: E_{k+1,i}(X) \to E_{k+1,i}(gX)$. We must show that this map is linear with respect to the vector space structures. Represent arbitrary elements of $E_{k+1,i}(X)$ as X^j_γ and Y^j_γ with $X^j_{\gamma'} = Y^j_{\gamma'}$ for all γ' of *type* 1. Let $gX^j_\gamma = \bar{X}^j_\gamma$ and $gY^j_\gamma = \bar{Y}^j_\gamma$. Finally, let $X^j_\gamma + Y^j_\gamma = W^j_\gamma$ (according to the addition rule given above), $\bar{X}^j_\gamma + \bar{Y}^j_\gamma = Z^j_\gamma$, and $gW^j_\gamma = \bar{W}^j_\gamma$.

We must show that $Z^j_\gamma = \bar{W}^j_\gamma$ for all γ and j. If γ is of *type* 1, this is obvious. Suppose then that γ is of type 2. The transformation law (formula (1.26)) says that

$$\bar{W}^j_\gamma = \sum_{\gamma_{i_1} \cup \ldots \cup \gamma_{i_r} = \gamma} \frac{\partial^r g^j}{\partial x^{k_1} \ldots \partial x^{k_r}} W^{k_1}_{\gamma_{i_1}} \ldots W^{k_r}_{\gamma_{i_r}}.$$

Now *precisely one* of the γ_{i_s} will be of type 2 in each join decomposition of γ, and for all of the others we will have

$$W^{k_t}_{\gamma_{i_t}} = X^{k_t}_{\gamma_{i_t}}.$$

Meanwhile, for the γ_{i_s} of type 2,

$$W^{k_s}_{\gamma_{i_s}} = X^{k_s}_{\gamma_{i_s}} + Y^{k_s}_{\gamma_{i_s}}.$$

Linearity of the rth partial derivative with respect to each variable gives the result on substitution. Scalar multiplication is handled similarly.

Observation 1.26. The natural projections

$$\begin{array}{c} T_{k+1}(M) \\ \big\downarrow D^i(k+1) \\ T_k(M) \end{array}$$

are *vector bundle projections* with fiber isomorphic to the intrinsic space $E_{k+1,i}(m)$. For $X \in T_{k+1}(M)$ denote $D^i(k+1)^{-1}[D^i(k+1)(X)]$ by $T_{k+1,i}(X)$ for $1 \leqslant i \leqslant k+1$.

Proof. For X^i_γ and Y^i_γ as above in $E_{k+1,i}(X)$, define

$$[\phi_{k+1}, X^i_\gamma] + [\phi_{k+1}, Y^i_\gamma] = [\phi_{k+1}, W^i_\gamma] \qquad \text{in } T_{k+1,i}(X) \tag{1.35}$$

for W^i_γ as defined above. Our earlier remarks show that this definition is unambiguous and it imposes an Abelian group structure on $T_{k+1,i}(X)$ which is 'intrinsic' in the sense that it is independent of frame. Scalar multiplication:

$$a[\phi_{k+1}, X^i_\gamma] = [\phi_{k+1}, a \cdot X^i_\gamma] \tag{1.36}$$

is likewise well-defined, if we give $a \cdot X^i_\gamma$ the meaning that it has above. We shall see later that these vector bundles have an interpretation as *tangent bundles.* ■

Example 1: $D^2(2)$. Let $[\phi_1, (X^i)]$ be a 1-sector $\bar{A}_1$. Let $[\phi_2, (X^i$——Z^r——$Y^j)] = \bar{B}_2$, and $[\phi_2, (X^i$——W^r——$U^j)] = \bar{C}_2$ be 2-sectors in the fiber over $\bar{A}_1$. Then we define $\bar{B}_2 + \bar{C}_2$ to be the 2-sector $[\phi_2, (X^i$——$Z^r + W^r$——$Y^j + U^j]$. Now if $\phi_2 = \theta_2 g_2$ for coordinate change germ g, we must compute $g_2(X^i$——$Z^r + W^r$——$Y^j + U^j)$ and according to (1.28) this is:

$$\left(\frac{\partial g^i}{\partial x^a} X^a \text{——} \frac{\partial g^r}{\partial x^b}(Z^b + W^b) + \frac{\partial^2 g^r}{\partial x^a \partial x^s} X^a (Y^s + U^s) \text{——} \frac{\partial g^j}{\partial x^s}(Y^s + U^s)\right),$$

and this is obviously equal to the second component of the sum of

$$\left[\theta_2, \left(\frac{\partial g^i}{\partial x^a} X^a \text{——} \frac{\partial g^r}{\partial x^b} Z^b + \frac{\partial^2 g^r}{\partial x^a \partial x^s} X^a Y^s \text{——} \frac{\partial g^j}{\partial x^s} Y^s\right)\right]$$

and

$$\left[\theta_2, \left(\frac{\partial g^i}{\partial x^a} X^a \text{——} \frac{\partial g^r}{\partial x^b} W^b + \frac{\partial^2 g^r}{\partial x^a \partial x^s} X^a U^s \text{——} \frac{\partial g^j}{\partial x^s} Y^s\right)\right].$$

Example 2: $D^3(3)$. Let $\bar{A}_2 = [\phi_2, (X^i$——$Z^r$——$Y^j)]$, and suppose we have two 3-sectors in the fiber over $\bar{A}_2$:

$$\begin{array}{c} [\phi_3, (X^i \text{——} Z^r \text{——} Y^j)] = \bar{G}_3 \\ \diagdown \qquad\qquad \diagup \\ B^s \quad E^u \quad C^t \\ \diagdown \quad \diagup \\ W^k \end{array}$$

and

$$\begin{array}{c} [\phi_3, (X^i \text{——} Z^r \text{——} Y^j)] = \bar{H}_3. \\ \diagdown \qquad\qquad \diagup \\ \bar{B}^s \quad \bar{E}^u \quad \bar{C}^t \\ \diagdown \quad \diagup \\ \bar{W}^k \end{array}$$

Then according to our definition, $\bar{G}_3 + \bar{H}_3$ is the sector

$$\begin{array}{ccccc} [\phi_3, (X^i & \text{———} & Z^r & \text{———} & Y^j)] \\ \diagdown & & & & \diagup \\ B^s + \bar{B}^s & & E^u + \bar{E}^u & & C^t + \bar{C}^t \\ & \diagdown & & \diagup & \\ & & W^k + \bar{W}^k & & \end{array}$$

following the scheme for locating faces given in (1.29).

Again, if $\phi_3 g_3 = \theta_3$ for frame change g_3, then (1.29) yields

$$\begin{array}{ccccc} g_3(X^i & \text{———} & Z^r & \text{———} & Y^j) \\ \diagdown & & & & \diagup \\ B^s + \bar{B}^s & & E^u + \bar{E}^u & & C^t + \bar{C}^t \\ & \diagdown & & \diagup & \\ & & W^k + \bar{W}^k & & \end{array}$$

has the following faces:

100:

$$\frac{\partial g^i}{\partial x^a} X^a$$

110:

$$\frac{\partial g^r}{\partial x^b} Z^b + \frac{\partial^2 g^r}{\partial x^a \partial x^c} X^a Y^c$$

010:

$$\frac{\partial g^j}{\partial x^c} Y^c$$

101:

$$\frac{\partial g^s}{\partial x^d} (B^d + \bar{B}^d) + \frac{\partial^2 g^s}{\partial x^a \partial x^e} X^a (W^e + \bar{W}^e)$$

001:

$$\frac{\partial g^k}{\partial x^e} (W^e + \bar{W}^e)$$

011:

$$\frac{\partial g^t}{\partial x^f} (C^f + \bar{C}^f) + \frac{\partial^2 g^t}{\partial x^c \partial x^e} Y^c (W^e + \bar{W}^e)$$

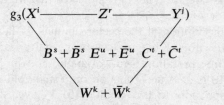

111:

$$\frac{\partial^3 g^u}{\partial x^a \partial x^c \partial x^e} X^a Y^c (W^e + \bar{W}^e) + \frac{\partial g^u}{\partial x^g}(E^g + \bar{E}^g) + \frac{\partial^2 g^u}{\partial x^a \partial x^f} X^a (C^f + \bar{C}^f)$$

$$+ \frac{\partial^2 g^u}{\partial x^c \partial x^d} Y^c (B^d + \bar{B}^d) + \frac{\partial^2 g^u}{\partial x^e \partial x^b} Z^b (W^e + \bar{W}^e).$$

The components of the sum of the transforms of $\bar{G}_3$ and $\bar{H}_3$ will easily be seen to equal those of the element of E_3 with these faces.

It will be convenient to denote a general k-sector, say an element of $T_k(M)$, at $x_0 \in M$ by a symbol like $\bar{A}_k(x_0)$. The 'bar' indicates that we are considering an F-related class for $F = \mathbb{R}^k$.

This concludes the *algebraic* description of the sector bundles. In later chapters, we shall be concerned with the use of sectors in analysis and geometry. They have several interpretations: for example, it will be possible to *differentiate* a smooth real-valued function on a manifold M with respect to a sector. The appendix to this chapter (§1.5) gives the interpretation of sectors as elements of $T_1[T_1[\ldots T_1(M)\ldots]]$. From another point of view, they are the 'elements' in the classical sense of sets of commuting flows, or multidimensional dynamics. Finally, such local notions as 'variation', 'parallel translation', and 'covariant derivative' can be conveniently formulated in terms of them.

§1.5. Appendix: Some natural isomorphisms

The material in this appendix is provided both for the sake of completeness and as a reference for some of the constructions which appear later in the book. It will be shown in the theorem that follows in precisely which sense the sector bundles can be thought of as iterated tangent bundles. It would not be necessary to focus on the *details* of the proof on a first reading; a general understanding of its meaning should be sufficient here. The reader will be referred back to this theorem from time to time in the course of later arguments, and can safely defer grappling with the technicalities until she/he needs them.

For $h \leqslant k$ there is an 'equivariant' projection $e_{k,h} : E_k \to E_h$ defined by iteration of certain side operators:

$$e_{k,h} = d^{h+1} \circ d^{h+2} \circ \ldots d^{k-1} \circ d^k.$$

These correspond to certain *natural* fiber bundle maps $(q_{k,h}, \mathrm{id})$

$$\begin{array}{ccc} T_k(M) & \xrightarrow[q_{k,h}]{} & T_h(M) \\ {\scriptstyle q_k}\downarrow & & \downarrow{\scriptstyle q_h} \\ M & \xrightarrow[\mathrm{id}]{} & M \end{array}$$

with

$$q_{k,h} = D^{h+1}(h+1)\circ D^{h+2}(h+2)\circ \ldots \circ D^{k-1}(k-1)\circ D^{k}(k). \qquad (1.37)$$

Theorem 1.27. *There is a* natural *bundle isomorphism* (ν, id) *for* $k > h$, *such that*

$$\begin{array}{ccc} T_k(M) & \xrightarrow{\nu(M)} & T_{k-h}(T_h(M)) \\ {\scriptstyle q_{k,h}}\downarrow & & \downarrow{\scriptstyle q_{k-h}} \\ T_h(M) & \xrightarrow[\mathrm{id}]{} & T_h(M) \end{array}$$

Proof. We begin with the observation that E_k is isomorphic to $(E_h)_{k-h}\times E_{k-h}\times E_h$ where, if $[X, Y]$ denotes the vector space of maps from set X to vector space Y, then we are saying that:

$$[\Gamma(k-1), E] \cong [\Gamma(k-h-1), E]\times[\Gamma(h-1), E] \\ \times[\Gamma(k-h-1), [\Gamma(h-1), E]].$$

In fact, this is obvious for dimensional reasons. Our demonstration will exhibit a specific isomorphism. Let $\phi: O \to U \subseteq M$ be a frame on M. Then define the map $\tilde{\phi}^k: O\times E_k \to q_k^{-1}(U)$ by the rule:

$$\tilde{\phi}^k(z, X^i_\gamma) = [(\phi\circ t_z)_k, (X^i_\gamma)] = \overline{(\phi\circ t_z\circ(X^i_\gamma t^\gamma))_k}$$

(the upper bar means F-related class, $F = \mathbb{R}^k$). $T_k(M)$ has the smallest topology for which these maps $\tilde{\phi}^k$ are homeomorphisms onto open sets.

Now for frame ϕ fixed, define the following map from $O\times E_k \to O\times E_h$ $\mathrm{id}\times e_{k,h}$. The elements of $\Gamma(k-1)$ which may be written $\gamma = (\gamma_1, \gamma_2, \ldots, \gamma_h, \ldots, \gamma_k)$ fall into three categories:

1) faces of the form $(\gamma_1, \ldots, \gamma_h, 0, \ldots, 0)$;
2) faces of the form $(0, \ldots, 0, \gamma_{h+1}, \ldots, \gamma_k)$; and
3) faces of the form $\gamma' \cup \gamma''$ for γ' of type 1 and γ'' of type 2.

Now let E_h denote the vector space of dimension $m(2^h-1)$ with basis the vector-valued monomials $e_j\cdot t^{\gamma'}$ where e_j is one of the standard basis vectors for E, and $t^{\gamma'} = (t^1)^{\gamma'_1}\ldots(t^h)^{\gamma'_h}$ for γ' of type 1.

The map $\tilde{\phi}^h: O\times E_h \to q_h^{-1}(U)$ can thus be considered a *frame* for $T_h(M)$ for $O\times E_h$ an open subset of $\mathbb{R}^{(m2^h)}$ where this last Euclidean space has basis $\{e_1, \ldots, e_m\}\cup\{e_j\cdot t^{\gamma'} \mid \gamma' \text{ is of type 1 in } \Gamma(k-1)\}$. Call this space E'_h.

Now form the map $\tilde{\tilde{\phi}}^h: (O\times E_h)\times(E'_h)_{k-h} \to q_{k-h}^{-1}(q_h^{-1}(U)) \subset T_{k-h}(T_h(U))$ by

$$\tilde{\tilde{\phi}}^h: ((z, X^i_{\gamma'}), W^\alpha_{\gamma''}) \to \overline{(\tilde{\phi}^h\circ t_{(z, X^i_{\gamma'})}\circ(W^\alpha_{\gamma''}t^{\gamma''}))_{k-h}}$$

where the last expression has the following meaning: $(z, X^i_{\gamma'})$ is a general element of $O\times E_h$. $W^\alpha_{\gamma''}$ is the image of γ'' under a map from the $(k-h-1)$ simplex (which we represent as the set of faces γ'' of type 2) to

E'_h with the basis given above. Thus

$$\tilde{\tilde{\phi}}^h((z, X^i_{\gamma'}), W^\alpha_{\gamma''}) = [(\tilde{\phi}^h \circ t_{(z, X^i_{\gamma'})})_{k-h}, W^\alpha_{\gamma''}].$$

Here γ' represents a face of $\Gamma(h-1)$ and γ'' represents a face of $\Gamma(k-h-1)$. By *definition*, $\tilde{\tilde{\phi}}^h$ is a *homeomorphism* of $(O \times E_h) \times (E'_h)_{k-h}$ *onto* $q^{-1}_{k-h}(q^{-1}_h(U))$.

Now let μ be the following isomorphism of $E_h \times (E'_h)_{k-h} \to E_k$:

1) a basis vector for E_h: $e_i \cdot t^{\gamma'}$ corresponds to itself;
2) a basis vector for $(E'_h)_{k-h}$ of the form $e_i \cdot t^{\gamma''}$ corresponds to itself;
3) a basis vector for $(E'_h)_{k-h}$ of the form $(e_i \cdot t^{\gamma'}) \cdot t^{\gamma''}$ corresponds to $e_i \cdot t^{\gamma' \cup \gamma''}$.

It is easy to see that this is a vector space isomorphism, and $\mathrm{id} \times \mu$ is a homeomorphism from $O \times E_h \times (E'_h)_{k-h} \to O \times E_k$ with $(\mathrm{id} \times e_{k,h}) \circ (\mathrm{id} \times \mu)$ equal to the projection on $O \times E_h$.

In order to patch these maps $\mathrm{id} \times \mu$ together, we must study their dependence on frame ϕ. Suppose that ϕ_k is a frame-jet at $x_0 \in M$. An element of $T_k(x_0)$ can be represented as $[\phi_k, (X^i_{\gamma'})$——(Z^i_γ)——$(Y^i_{\gamma''})]$ (abusing notation slightly), where γ' is a type 1 face of $\Gamma(k-1)$ thought of also as a face of $\Gamma(h-1)$, γ'' is a type 2 face of $\Gamma(k-1)$ thought of also as a face of $\Gamma(k-h-1)$, γ is a face of type 3 of $\Gamma(k-1)$: $\gamma = \gamma' \cup \gamma''$, X^i, Y^i, Z^i belong to $E = \mathbb{R}^m$.

The transformation law for sectors presented in this form is the following: Suppose that $f : (M, x_0) \to (M, x_0)$ is an invertible germ at x_0, $\phi : (\mathbb{R}^m, 0) \to (M, x_0)$ and $\theta : (\mathbb{R}^m, 0) \to (M, x_0)$ are two frame-germs. As earlier, we let $\theta^{-1} \circ f \circ \phi = \bar{f}$. Then

$$[\phi_k, (X^i_{\gamma'})\text{——}(Z^i_\gamma)\text{——}(Y^i_{\gamma''})] = [\theta_k, (\bar{X}^i_{\gamma'})\text{——}(\bar{Z}^i_\gamma)\text{——}(\bar{Y}^i_{\gamma''})]$$

with (following formula (1.26)):

$$\bar{X}^i_{\gamma'} = \sum_{\gamma'_{i_1} \cup \gamma'_{i_2} \cup \ldots \cup \gamma'_{i_r} = \gamma'} \frac{\partial^r \bar{f}^i}{\partial x^{k_1} \ldots \partial x^{k_r}} X^{k_1}_{\gamma'_{i_1}} \ldots X^{k_r}_{\gamma'_{i_r}}$$

since two faces must be of type 1 if their join is to be of type 1.

Similarly,

$$\bar{Y}^i_{\gamma''} = \sum_{\gamma''_{i_1} \cup \gamma''_{i_2} \cup \ldots \cup \gamma''_{i_s} = \gamma''} \frac{\partial^s \bar{f}^i}{\partial x^{k_1} \ldots \partial x^{k_s}} Y^{k_1}_{\gamma''_{i_1}} \ldots Y^{k_s}_{\gamma''_{i_s}}.$$

$\bar{Z}^i_\gamma = \bar{Z}^i_{\gamma' \cup \gamma''}$ is rather more difficult to describe. A decomposition into joins of a face $\gamma' \cup \gamma''$ can be described by the following data:

1) a decomposition of type 1 faces $\gamma' = \gamma'_{i_1} \cup \ldots \cup \gamma'_{i_t} \cup \gamma'_{i_{t+1}} \cup \ldots \cup \gamma'_{i_R}$ (possibly $t = R$);
2) a decomposition of type 2 faces $\gamma'' = \gamma''_{i_1} \cup \ldots \cup \gamma''_{i_t} \cup \gamma''_{i_{t+1}} \cup \ldots \cup \gamma''_{i_S}$ (possibly $t = S$);

3) in the case $t \neq 0$, a *bijective correspondence* $\{\gamma'_{i_1}, \gamma'_{i_2}, \ldots, \gamma'_{i_t}\} \to \{\gamma''_{i_1}, \ldots, \gamma''_{i_t}\}$, between the special t element sets of faces. This uniquely gives:

$$\gamma' \cup \gamma'' = [(\gamma'_{i_1} \cup \gamma''_{i_1}) \cup \ldots \cup (\gamma'_{i_t} \cup \gamma''_{i_t})] \cup [\gamma'_{i_{t+1}} \cup \ldots \cup \gamma'_{i_R}] \cup [\gamma''_{i_{t+1}} \cup \ldots \cup \gamma''_{i_S}].$$

Thus for fixed $\gamma' \cup \gamma''$:

$$\bar{Z}^j_{\gamma' \cup \gamma''} = \sum_{\substack{\gamma' = \gamma'_{i_1} \cup \ldots \cup \gamma'_{i_t} \cup \ldots \cup \gamma'_{i_R} \\ \gamma'' = \gamma''_{i_1} \cup \ldots \cup \gamma''_{i_t} \cup \ldots \cup \gamma''_{i_S}}} \frac{\partial^{(R+S-t)} \bar{f}^j}{\partial x^{a_1} \ldots \partial x^{a_{R-t}} \partial y^{b_1} \ldots \partial y^{b_{S-t}} \partial z^{c_1} \ldots \partial z^{c_t}} (P)$$

where

$$(P) = (X^{a_1}_{\gamma'_{i_{t+1}}} \ldots X^{a_{R-t}}_{\gamma'_{i_R}} Y^{b_1}_{\gamma''_{i_{t+1}}} \ldots Y^{b_{S-t}}_{\gamma''_{i_S}} Z^{c_1}_{(\gamma'_{i_1} \cup \gamma''_{i_1})} \ldots Z^{c_t}_{(\gamma'_{i_t} \cup \gamma''_{i_t})}).$$

This follows (with some application) from the transformation law given in formula (1.26).

Now consider the frames $\tilde{\phi}^h : O \times E_h \to q_h^{-1}(U) \subset T_h(U)$ and $\tilde{\theta}^h : O \times E_h \to q_h^{-1}(U)$ induced by frames ϕ and $\theta : (\mathbb{R}^m, 0) \to (M, x_0)$. The composition of germs

$$(\tilde{\theta}^h)^{-1} \circ T_h(f) \circ (\tilde{\phi}^h) : [O \times E_h, (0, X^i_{\gamma'})] \to [O \times E_h, (0, \bar{X}^i_{\gamma'})]$$

will be denoted simply $\bar{g}$. Here, the pair $(0, X^i_{\gamma'})$ gives the components of an element of E'_h, the $m(2^h)$-dimensional vector space introduced earlier with basis $(e_i, e_i t^{\gamma'})$, $i = 1, 2, \ldots, m$, and γ' a type 1 face of $\Gamma(k-1)$.

In general, the pair $(z^i, X^i_{\gamma'})$ represents the element of E'_h: $z^i e_i + X^i_{\gamma'} e_i t^{\gamma'}$ invoking the summation convention.

Now let $[(z, X^i_{\gamma'}), W^\alpha_{\gamma''}]$ be an element of $(O \times E_h) \times (E'_h)_{k-h}$. The map $\mathrm{id} \times \mu$ corresponds the following element of $O \times E_k$ to this element. Distinguish the *two* types of index α: either α has the form $(j, 0)$ corresponding to basis vector e_j or α has the form (j, γ') corresponding to basis vector $e_j t^{\gamma'}$. Then denote $W^{(j,0)}_{\gamma''}$ as $Y^j_{\gamma''}$ (an element of $\mathbb{R}$), and denote $W^{(j,\gamma')}_{\gamma''}$ as $Z^j_{\gamma' \cup \gamma''}$ (an element of $\mathbb{R}$). Then

$$\mathrm{id} \times \mu : [(z, X^i_{\gamma'}), W^\alpha_{\gamma''}] \to [z; X^i_{\gamma'} \text{——} Z^i_{\gamma' \cup \gamma''} \text{——} Y^i_{\gamma''}]$$

where for the latter term we make use of the identifications introduced earlier in this discussion. For $z = 0$, we computed the effect of a coordinate change at the origin of O: $\bar{f}$ on the right-hand expression. This coordinate change gives rise to the coordinate change $\bar{g}$: $[O \times E_h, (0, X^i_{\gamma'})] \to [O \times E_h, (0, \bar{X}^i_{\gamma'})]$ which *does not* fix the base point. The latter coordinate change is computed with respect to frames $\tilde{\phi}^h$ and $\tilde{\theta}^h$.

For the *fixed* element $[(0, X^i_{\gamma'}), W^\alpha_{\gamma''}]$ we must now compute the effect of *left multiplication* of $W^\alpha_{\gamma''} \in (E'_h)_{k-h}$ by the $(k-h)$ jet $(t^{-1}_{(0, \bar{X}^i_{\gamma'})} \circ \bar{g} \circ t_{(0, X^i_{\gamma'})})_{k-h}$ in $G_{k-h}(m2^h)$. The latter is the $(k-h)$ jet at the origin of the germ $F:(O\times E_h,(0,0))\to(O\times E_h,(0,0))$

$$F^{(j,0)}(z,V^i_{\gamma'})=\bar f^i(z)$$

$$F^{(j,\bar\gamma')}(z,V^i_{\gamma'})=\sum_{\bar\gamma_{i_1}\cup\ldots\cup\bar\gamma'_{i_r}=\bar\gamma'}\frac{\delta^r\bar f^i}{\delta x^{k_1}\ldots\delta x^{k_r}}_{|z}(X^{k_1}_{\bar\gamma'_{i_1}}+V^{k_1}_{\bar\gamma'_{i_1}})\ldots(X^{k_r}_{\bar\gamma'_{i_r}}+V^{k_r}_{\bar\gamma'_{i_r}})$$
$$-\sum_{\bar\gamma'_{i_1}\cup\ldots\cup\bar\gamma'_{i_r}=\bar\gamma'}\frac{\delta^r\bar f^i}{\delta x^{k_1}\ldots\delta x^{k_r}}_{|0}X^{k_1}_{\bar\gamma'_{i_1}}\ldots X^{k_r}_{\bar\gamma'_{i_r}}$$

This will give an element $\bar W^\beta_{\gamma''}$ of $(E'_h)_{k-h}$ and we will want to show that

$$(\mathrm{id}\times\mu)[(0,\bar X^i_{\gamma'}),\bar W^\beta_{\gamma''})]=[0;\bar X^i_{\gamma'}\text{——}\bar Z^i_{\gamma'\cup\gamma''}\text{——}\bar Y^i_{\gamma''}].$$

Now $\bar W^\beta_{\gamma''}$ (for $\beta=(j,0)$ or $(j,\bar\gamma')$) is equal to

$$\sum_{\gamma''_{i_1}\cup\gamma''_{i_2}\cup\ldots\cup\gamma''_{i_s}=\gamma''}\frac{\partial^s F^\beta}{\partial u^{\alpha_1}\ldots\partial u^{\alpha_s}}W^{\alpha_1}_{\gamma''_{i_1}}\ldots W^{\alpha_s}_{\gamma''_{i_s}}$$

where u^α is a variable in E'_h, and α is of the form $(j,0)$ or (j,γ').

Case 1: $\beta=(j,0)$, $F^\beta(z,V^i_{\gamma'})=\bar f^i(z)$. In this case, all α_i are of the form $(i,0)$ since other partial derivatives yield 0. It is easy to see that $\bar W^\beta_{\gamma''}=\bar Y^i_{\gamma''}$.

Case 2: $\beta=(j,\bar\gamma')$. Here we must show that

$$\bar Z^i_{\bar\gamma'\cup\gamma''}=\sum_{\gamma''_{i_1}\cup\ldots\cup\gamma''_{i_s}=\gamma''}\frac{\partial^s F^{(j,\bar\gamma')}}{\partial u^{\alpha_1}\ldots\partial u^{\alpha_s}}W^{\alpha_1}_{\gamma''_{i_1}}\ldots W^{\alpha_s}_{\gamma''_{i_s}}$$

where

$$F^{(j,\bar\gamma')}(z,V^i_{\gamma'})=\sum_{\bar\gamma'_{i_1}\cup\ldots\cup\bar\gamma'_{i_r}=\bar\gamma'}\frac{\delta^r\bar f^i}{\delta x^{k_1}\ldots\delta x^{k_r}}_{|z}(X^{k_1}_{\bar\gamma_{i_1}}+V^{k_1}_{\bar\gamma_{i_1}})\ldots(X^{k_r}_{\bar\gamma_{i_r}}+V^{k_r}_{\bar\gamma_{i_r}})$$
$$-\sum_{\bar\gamma'_{i_1}\cup\ldots\cup\bar\gamma'_{i_r}=\bar\gamma'}\frac{\delta^r\bar f^i}{\delta x^{k_1}\ldots\delta x^{k_r}}_{|0}X^{k_1}_{\bar\gamma'_{i_1}}\ldots X^{k_r}_{\bar\gamma'_{i_r}}$$

For $\bar\gamma'$ and γ'' fixed, there will be for each j a single non-zero term for each pair of join decompositions of the form:

$$\bar\gamma'=\bar\gamma'_{i_1}\cup\ldots\cup\bar\gamma'_{i_t}\cup\ldots\cup\bar\gamma'_{i_r}$$
$$\gamma''=\gamma''_{i_1}\cup\ldots\cup\gamma''_{i_t}\cup\ldots\cup\gamma''_{i_s}$$

with a *bijective correspondence* between the special t element sets of faces. Here the $\alpha_1,\alpha_2,\ldots,\alpha_t$ are chosen so that $\alpha_n=(k_n,\bar\gamma'_{i_n})$ where $\bar\gamma'_{i_n}$ corresponds to γ''_{i_n}. The remaining α_p are of the form $(i_p,0)$, and *their* order is unimportant. A direct calculation of the partial derivative with respect to all such choices of join decompositions and bijective correspondences gives the expression obtained earlier for $\bar Z^i_{\bar\gamma'\cup\gamma''}$, for j, $\bar\gamma'$ and γ'' fixed.

The previous calculation shows that the maps $\mathrm{id}\times\mu : O\times E_h\times (E'_h)_{k-h}\to O\times E_k$ can be 'patched together' to give an isomorphism of fiber bundles for $k>h$:

$$\begin{array}{ccc} T_k(M) & \xrightarrow{\nu(M)} & T_{k-h}(T_h(M)) \\ {\scriptstyle q_{k,h}}\downarrow & & \downarrow{\scriptstyle q_{k-h}} \\ T_h(M) & \xrightarrow[\mathrm{id}]{} & T_h(M) \end{array}$$

The local maps on frames $\tilde{\phi}^h$, $\tilde{\tilde{\phi}}^h$, and $\tilde{\phi}^k$ are defined in terms of the original choice of a frame ϕ on M. The calculation shows that the map $\nu(M)$ is independent of that choice.

Finally, naturality of (ν, id) is formulated in the following way. If $f: M\to N$ is smooth, the diagram below commutes:

$$\begin{array}{ccc} T_k(M) & \xrightarrow{\nu(M)} & T_{k-h}(T_h(M)) \\ {\scriptstyle T_k(f)}\downarrow & & \downarrow{\scriptstyle T_{k-h}(T_h(f))} \\ T_k(N) & \xrightarrow[\nu(N)]{} & T_{k-h}(T_h(N)) \end{array}$$

We omit the proof of this last fact. It requires a straightforward modification of the argument just given. ■

Exercises

17. Show directly that there is a *natural* isomorphism $T_2(M)\to T_1[T_1(M)]$.

18. Interpret the intrinsic spaces $T_{2,1}(X)\subset T_2(M)$ in terms of the above isomorphism (for $X\in T_2(M)$ at $x_0\in M$).

19. Show that there are in general $k!$ *natural* ways to identify $T_k(M)$ with $T_1[T_1 \ldots T_1(M)\ldots]]$ (k times).

20. Prove 'naturality' of (ν, id) as formulated above.

21. Give a geometric interpretation of the symmetric group $(S(2))$ action on $T_1[T_1(M)]$ via its action on $T_2(M)$.

2

Brackets of tangent sectors

§2.1. *K*-sector bundles

The sector bundles provide the 'simplices' for the local objects, K-sectors and *K-sector bundles* which we shall construct in this chapter. Suppose that M is a smooth manifold and $x_0 \in M$. Let $T_k(x_0)$ denote $q_k^{-1}(x_0) \subset T_k(M)$, and let $\bar{A}_k(x_0)$ denote a general element of $T_k(x_0)$.

Definition **2.1.** The $(k-1)$ simplex $\Gamma(k-1)$ is the collection of non-empty subsets of $\{1, 2, \ldots, k\}$ (Definition 1.20). A *subcomplex* of $\Gamma(k-1)$, that is $K \subset \Gamma(k-1)$, is a collection of non-empty subsets of $\{1, 2, \ldots, k\}$ with the property that if $C \in K$, $C' \neq \varnothing$ and $C' \subset C$, then $C' \in K$.

If K is a subcomplex of $\Gamma(k-1)$ the *faces* of K are the sets $C \in K$; a *maximal face* is a face which is not properly contained in any other face of K.

An *orientation* for a subcomplex K of $\Gamma(k-1)$ is an *ordering* for the vertices of $\Gamma(k-1)$. This induces for each *maximal* $(i-1)$ face C_i a map $\Gamma(i-1) \to \Gamma(k-1)$ by extending the unique strictly increasing map $\{1, 2, \ldots, i\} \to \{\text{vertices of } \Gamma(k-1) \text{ with their ordering}\}$. The *standard* ordering of the vertices of $\Gamma(k-1)$ is given by the arrangement $\{1, 2, \ldots, k\}$.

Finally, a subcomplex K of $\Gamma(k-1)$ will be assumed to contain all vertices of $\Gamma(k-1)$; an *oriented* subcomplex of $\Gamma(k-1)$ is entirely specified by the data $[C_1, C_2, \ldots, C_r; \leqslant]$, the *maximal* faces C_i together with the ordering on the vertices: when the ordering is the standard one, we shall not mention it explicitly. ■

This definition is entirely analogous to the topological case. Notationally we represent the 0, 1, and 2 simplices in the following way with standard orderings:

0-*simplex* (1)

1-*simplex* 1——(12)——2

2-*simplex*

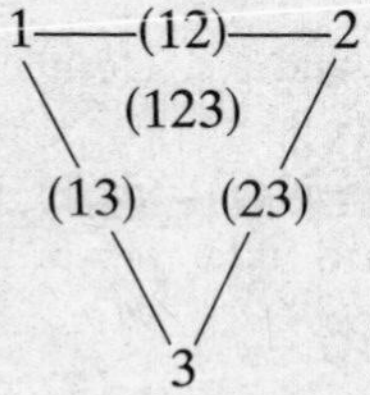

(We later introduce a similar notation for the 3-simplex.)

Thus we may represent subcomplexes of the 1-simplex with such symbols as:

*   * or *——*——*

(which respectively *omit* and *contain* the 1-face).
Also some subcomplexes of the 2-simplex can be represented:

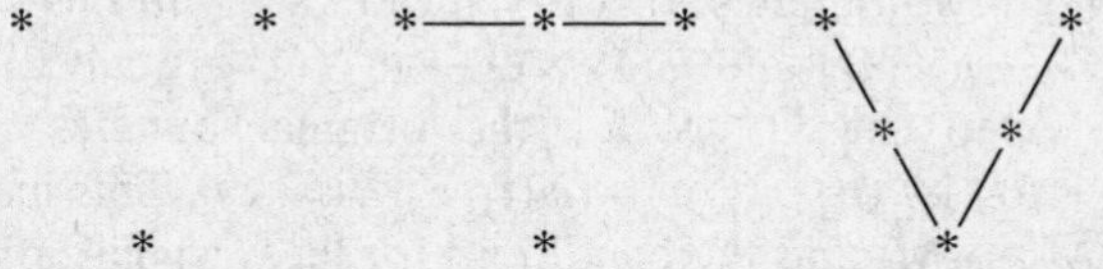

and so forth.

Now suppose that K is a subcomplex of $\Gamma(k-1)$ oriented with the standard ordering. Denote it as $[C_1, \ldots, C_r]$. For each maximal face C_i there is an inclusion (Definition 2.1) $j_i : \Gamma(c_i - 1) \to \Gamma(k-1)$ where C_i is a $(c_i - 1)$ face. The 'adjoint' of this map is a projection $j_i^* : E_k \to E_{c_i}$. This map is $G_k(m)$ equivariant for $\dim E = m$ and it gives rise to a natural fiber bundle map $(D(C_i), \mathrm{id})$

$$\begin{array}{ccc} T_k(M) & \xrightarrow[D(C_i)]{} & T_{c_i}(M) \\ {\scriptstyle q_k}\downarrow & & \downarrow{\scriptstyle q_{c_i}} \\ M & \xrightarrow[\mathrm{id}]{} & M \end{array}$$

Now for a smooth manifold M we want to construct a fiber bundle on M which is associated in certain natural ways with the subcomplex K of $\Gamma(k-1)$. This fiber bundle will be the *bundle of K-sectors* determined by K. And in order to construct it we make use of the *Whitney product* of fiber bundles which we now describe.

Definition 2.2. Suppose that

$$\begin{array}{c} M_1 \\ \downarrow{\scriptstyle p_1} \\ N \end{array} \quad \text{and} \quad \begin{array}{c} M_2 \\ \downarrow{\scriptstyle p_2} \\ N \end{array}$$

are fiber bundles, with respective fibers F_1 and F_2. Define $M_1 \times_N M_2$ to be the subset of $M_1 \times M_2 : \{(x, y) \mid p_1(x) = p_2(y)\}$.

Map $M_1 \times_N M_2 \to N$ by the rule $(x, y) \to p_1(x) = p_2(y)$ and call that mapping $p_1 \times_N p_2$. Also define $p_1^*(x, y) = y$ and $p_2^*(x, y) = x$. The following diagram commutes:

$$\begin{array}{ccc} M_1 \times_N M_2 & \xrightarrow{p_1^*} & M_2 \\ {\scriptstyle p_2^*}\downarrow & \overset{p_1 \times_N p_2}{\searrow} & \downarrow{\scriptstyle p_2} \\ M_1 & \xrightarrow[p_1]{} & N \end{array}$$

Each mapping is the projection of a fiber bundle: p_1^* has fiber F_1, p_2^* has fiber F_2, and $p_1 \times_N p_2$ has fiber $F_1 \times F_2$. We call the bundle

$$\begin{array}{c} M_1 \times_N M_2 \\ \downarrow{\scriptstyle p_1 \times_N p_2} \\ N \end{array}$$

the *product* of the original two bundles. It is trivialized locally in this way: Suppose that $\phi : F_1 \times O \to p_1^{-1}(O)$ and $\theta : F_2 \times O \to p_2^{-1}(O)$ are trivializations over the *same* open set $O \subset N$ for the original bundles. Let $\psi : (F_1 \times F_2) \times O \to M_1 \times M_2$ by $\psi(f_1, f_2; z) = (\phi(f_1, z), \theta(f_2, z))$. This image is easily seen to be in $M_1 \times_N M_2$ and the conditions for local trivializations are easily checked.

If p_1 is the projection of a vector bundle, then so is p_1^*. Finally the Whitney product

$$\begin{array}{c} M_1 \times_N M_2 \\ \downarrow{\scriptstyle p_1 \times_N p_2} \\ N \end{array}$$

is a 'product' in the category of fiber bundles over N in the following sense: Suppose that

$$\begin{array}{c} M \\ \downarrow{\scriptstyle p} \\ N \end{array}$$

is a fiber bundle and there are fiber bundle maps (q_1, id), (q_2, id)

$$\begin{array}{ccc} M & \xrightarrow[q_1]{} & M_1 \\ {\scriptstyle p}\downarrow & & \downarrow{\scriptstyle p_1} \\ N & \xrightarrow[\text{id}]{} & N \end{array} \quad \text{and} \quad \begin{array}{ccc} M & \xrightarrow[q_2]{} & M_2 \\ {\scriptstyle p}\downarrow & & \downarrow{\scriptstyle p_2} \\ N & \xrightarrow[\text{id}]{} & N \end{array}$$

Then there is a *unique* fiber bundle map (F, id)

$$\begin{array}{ccc} M & \xrightarrow{F} & M_1 \times_N M_2 \\ p\downarrow & & \downarrow p_1 \times_N p_2 \\ N & \xrightarrow[\mathrm{id}]{} & N \end{array}$$

such that $p_1^* \circ F = q_1$ and $p_2^* \circ F = q_2$. ■

Exercises

1. Prove the last assertion in the previous definition: that the Whitney product is a 'product' in the above category-theoretic sense.
2. Prove that if, in the notation of the definition above, p_1 is a vector bundle projection, then p_1^* is also a vector bundle projection. What is the fiber?

Definition 2.3. Suppose $K = [C_1, C_2, \ldots, C_r; \leqslant]$ is an oriented subcomplex of $\Gamma(k-1)$. For each C_i a $(c_i - 1)$ face, let the natural bundle mappings

$$\begin{array}{ccc} T_k(M) & \xrightarrow[D(C_i)]{} & T_{c_i}(M) \\ q_k\downarrow & & \downarrow q_{c_i} \\ M & \xrightarrow[\mathrm{id}]{} & M \end{array}$$

be constructed. Let

$$\begin{array}{c} X_M[T_{c_i}(M)] \\ \downarrow x_m(q_{c_i}) \\ M \end{array} \qquad 1 \leqslant i \leqslant r$$

denote the Whitney product of the bundles

$$\begin{array}{c} T_{c_i}(M) \\ \downarrow q_{c_i} \\ M \end{array} \qquad \text{with} \qquad q_{c_i} \circ Q_{c_i} = X_M(q_{c_i})$$

for all i. There is then (from the universal property of Whitney products) a unique fiber bundle mapping (F, id)

$$\begin{array}{ccc} T_k(M) & \xrightarrow[F]{} & X_M[T_{c_i}(M)] \\ q_k\downarrow & & \downarrow X_M(q_{c_i}) \\ M & \xrightarrow[\mathrm{id}]{} & M \end{array}$$

such that $D(C_i) = Q_{c_i} \circ F$ for all i.

Now the image of F gives a *sub-bundle*

$$\begin{array}{c}(\operatorname{im} F)\\ \big\downarrow{\scriptstyle \chi_M(q_{c_i})_{|\operatorname{im}(F)}}\\ M\end{array}$$

of the Whitney product. Denote it

$$\begin{array}{c}T_k[M; K]\\ \big\downarrow{\scriptstyle q_k}\\ M\end{array}$$

and call it the K-complex bundle of tangent sectors, or the *K-sector bundle.*

Elements of the K-sector bundle

$$\begin{array}{c}T_k[M; K]\\ \big\downarrow{\scriptstyle q_k}\\ M\end{array}$$

say at a point $x_0 \in M$, can be represented as $[\phi_k, X^i_\gamma]$ with the restriction $\gamma \in K$, for ϕ a frame-germ at x_0, and $X^i_\gamma \in E$. The transformation laws apply *mutatis mutandis* to such symbols and their interpretation as equivariant elements of the appropriate product bundle will be left as an exercise. Similarly, for $\phi: V \to U \subset M$ a local frame, a local cross section of the K-sector bundle (a K-sector *field*) will be denoted $[(\phi \circ t_z)_k, X^i_\gamma(z)]$ or simply $\tilde{X}^i_\gamma(z)$ when the frame is understood.

Finally, in the low dimensions, a K-sector at a point may be represented in components

$$[\phi_2, (X^i) \qquad (Y^j)] \qquad \text{or} \qquad \begin{array}{c}[\phi_3, X^i \text{——} A^r \text{——} Y^j]\\ \\ \\ (Z^k)\end{array}$$

and so on, where, unless mention is made to the contrary, the standard ordering is in force. ■

Exercises

3. Let K be a complex in $\Gamma(k-1)$, $K = [C_1, C_2, \ldots, C_r]$. Let $j_i: \Gamma(c_i - 1) \to \Gamma(k-1)$ be as in the opening discussion of this chapter, and let $j_i^*: E_k \to E_{c_i}$ be the adjoints. Let $E_k(K)$ be the vector space in $\underset{i}{X} j^*_i$, $\underset{i}{X} j^*_i: E_k \to \underset{i}{X} E_{c_i}$. Thus, $E_k(K) \subset \underset{i}{X} E_{c_i}$. Define an action of $G_k(m)$ on $E_k(K)$ with respect to which $T_k[M; K]$ is the quotient of $F_k(M) \times E_k(K)$.

4. Construct $T_2[M; K]$ where K is the complex (* *) in $\Gamma(1)$, and describe in terms of $T_1(M)$.

The association for a given complex $K \subset \Gamma(k-1)$

$$\begin{array}{ccc} M \mapsto T_k[M, K] & \text{and} & f \mapsto T_k[f, K] \\ \downarrow & & \\ M & & \end{array}$$

gives in the usual way a covariant functor from the category of smooth manifolds and maps to the category of fiber bundles and fiber bundle maps.

Now let K be the subcomplex (* *) of $\Gamma(1)$. Elements of $T_2[M; K]$ have the form $[\phi_2, (X^i) \quad (Y^j)]$ for ϕ_2 a frame-jet at $x_0 \in M$ and they transform according to the following rule: if $\theta_2 g_2 = \phi_2$ then

$$[\phi_2, (X^i) \qquad (Y^j)] = \left[\theta_2, \left(\frac{\partial g^i}{\partial x^a} X^a\right) \qquad \left(\frac{\partial g^j}{\partial x^b} Y^b\right)\right].$$

In this case, we may *identify*

$$\begin{array}{ccc} T_2[M; K] & \text{with} & T_1(M) \times_M T_1(M) \\ \downarrow{\scriptstyle q_2} & & \downarrow{\scriptstyle q_1 \times_M q_1} \\ M & & M \end{array}$$

(the product of the tangent bundle with itself). For simplicity denote the total space of this last bundle $T_1^2(M)$ and represent the elements $[\phi_1, (X^i, Y^j)] \cong [\phi_2, (X^i) \qquad (Y^j)]$.

Finally, the projection

$$\begin{array}{c} T_2(M) \\ \downarrow{\scriptstyle F} \\ T_2[M; K] \end{array}$$

will be denoted the 2-*boundary transformation* and will be represented (via the identifications above)

$$\begin{array}{c} T_2(M) \\ \downarrow{\scriptstyle D(2)} \\ T_1^2(M) \end{array}$$

with

$$D(2)[\phi_2, X^i \text{——} A^r \text{——} Y^j] = [\phi_1, (X^i, Y^j)]. \tag{2.1}$$

We give the 2-boundary transformation a special notation because it will appear in a number of different contexts. While $D(2)$ is not a vector bundle projection, it is the projection of an especially simple type of bundle. We elaborate with the following observation.

Observation 2.4. The 2-boundary transformation

$$\begin{array}{c} T_2(M) \\ \downarrow D(2) \\ T_1^2(M) \end{array}$$

is a fiber bundle projection with fiber $E=\mathbb{R}^m$. Further, it is a *natural transformation* in the sense that if $f:M\to N$ is smooth, then the diagram below commutes:

$$\begin{array}{ccc} T_2(M) & \xrightarrow{T_2(f)} & T_2(N) \\ {\scriptstyle D(2)}\downarrow & & \downarrow{\scriptstyle D(2)} \\ T_1^2(M) & \xrightarrow[T_1^2(f)]{} & T_1^2(N) \end{array}$$

Proof. This is standard. We give some details of construction here because they will be useful in the immediate sequel.

Choose a frame $\phi:O\to U\subset M$. We use this to construct a frame $\tilde{\phi}$ for $T_1^2(M)$. Thus, let $\tilde{\phi}:O\times(E\times E)\to T_1^2(M)$ by $\tilde{\phi}(z,(X^i,Y^j))=[(\phi\circ t_z)_1,(X^i,Y^j)]$. This map is smooth.

In a similar way, define frame $\tilde{\tilde{\phi}}:O\times E^3\to T_2(M)$ by $\tilde{\tilde{\phi}}(z,X^i_\gamma)=[(\phi\circ t_z)_2,X^i_\gamma]$. It is clear that the following diagram commutes:

$$\begin{array}{ccc} O\times E^3 & \xrightarrow{\tilde{\tilde{\phi}}} & q_2^{-1}(U) \\ {\scriptstyle \mathrm{id}\times p}\downarrow & & \downarrow{\scriptstyle D(2)} \\ O\times(E\times E) & \xrightarrow{\tilde{\phi}} & (q_1^2)^{-1}(U) \end{array}$$

where $p(X^i$——A^r——$Y^j)=(X^i,Y^j)$. This guarantees that $D(2)$ is a projection of a fiber bundle with fiber E. Naturality of $D(2)$ is immediate. ■

Unfortunately, $D(2)$ is *not* the projection of a vector bundle. To see this, we look at the maps $\tilde{\theta}^{-1}\circ\tilde{\phi}$ and $\tilde{\tilde{\theta}}^{-1}\circ\tilde{\tilde{\phi}}$ induced by another frame $\theta:O'\to V\subset M$. Suppose that $\phi(z)=\theta(w)$. Then

$$\tilde{\theta}^{-1}\circ\tilde{\phi}(z,(X^i,Y^j))=\left(w,\left(\frac{\partial g^i}{\partial x^a}X^a,\frac{\partial g^j}{\partial x^b}Y^b\right)\right)$$

where $g=t_w^{-1}\circ\theta^{-1}\circ\phi\circ t_z$. Now

$$\tilde{\tilde{\theta}}^{-1}\circ\tilde{\tilde{\phi}}(z,X^i\text{——}A^r\text{——}Y^j)$$

$$=\left(w,\left(\frac{\partial g^i}{\partial x^a}X^a\text{——}\frac{\partial g^r}{\partial x^c}A^c+\frac{\partial^2 g^r}{\partial x^a\partial x^b}X^aY^b\text{——}\frac{\partial g^j}{\partial x^b}Y^b\right)\right).$$

The 'fiber' map takes

$$A^r \to \frac{\partial g^r A^c}{\partial x^c} + \frac{\partial^2 g^r}{\partial x^a \partial x^b} X^a Y^b$$

and this is an *affine map* from E to E, not a linear map in general. In order to see the structure of the bundle whose projection is $D(2)$, we make the following definitions which introduce a generalization of the notion of affine bundle and lead to an interesting analog to the notion of 'short exact sequence' of vector bundles.

If F is an affine space over E, for E a finite-dimensional vector space, recall that an *affine bundle with fiber F* is a fiber bundle

$$\begin{array}{c} M \\ \downarrow{\scriptstyle p} \\ N \end{array}$$

with the property that for frames $\phi : O \to U \subset N$, and $\theta : O' \to V \subset N$, there are trivializations $\tilde{\phi} : F \times O \to p^{-1}(U)$ and $\tilde{\theta} : F \times O' \to p^{-1}(V)$ such that if $\phi(z) = \theta(w)$ then $\tilde{\theta}^{-1} \circ \tilde{\phi}(\gamma, z) = (\delta\gamma, w)$ for δ fixed for each fiber but smoothly varying with z.

It amounts to the same thing to say that each fiber $p^{-1}(x_0)$ is an affine space over E, and that for some set of trivializations over frames $\tilde{\phi} : F \times O \to p^{-1}(U)$ the restriction of $\tilde{\phi}$ to each fiber is a *translation*. (That is, for each $z \in O$, there is associated $y \in p^{-1}(\phi(z))$ varying smoothly with z such that for some γ_0 fixed in F, $\tilde{\phi}(\delta\gamma_0, z) = \delta y$ for $\delta \in E$.) The translations are a distinguished set of affine maps between affine spaces: a *pair* (A, f), where $A : F \to F'$ is a mapping between affine spaces over E and $f : E \to E$ is a homomorphism, defines an *affine map* if $A(\delta\gamma) = f(\delta)A(\gamma)$ for $\delta \in E$ and $\gamma \in F$. The translations are the affine maps for which $f = \text{id}$.

Now suppose that $p : M \to N$ is a fiber bundle with fiber F. Suppose also that

$$\begin{array}{c} V \\ \downarrow{\scriptstyle \pi} \\ N \end{array}$$

is a *vector bundle* with fiber E. Form the bundle

$$\begin{array}{c} V \times_N M \\ \downarrow{\scriptstyle \pi \times_N p} \\ N \end{array}$$

Now the idea is to define for *each* fiber $p^{-1}(x_0)$ an *affine structure* over the fiber vector space $\pi^{-1}(x_0)$. This is the direction in which the notion of affine bundle must be generalized.

Definition 2.5. Suppose p and π are as above and the product bundle with projection $\pi \times_N p$ is constructed. Suppose also that there is a smooth bundle mapping (λ, id):

$$\begin{array}{ccc} V \times_N M & \xrightarrow{\lambda} & M \\ {\scriptstyle \pi\times_N p}\downarrow & & \downarrow{\scriptstyle p} \\ N & \xrightarrow{\text{id}} & N \end{array}$$

satisfying the following conditions:

1) $\lambda(v+w, y) = \lambda(v, \lambda(w, y))$ for $v, w \in \tau^{-1}(x)$, $y \in p^{-1}(x)$ (note that there is a well defined addition in each fiber of π) for each $x \in N$.
2) $\lambda(0, y) = y$ for $y \in p^{-1}(x)$ for each $x \in N$.
3) for *each* pair of points y_1, y_2 in $p^{-1}(x)$ there is a unique v in $\pi^{-1}(x)$ such that $\lambda(v, y_1) = y_2$; this is so for each x in N.

Under these conditions, we say the fiber bundle p is an *affine bundle* over the vector bundle π. The case where p is an affine bundle over the vector space E corresponds to the situation where π is the projection of the *product bundle* $V = E \times N$. ■

According to this definition, each fiber space $p^{-1}(x)$ is an affine space over the corresponding fiber vector space $\pi^{-1}(x)$. Now, in order to use this definition, we give the following sufficient conditions that p be an affine bundle over vector bundle π: these conditions are analogous to those restated at the beginning of this discussion for a fiber bundle to be an affine bundle over a vector space E.

Proposition 2.6. Suppose that

$$\begin{array}{c} M \\ \downarrow{\scriptstyle p} \\ N \end{array}$$

is a fiber bundle with fiber F and that

$$\begin{array}{c} V \\ \downarrow{\scriptstyle \pi} \\ N \end{array}$$

is a vector bundle with fiber E. Then *p is an affine bundle over π* if the following conditions are met. There are sets of trivializations $\tilde{\phi}: F \times O \to p^{-1}(O)$ and $\tilde{\tilde{\phi}}: E \times O \to \pi^{-1}(O)$ over some family of open sets O covering N (simultaneous trivializations) such that for each O in the covering family there is a smooth fiber-preserving map $\lambda_O: (E \times F) \times O \to F \times O$ such that for each $z \in O$, $\lambda_O: (E \times F) \times z \to F \times z$ gives $F \times z$ the structure of affine space over E (Definition 1.18) where $\lambda_O(\delta, (f, z)) \equiv \delta(f, z)$ defines the action for $\delta \in E$.

Finally, these maps λ_O satisfy the following consistency conditions: Suppose that O', O are sets in the covering family and that $\tilde{\theta}: F \times O' \to p^{-1}(O')$ and $\tilde{\tilde{\theta}}: E \times O' \to \pi^{-1}(O')$ are the corresponding trivializations for O', and suppose that $O \cap O' = O''$. The map $(E \times F) \times O \to (E \times O) \times (F \times O)$ taking $((v, f), z) \to ((v, z), (f, z))$ will be called h_O. Then the following diagram should commute:

$$\begin{array}{ccccc} (E \times O'') \times (F \times O'') & \xleftarrow{h_O} & (E \times F) \times O'' & \xrightarrow{\lambda_O} & F \times O \\ \big\downarrow{\scriptstyle (\tilde{\tilde{\theta}}^{-1} \circ \tilde{\tilde{\phi}}) \times (\tilde{\theta}^{-1} \circ \tilde{\phi})} & & & & \big\downarrow{\scriptstyle \tilde{\theta}^{-1} \circ \tilde{\phi}} \\ (E \times O'') \times (F \times O'') & \xrightarrow[h_{O'}^{-1}]{} & (E \times F) \times O'' & \xrightarrow[\lambda_{O'}]{} & F \times O' \end{array}$$

Proof. The statement of the proposition is more difficult than the proof. We define $\lambda : V \times_N M \to M$ in the obvious way. A typical trivialization for

$$\begin{array}{c} V \times_N M \\ \downarrow \\ N \end{array}$$

over $O \subset N$ is given by a map such as

$$(\tilde{\tilde{\phi}} \times \tilde{\phi}) \circ h_O : (E \times F) \times O \to (\pi \times_N p)^{-1}(O).$$

The 'consistency condition' given by the diagram above is the standard condition for defining a bundle map by specifying it on the local products given by pairs of trivializations.

What is perhaps more useful than the details of this argument will be an *interpretation* of these conditions. The map $\tilde{\theta}^{-1} \circ \tilde{\phi} : F \times O'' \to F \times O''$ will be called $(f, z) \mapsto (A_z(f), z)$ for $z \in O''$. Also suppose the map $\tilde{\tilde{\theta}}^{-1} \circ \tilde{\tilde{\phi}} : E \times O'' \to E \times O''$ has the form $(v, z) \mapsto (\gamma_z(v), z)$ for γ_z a *linear* map from $E \to E$ depending smoothly on z. Then the composition $h_{0'}^{-1} \circ (\tilde{\tilde{\theta}}^{1} \circ \tilde{\tilde{\phi}}) \times (\tilde{\theta}^{-1} \circ \tilde{\phi}) \circ h_O$ can be written in the following simple form: $(v, f; z) \to (\gamma_z(v), A_z(f); z)$. Commutativity of the diagram then means that

$$\gamma_z(v) \cdot A_z(f) = A_z(v \cdot f) \tag{2.2}$$

where the left-hand side has action computed with respect to the affine structure given by $\lambda_{O'}$ and the right-hand side has action computed with respect to λ_O.

Equation (2.2) expresses the fact that for each $z \in O''$, the map $A_z : F \to F$ gives an *affine* map of the form (A_z, γ_z). ■

We are now in a position to elaborate the nature of the 2-boundary transformation $D(2): T_2(M) \to T_1^2(M)$. Recall that it is a fiber bundle projection with fiber $E = \mathbb{R}^m$ (where $m = \dim M$). As a preliminary we consider the pair of bundles

$$\begin{array}{c} T_1(M) \\ \downarrow{\scriptstyle q_1} \\ M \end{array} \quad \text{and} \quad \begin{array}{c} T_1^2(M) \\ \downarrow{\scriptstyle q_1(2)} \\ M \end{array}$$

There is an obvious isomorphism of bundles from

$$\begin{array}{c} T_1^3(M) \\ \downarrow{\scriptstyle q_1(3)} \\ M \end{array}$$

to the *product* bundle

$$\begin{array}{c} T_1(M) \times_M T_1^2(M) \\ \downarrow{\scriptstyle q_1 \times_M q_1(2)} \\ M \end{array}$$

and we shall *identify* these bundles with each other via this natural isomorphism.

According to our discussion of products (Definition 2.2) there is a *vector bundle* projection

$$\begin{array}{c} T_1^3(M) \\ \downarrow{\scriptstyle q_1^*} \\ T_1^2(M) \end{array}$$

which associates to each ordered triple of tangent vectors the last two vectors

$$[\phi_1, (W^r, X^i, Y^j)] \underset{q_1^*}{\rightarrow} [\phi_1, (X^i, Y^j)].$$

We may now state our next theorem.

Theorem 2.7. *The fiber bundle*

$$\begin{array}{c} T_2(M) \\ \downarrow{\scriptstyle D(2)} \\ T_1^2(M) \end{array}$$

is an affine bundle *over the* vector bundle

$$\begin{array}{c} T_1^3(M) \\ \downarrow{\scriptstyle q_1^*} \\ T_1^2(M) \end{array}$$

In fact, for frame-jet ϕ_2 at x_0 in M, a typical element of $T_1^2(M)$ at x_0 may be represented $[\phi_1, (X^i, Y^j)]$; a typical element of $(q_1^*)^{-1}([\phi_1, (X^i, Y^j)])$

may be represented $[\phi_1, (W^r, X^i, Y^j)]$ *(so the fiber for* q_1^* *is* $E = \mathbb{R}^m$ *for m the dimension of M) and a typical element of* $D(2)^{-1}([\phi_1, (X^i, Y^j)])$ *may be represented* $[\phi_2, (X^i \text{——} Z^r \text{——} Y^j)]$. *The action of* $(q_1^*)^{-1}([\phi_1, (X^i, Y^j)])$ *on* $D(2)^{-1}([\phi_1, (X^i, Y^j)])$ *is given by*

$$[\phi_1, (W^r, X^i, Y^j)] \cdot [\phi_2, X^i \text{——} Z^r \text{——} Y^j] \\ = [\phi_2, X^i \text{——} Z^r + W^r \text{——} Y^j]. \quad (2.3)$$

Proof. Formula (2.3) can be *immediately* shown to be well-defined using the transformation rules for 1- and 2-sectors. The 'invariance' of that rule is the algebraic goal we are after at this stage of the development of the calculus. What follows is the topological justification for framing that rule in terms of affine bundles. (This is not simply an indulgence; this setting gives a very natural generalization of the notion of *short exact sequence of vector bundles* to the case of bundles such as $T_2(M)$ which have no natural linear structure in the fiber. We shall eventually formulate the 'covariant derivative' as a tensor associated to a 2-sector by means of a 'splitting' of $D(2)$ induced by the affine connection associated to a metric.)

We refer the reader to the proof of Observation 2.4 for a picture of the setup.

Now define a family of frames for $M, \phi: O \to U \subset M$, and let $\tilde{\phi}: O \times (E \times E) \to T_1^2(M)$ be the corresponding family of *frames* for the manifold $T_1^2(M) \cdot \tilde{\phi}(z, (X^i, Y^j)) = [(\phi \circ t_z)_1, (X^i, Y^j)]$. Then we have a covering of $T_1^2(M)$ by open sets, the images of the $\tilde{\phi}$.

For simplicity, the image of $\tilde{\phi}$ in $T_1^2(M)$ we shall call O'. Next define a trivialization of $T_2(M)$ over O' by the map $\phi': (O \times E \times E) \times E \to D(2)^{-1}(O')$ by the rule

$$\phi'((z, X^i, Y^j), Z^r) = [(\phi \circ t_z)_2, (X^i \text{——} Z^r \text{——} Y^j)].$$

Similarly, define a trivialization for $T_1^3(M)$ over $O' \cdot \phi'': (O \times E \times E) \times E \to (q_1^*)^{-1}(O')$ by $\phi'': ((z, X^i, Y^j), W^r) \to [(\phi \circ t_z)_1, (W^r, X^i, Y^j)]$. Now define $\lambda_{O'}: (O \times E \times E) \times E \times E \to (O \times E \times E) \times E$ by

$$\lambda_{O'}: (z, X^i, Y^j; Z^r, W^r) \to (z, X^i, Y^j; Z^r + W^r).$$

This makes each fiber of ϕ' an affine space over the corresponding fiber of ϕ'' for each point (z, X^i, Y^j) in $O \times E \times E$. Such a map $\lambda_{O'}$ is defined for each O'.

We have only to check the 'consistency conditions' (Proposition 2.6). Thus, suppose that $\theta: \bar{O} \to V$ is another frame on M, and that $\tilde{\theta}: \bar{O} \times E \times E \to T_1^2(M)$ is the corresponding frame for $T_1^2(M)$. Denote the image of $\bar{O}$ by $\bar{O}'$. Also, let θ' and θ'' be the corresponding maps for θ, and let $\lambda_{\bar{O}'}: (\bar{O} \times E \times E) \times E \times E \to (\bar{O} \times E \times E) \times E$ by the rule

$$(w, X^i, Y^j; Z^r, W^r) \to (w, X^i, Y^j; Z^r + W^r).$$

Finally, suppose $\phi(z)=\theta(w)$, and (z, X^i, Y^j) is fixed in $O\times E\times E$. Consider

$$(\theta')^{-1}\circ\phi'\circ\lambda_{O'}(z, X^i, Y^j; Z^r, W^r)$$
$$=\left(w, \frac{\partial g^i}{\partial x^a}X^a, \frac{\partial g^j}{\partial x^b}Y^b; \frac{\partial g^r}{\partial x^c}(Z^c+W^c)+\frac{\partial^2 g^r}{\partial x^a \partial x^b}X^aY^b\right)$$

where g is the germ of $t_w^{-1}\circ\theta^{-1}\circ\phi\circ t_z$.

This is easily seen to be equal to

$$\lambda_{\bar{O}'}\left(w, \frac{\partial g^i}{\partial x^a}X^a, \frac{\partial g^j}{\partial x^b}Y^b; \frac{\partial g^r}{\partial x^c}Z^c+\frac{\partial^2 g^r}{\partial x^a \partial x^b}X^aY^b, \frac{\partial g^r}{\partial x^c}W^c\right)$$

Thus the proof is complete in light of Proposition 2.6. ■

For calculations, the useful formula is (2.3). We think of $D(2)$ as a 'boundary' operator. This will be generalized shortly to the case of higher sectors, and a theorem analogous to Theorem 2.7 will be stated there. But this much is already clear: if $\bar{A}_2$ and $\bar{B}_2$ are sectors with the *same* boundary then there is a *unique* tangent vector (with source the same point as for the sector) which gives rise to a translation taking the first to the second. By this process, pairs of 2-sectors with the same boundary determine *tensors*. The local version of the Lie bracket of vector fields will be a notable example; in the case of 3-sectors, the Riemann curvature tensor will arise in this fashion.

Exercise

5. Let C be the subcomplex of $\Gamma(2)$ consisting of all faces but the 2-face. Let

$$\begin{array}{c} T_2^3(M) \\ \downarrow{\scriptstyle q_2(3)} \\ M \end{array}$$

be the product of

$$\begin{array}{c} T_2(M) \\ \downarrow{\scriptstyle q_2} \\ M \end{array}$$

with itself three times. Give an explicit description of the map $F: T_3(M)\to T_2^3(M)$ (C has three maximal 1-faces). Describe the fiber of the bundle

$$\begin{array}{c} T_3[M; C] \\ \downarrow{\scriptstyle q_3} \\ M \end{array}$$

The last topics to be discussed in this section will be the generalization of the 2-boundary operator to the case of higher-order sectors, and the derived operation which we shall call the *bracket* of sectors with the same boundary.

Let k be fixed, and suppose C is the subcomplex of $\Gamma(k-1)$ consisting of all faces but the $(k-1)$ face. As in the previous exercise we observe that there are k maximal $(k-1)$ faces in C. Denoting the product of

$$\begin{array}{c} T_{k-1}(M) \\ \downarrow q_{k-1} \\ M \end{array}$$

with itself k times as

$$\begin{array}{c} T_{k-1}^{k}(M) \\ \downarrow q_{k-1}(k) \\ M \end{array}$$

we may represent elements of

$$T_k[M; C]:[\phi_k, (X^{i_1}_{\gamma_{i_1}}, X^{i_2}_{\gamma_{i_2}}, \ldots, X^{i_k}_{\gamma_{jk}})]$$

for the $\gamma_j \in \Gamma(k-2)$ and subject to the condition that they be in image(F). We call

$$\begin{array}{c} T_k[M; C] \\ \downarrow q_k \\ M \end{array}$$

the *bundle of k-boundaries* and denote it simply

$$\begin{array}{c} B_k(M) \\ \downarrow q_{k-1}(k) \\ M \end{array}$$

and the map $F: T_k(M) \to B_k(M)$ we call $D(k)$ in analogy with the 2-dimensional case.

We are interested then in the *fiber bundle projection*

$$\begin{array}{c} T_k(M) \\ \downarrow D(k) \\ B_k(M) \end{array}$$

which we shall refer to as the *k-boundary operator*. It is clear that the fiber of this bundle is $E = \mathbb{R}^m$, but that $D(k)$ is *not* the projection of a vector bundle. Considering the two bundles

$$\begin{array}{ccc} B_k(M) & & T_1(M) \\ \downarrow{\scriptstyle q_{k-1}(k)} & \text{and} & \downarrow{\scriptstyle q_1} \\ M & & M \end{array}$$

we may again form the *product bundle*

$$\begin{array}{c} B_k \times_M T_1 \\ \downarrow{\scriptstyle q_{k-1}(k) \times_M q_1} \\ M \end{array}$$

Then

$$\begin{array}{c} B_k \times_M T_1 \\ \downarrow{\scriptstyle q_1^*} \\ B_k(M) \end{array}$$

is a *vector bundle* with fiber $E = \mathbb{R}^m$. Now we have the following extension of Theorem 2.7.

Theorem 2.8. 1) *The fiber bundle*

$$\begin{array}{c} T_k(M) \\ \downarrow{\scriptstyle D(k)} \\ B_k(M) \end{array}$$

is an affine bundle *over the* vector bundle

$$\begin{array}{c} B_k \times_M T_1 \\ \downarrow{\scriptstyle q_1^*} \\ B_k(M) \end{array}$$

2) *This affine bundle structure is* natural *in a sense to be made precise below,*

3) *Finally,* any *smooth cross section* $s : B_k(M) \to T_k(M)$ *gives a 'choice of origin' in each fiber for* $D(k)$, *thus establishing a* fiber bundle isomorphism

$$\begin{array}{ccc} B_k \times_M T_1 & \to & T_k(M) \\ \downarrow{\scriptstyle q_1^*} & & \downarrow{\scriptstyle D(k)} \\ B_k(M) & \to & B_k(M) \end{array}$$

Via this isomorphism the fibers of $D(k)$ *are endowed with linear structure and* $D(k)$ *becomes the projection of a* vector bundle.

Proof. For the first sentence, the topological details are easily handled following Theorem 2.7 once the action of fibers of q_1^* on corresponding fibers of $D(k)$ is established.

Suppose ϕ is a frame-germ at x_0 in M. Let us choose an element

$$[\phi_k, (X^{i_1}_{\gamma_{i_1}}, X^{i_2}_{\gamma_{i_2}}, \ldots, X^{i_k}_{\gamma_{i_k}})] \times [\phi_1, W^i]$$

of $B_k \times_M T_1$ over the element $[\phi_k, (X^{i_1}_{\gamma_{i_1}}, X^{i_2}_{\gamma_{i_2}}, \ldots, X^{i_k}_{\gamma_{i_k}})]$ of $B_k(M)$. Also suppose that $[\phi_k, Y^i_{\gamma'}] \in T_k(M)$ is in the fiber over this element of $B_k(M)$.

Then define

$$([\phi_k, (X^{i_1}_{\gamma_{i_1}}, X^{i_2}_{\gamma_{i_2}}, \ldots, X^{i_k}_{\gamma_{i_k}})] \times [\phi_1, W^i]) \cdot [\phi_k, Y^i_{\gamma'}] = [\phi_k, Z^i_{\gamma'}]$$

where

$$Y^i_{\gamma'} = Z^i_{\gamma'}, \qquad \text{if } \gamma' \text{ is } \textit{not} \text{ the } (k-1) \text{ face},$$
$$Y^i_{\gamma''} + W^i = Z^i_{\gamma''}, \qquad \text{for } \gamma'' \text{ the } (k-1) \text{ face}.$$

We must show that if θ_k is a k-frame at x_0 and if $\phi_k = \theta_k g_k$ for $g_k \in G_k(m)$, then the action computed with respect to θ_k gives the same result. Computing, we have (abusing the notation for group action on components)

$$([\theta_k, (g_{k-1} X^{i_1}_{\gamma_{i_1}}, \ldots, g_{k-1} X^{i_k}_{\gamma_{i_k}})] \times [\theta_1, g_1 W^i)]) \cdot [\theta_k, g_k Y^i_{\gamma'}] = [\theta_k, U^i_{\gamma'}].$$

Now

$$U^i_{\gamma'} = g_k Y^i_{\gamma'} = g_k Z^i_{\gamma'}, \qquad \text{if } \gamma' \text{ is not the } (k-1) \text{ face},$$
$$U^i_{\gamma''} = g_1 W^i + g_k Y^i_{\gamma''}, \qquad \text{for } \gamma'' \text{ the } (k-1) \text{ face},$$
$$g_k Y^i_{\gamma''} = g_1 Y^i_{\gamma''} + (\text{multilinear expressions in } Y^i_{\gamma'}) \quad \text{for lower-order faces } \gamma'.$$

It is easy to see that

$$g_k Z^i_{\gamma''} = g_1(W^i + Y^i_{\gamma''}) + (\text{the } \textit{same} \text{ multilinear terms}).$$

The reader is referred to equation (1.26) for verification.

For the second statement, we need the following notion to formulate naturality. Suppose that

$$\begin{array}{ccc} M_1 & & M_2 \\ \downarrow{\scriptstyle p_1} & \text{and} & \downarrow{\scriptstyle p_2} \\ N & & N \end{array}$$

are fiber bundles and that

$$\begin{array}{ccc} M'_1 & & M'_2 \\ \downarrow{\scriptstyle p'_1} & \text{and} & \downarrow{\scriptstyle p'_2} \\ N' & & N' \end{array}$$

are also fiber bundles. Suppose that $f: N \to N'$ is a smooth map and that $F_1: M_1 \to M'_1$ and $F_2: M_2 \to M'_2$ are smooth maps such that (F_1, f) and

(F_2, f) are fiber bundle maps. There is then induced a mapping of fiber bundles:

$$\begin{array}{ccc} M_1 \times_N M_2 & \xrightarrow[F_1 \times_f F_2]{} & M_1' \times_{N'} M_2' \\ {\scriptstyle p_1 \times_N p_2} \downarrow & & \downarrow {\scriptstyle p_1' \times_{N'} p_2'} \\ N & \xrightarrow[f]{} & N' \end{array}$$

where $(F_1 \times_f F_2)(x, y) = (F_1(x), F_2(y))$.

In this notation, the affine bundle structure for $T_k(M)$ over $B_k(M) \times_M T_1(M)$ is expressed as a certain bundle mapping $(\lambda(M), \mathrm{id})$

$$\begin{array}{ccc} [B_k(M) \times_M T_1(M)] \times_{B_k(M)} T_k(M) & \xrightarrow[\lambda(M)]{} & T_k(M) \\ \downarrow & & \downarrow {\scriptstyle D(k)} \\ B_k(M) & \xrightarrow{\mathrm{id}} & B_k(M) \end{array}$$

satisfying conditions 1)–3) of Definition 2.5.

Now if $f: M \to N$ is a smooth map of manifolds, it induces a bundle map

$$\begin{array}{ccc} B_k(M) \times_M T_1(M) & \xrightarrow[B_k(f) \times_f T_1(f)]{} & B_k(N) \times_N T_1(N) \\ \downarrow & & \downarrow \\ M & \xrightarrow[f]{} & N \end{array}$$

Next, there is a *bundle map* $([B_k(f) \times_f T_1(f)] \times_{B_x(f)} T_k(f), B_k(f))$:

$$\begin{array}{ccc} [B_k(M) \times_M T_1(M)] \times_{B_k(M)} T_k(M) & \to & [B_k(N) \times_N T_1(N)] \times_{B_k(N)} T_k(N) \\ \downarrow & & \downarrow \\ B_k(M) & \xrightarrow[B_k(f)]{} & B_k(N) \end{array}$$

Naturality of the affine structure means commutativity of the diagram:

$$\begin{array}{ccc} [B_k(M) \times_M T_1(M)] \times_{B_k(M)} T_k(M) & \xrightarrow{\lambda(M)} & T_k(M) \\ \downarrow {\scriptstyle [B_k(f) \times_f T_1(f)] \times_{B_k(f)} T_k(f)} & & \downarrow {\scriptstyle T_k(f)} \\ [B_k(N) \times_N T_1(N)] \times_{B_k(N)} T_k(N) & \xrightarrow[\lambda(N)]{} & T_k(N) \end{array}$$

A straightforward modification of the argument given above for coordinate changes (replace g_k with the k-jet of $\theta^{-1} \circ f \circ \phi$) guarantees that this diagram is commutative.

Now, for the third statement, suppose that $s: B_k(M) \to T_k(M)$ is a smooth section. Let $z \in B_k$, $s(z) \in T_k$, then map $(q_1^*)^{-1}(z) \to D(k)^{-1}(z)$ by the prescription $(z, \bar{A}_1) \to (z, \bar{A}_1)s(z)$. According to Definition 2.5, this is a bijection, and $(z, 0) \cdot s(z) = s(z)$. ■

This theorem leads naturally to the notion of the *bracket* of sectors. Suppose that $\bar{A}_k(x_0)$ and $\bar{B}_k(x_0)$ satisfy $D(k)\bar{A}_k = D(k)\bar{B}_k$. Under these conditions we shall say that $\bar{A}_k$ and $\bar{B}_k$ *have the same boundary.*

Definition 2.9. Let M be a smooth manifold of dimension m, and let $\bar{A}_k(x_0)$ and $\bar{B}_k(x_0)$ be k-sectors at x_0 with the same boundary. There is a *unique* tangent vector $[\phi_1, W^i]$ such that

$$((D(k)\bar{A}_k)\times[\phi_1, W^i])\cdot\bar{A}_k=\bar{B}_k$$

for ϕ_k some k-frame at x_0. Call the tangent vector $[\phi_1, W^i]$ at x_0 the *bracket* $[\bar{A}_k;\bar{B}_k]$.

Clearly $[\bar{A}_k;\bar{B}_k]=-[\bar{B}_k;\bar{A}_k]$. The bracket can be computed with the following simple rule. Let ϕ_k be a k-frame at x_0, and suppose that $\bar{A}_k=[\phi_k, X^i_\gamma]$ and $\bar{B}_k=[\phi_k, Y^i_\gamma]$. Let γ'' denote the $(k-1)$ face of the $(k-1)$ simplex. Under these conditions, we have

$$[\bar{A}_k;\bar{B}_k]=[\phi_1, Y^i_{\gamma''}-X^i_{\gamma''}]. \tag{2.4}$$

The bracket is *natural* in the following sense. If $f:M\to N$ is a smooth map then

$$[T_k(f)(\bar{A}_k);T_k(f)(\bar{B}_k)]=T_1(f)([\bar{A}_k;\bar{B}_k]). \tag{2.5}$$

This follows at once from the naturality of the affine bundle structure given in Theorem 2.8. ■

Exercises

6. Prove directly that the bracket of 3-sectors with the same boundary is natural in the sense of the above definition.

7. Let $\gamma:\mathbb{R}\to\mathbb{R}^3$ be a smooth curve with non-vanishing tangent vector. With the Euclidean metric in $\mathbb{R}^3$ use this γ to construct a section of the bundle

$$\begin{array}{c} T_2(\mathbb{R}) \\ \downarrow D(2) \\ T_1^2(\mathbb{R}) \end{array}$$

[*Hint:* arc length parametrization.]

§2.2. Local fields and variations

In this section, certain classical constructions will be reinterpreted in terms of sectors. In particular, the bracket operation will appear in one of its most well-known and natural contexts: as the Lie bracket of a pair of local vector fields. Via an operation which we shall discuss in some detail here, a smooth field of tangent sectors on a manifold M gives rise to a smooth field of tangent sectors on the tangent bundle $T_1(M)$. This operation is called *promotion* and will have some useful analytic and geometric interpretations, in particular when done in conjunction with bracket operations.

The purpose of the section is twofold. On the one hand, it can serve as an aid to the intuition in working with sectors. The heavily *algebraic* presentation of these invariants which preceded hardly exhausted their properties. The real interest of sectors is *geometrical* and it is to the development of some of their geometric properties that the central part of this book is addressed. Thus, examples from analysis and geometry will be indispensable. On the other hand, sectors and K-sector bundles will be the local elements in a calculus which can be used to give a unified account of some of the basic constructions in Riemannian geometry. These constructions are cast in a somewhat new light from the point of view of this book and the examples from this section will give a first indication of the direction and the scope of the calculus.

We must defer until Chapter 3 the real springboard from which this calculus takes its own characteristic development. Just as vector fields are dualized by and can be 'contracted with' differential forms, sector fields are dualized by and contracted with *sector forms*. It will be necessary to return later to some of the constructions and examples of this section to reinterpret and extend them in the light of their relationship with sector forms.

Let us begin with a simple example of the bracket operation which derives from the observation that the 'second derivative' is a tensor when the first derivative is 0.

Suppose that M and N are smooth manifolds and $f:(M, x_0)\to(N, y_0)$ is a smooth germ. Suppose also that the *derivative* of f at x_0 is 0, that is $T_1(f): T_{x_0}(M)\to T_{y_0}(N)$ is the zero map. Now consider the map $T_2(f): q_2^{-1}(x_0)\to q_2^{-1}(y_0)$. Let ϕ be a frame-germ at x_0 and θ a frame-germ at y_0, and let $\bar{f}=\theta^{-1}\circ f\circ\phi$. Then the action of $T_2(f)$ is given by

$$T_2(f):[\phi_2, X^i\!-\!\!-\!Z^i\!-\!\!-\!Y^i]\to\left[\theta_2, 0\!-\!\!-\!\frac{\partial^2\bar{f}^i}{\partial x^r\partial x^s}X^rY^s\!-\!\!-\!0\right]$$

since $\partial\bar{f}^i/\partial x^j=0$. It is easy to see that, for this map, $T_2(f)$ has constant value on the fibers of $D(2): T_2(M)\to T_1^2(M)$ over x_0 and therefore *passes to quotients* giving a map $q_1(2)^{-1}(x_0)\to D(2)^{-1}[\theta_2, (0,0)]$.

Now define the *zero* 2-sector at y_0 to be the sector $[\theta_2, 0\!-\!\!-\!0\!-\!\!-\!0]$ and call it $\bar{O}_2(y_0)$. Its components are the same in every frame. It allow us to establish an isomorphism between $D(2)^{-1}[\theta_2, (0,0)]$ and $T_{y_0}(N)$ by taking $\bar{A}_2(y_0)$ to $[\bar{O}_2(y_0); \bar{A}_2(y_0)]$.

We shall always identify the set of 2-sectors with 'zero boundary' at a point with the tangent space at that point via this isomorphism.

Therefore f induces a map from the fiber of $T_1^2(M)$ over x_0 (pairs of tangent vectors at x_0) to the tangent space of N at y_0. This map is called the *Hessian* of f. We see from the formula given for calculating it in frames that it is *symmetric* and *bilinear*.

It was important that the derivative of f vanished at x_0 in order to make this definition work; however, a slight generalization is possible if we are willing to settle for a weaker invariant associated to f. Suppose that $K \subset T_{x_0}(M)$ is the *kernel* of the derivative of f at x_0. Abusing language, write $K \times K \subset q_1(2)^{-1}(x_0)$. Then

$$T_2(f): D(2)^{-1}(K \times K) \to D(2)^{-1}[\theta_2, (0, 0)] \equiv T_{y_0}(N)$$

by the rule

$$T_2(f): [\phi_2, X^i —— Z^i —— Y^i]$$
$$\to \left[\theta_2, 0 —— \frac{\partial \bar{f}^i}{\partial x^j} Z^j + \frac{\partial^2 \bar{f}^i}{\partial x^r \partial x^s} X^r Y^s —— 0\right].$$

'Passing to quotients' leads to difficulties: since Z^j is arbitrary, the image is only well-defined up to the *quotient* of $T_{y_0}(N)$ by $\mathrm{im}(T_{x_0}(f))$. Thus, *arbitrary* $f: (M, x_0) \to (N, y_0)$ produces a *bilinear* and *symmetric* map defined on pairs of tangent vectors in the *kernel* of $T_{x_0}(f)$ and taking values in the *cokernel* of $T_{x_0}(f)$ (e.g. Smale [19]).

By this procedure, we may easily associate to a map $(M, x_0) \to (N, y_0)$ with 0 $(k-1)$ *jet* a *k-linear* map from $T_1^k(M)$ over x_0 to $T_{y_0}(N)$.

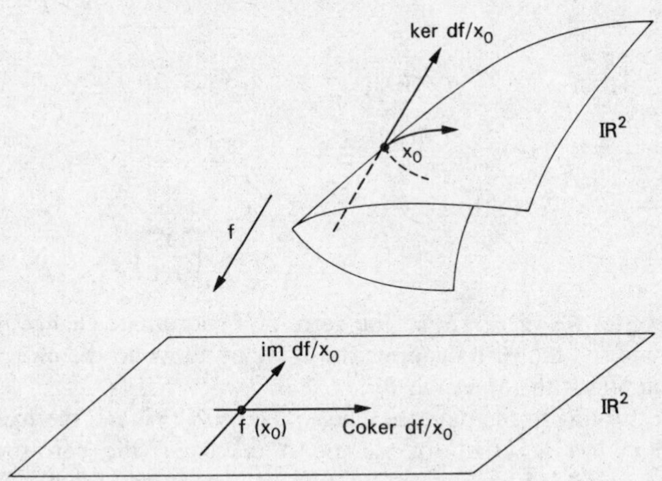

Fig. 2.1. Setup for generalized Hessian at x_0 of $f: \mathbb{R}^2 \to \mathbb{R}^2$ with $df_{|x_0}$ of rank 1

The next example will be developed and expanded in some detail in the sequel. Here, we present the simplest case. Recall that if $\phi: O \to U \subset M$ is a frame, then a local vector field (over U) may be represented as $[(\phi \circ t_z)_1, Y^i(z)]$ or, if reference to ϕ is suppressed, as $Y^i(z)\, \delta/\delta x^i$, $z \in O$.

Suppose that $\phi(z_0) = x_0 \in U$ and let $X^i(z_0)\, \partial/\partial x^i$ be a tangent vector at x_0. For simplicity, represent the former local vector field by $\tilde{Y}^i(z)$ to indicate that it is defined on U (although $z \in O$) and the latter tangent vector $X^i(z_0)$.

Definition 2.10. If $\phi: O \to U \subset M$ is a frame on a smooth manifold M, and if $\phi(z_0) = x_0$, let $X^i(z_0)$ be a tangent vector at x_0 and let $\tilde{Y}^j(z)$ be a *local vector field* on U.

Define the *suspension* of $\tilde{Y}^j(z)$ by $X^i(z_0)$: $X^i(z_0) \cup \tilde{Y}^j(z)$ to be the following 2-sector at x_0:

$$\left[(\phi \circ t_{z_0})_2, X^i(z_0) \text{---} \frac{\partial Y^r}{\partial x^a}_{|z_0} X^a(z_0) \text{---} Y^j(z_0)\right]. \tag{2.6.}$$

Of course, we must show that this is well-defined. Suppose that $\theta: O' \to V \subset M$ is another frame, $\theta(u_0) = \phi(z_0) = x_0$, $\theta(u) = \phi(z)$ for u in some neighborhood of u_0 and z in the corresponding neighborhood of z_0. Then we may write for $g(z) = \theta^{-1} \circ \phi(z)$:

$$[(\phi \circ t_{z_0})_1, X^i(z_0)] = \left[(\theta \circ t_{u_0})_1, \frac{\partial g^i}{\partial x^a}_{|g^{-1}(u_0)} \cdot X^a(g^{-1}(u_0))\right],$$

$$[(\phi \circ t_z)_1, Y^j(z)] = \left[(\theta \circ t_u)_1, \frac{\partial g^j}{\partial x^b}_{|g^{-1}(u)} \cdot Y^b(g^{-1}(u))\right].$$

Calculating

$$\frac{\partial}{\partial u^k}_{|u_0} \left(\frac{\partial g^j}{\partial x^b}_{|g^{-1}(u)} \cdot Y^b(g^{-1}(u))\right)\left(\frac{\partial g^k}{\partial x^s}_{|g^{-1}(u_0)} \cdot X^s(g^{-1}(u_0))\right),$$

we have

$$\frac{\partial^2 g^j}{\partial x^b \partial x^s}_{|g^{-1}(u_0)} Y^b(g^{-1}(u_0)) X^s(g^{-1}(u_0)) + \frac{\partial g^j}{\partial x^b}_{|g^{-1}(u_0)} \left(\frac{\partial Y^b}{\partial x^s}\right)_{|g^{-1}(u_0)} \cdot X^s(g^{-1}(u_0)).$$

Now if we let $t_{u_0}^{-1} \circ g(z_0) \circ t_{z_0}$ be the germ of a coordinate change in $(E, 0)$, it is easy to see that it transforms the 2-sector above to the one obtained by calculating with respect to θ.

Now returning to the pair, the *tangent vector* $X^i(z_0)$ and the *local vector field* $\tilde{Y}^j(z)$, we see that for z_0 and $\tilde{Y}^j(z)$ fixed, the correspondence $X^i(z_0) \mapsto X^i(z_0) \cup \tilde{Y}^j(z)$ gives a *linear* mapping from $T_{x_0}(M) \to T_{2,1}[X^i(z_0) \cup \tilde{Y}^j(z)]$ (see Observation 1.26), the image space being the 2-sectors at x_0 with 1-side $Y^j(z_0)$. ■

In the case that $\tilde{X}^i(z)$ is a local vector field and $Y^j(z_0)$ a tangent vector at x_0, the symbol $\tilde{X}^i(z) \cup Y^j(z_0)$ will denote the 2-sector

$$\left[(\phi \circ t_{z_0})_2, X^i(z_0) \text{---} \frac{\partial X^r}{\partial x^a}\bigg|_{z_0} Y^a(z_0) \text{---} Y^j(z_0)\right].$$

This is read: the suspension of $\tilde{X}^i(z)$ by $Y^j(z_0)$. The 'tilde' will be used notationally to indicate a local field.

This construction will be extended in the following way. Suppose that we are given a frame $\phi: O \to U \subset M$ and let $\bar{A}_k(x)$ be a local k-sector field over U: $[(\phi \circ t_z)_k, X^i_\gamma(z)]$, with $\phi(z) = x$. Denote the field $\bar{A}_k(x)$ in components $\tilde{X}^i_\gamma(z)$. Suppose also that $W^i(z_0)$ is a tangent vector at $x_0 = \phi(z_0)$: $[(\phi \circ t_{z_0})_1, W^i(z_0)]$.

Definition 2.11. Under the above conditions, define the suspension of $\tilde{X}^i_\gamma(z)$ by $W^i(z_0)$, *i.e.* $\tilde{X}^i_\gamma(z) \cup W^i(z_0)$, to be the following $(k+1)$ sector at x_0. Recalling the notation of Theorem 1.27, let γ' denote a general face of $\Gamma(k)$ of the form $(\gamma'_1, \gamma'_2, \ldots, \gamma'_k; 0)$ and let γ'' denote the $(k+1)$ vertex $(0, 0, \ldots, 0; 1)$ of $\Gamma(k)$. Represent elements of E_{k+1} by $(U^i_{\gamma'} \text{---} Z^i_{\gamma' \cup \gamma''} \text{---} V^i_{\gamma''})$. Then

$$\tilde{X}^i_\gamma(z) \cup W^i(z_0) = [(\phi \circ t_z)_{k+1}, U^i_{\gamma'} \text{---} Z^i_{\gamma' \cup \gamma''} \text{---} V^i_{\gamma''}]$$

with $U^i_{\gamma'} = X^i_\gamma(z_0)$ for $\gamma = (\gamma_1, \gamma_2, \ldots, \gamma_k)$ identified with $\gamma' = (\gamma_1, \gamma_2, \ldots, \gamma_k; 0)$. Also $V^i_{\gamma''} = W^i$. Finally, $Z^i_{\gamma' \cup \gamma''}$ is defined by

$$Z^i_{\gamma' \cup \gamma''} = \frac{\partial X^i_\gamma}{\partial x^j}\bigg|_{z_0} W^j$$

(with γ identified with γ' as above). A simple argument using the transformation law for sectors represented in this form shows that this sector is well-defined.

In a similar way, let $W^i(z_0) \cup \tilde{X}^i_\gamma(z)$ denote the sector $\sigma^k[\tilde{X}^i_\gamma(z) \cup W^i(z_0)]$ for σ the shift in $S(k+1)$. ■

The suspension operator has the following useful interpretation.

Definition 2.12. Suppose for frame $\phi: O \to U \subset M$ that $\bar{A}_r(x)$ is a local r-sector field on U with components $\tilde{X}^i_\gamma(z)$ $(\phi(z) = x)$. This gives a local cross section of the bundle

$$\begin{array}{c} T_r(T_1(U)) \\ \downarrow q_r \\ T_1(U) \end{array}$$

for $T_1(U) = q_1^{-1}(U) \subset T_1(M)$. This cross section is given by the rule

$$W^i(z_0) \to \nu_U[W^i(z_0) \cup \tilde{X}^i_\gamma(z)] \tag{2.7}$$

for ν_U the natural isomorphism $T_{r+1} \to T_r(T_1(U))$ given in Theorem 1.27. Call this cross section, which is itself a local r-sector field, the promotion of $\bar{A}_r(x)$ on U. We shall denote it $\pi \bar{A}_r(x)$. ■

A notion related to suspension and which has important application in the formulation of mechanical laws (least-action principle) and in the description of the solutions of differential equations (equation of variations) is the idea of a (local) *variation*. In order to formulate it, we enlarge the notion of a map-germ.

Suppose, in general, that A, B, A', B' are sets. Let $B \subset A$ and $B' \subset A'$ and suppose that $f: A \to A'$ in such a way that $f(B) \subset B'$. Say that $f:(A, B) \to (A', B')$. Now the *germ* of $f:(A, B) \to (A', B')$ is the equivalence class of maps taking $(A, B) \to (A', B')$ which agree with f on a *neighborhood* of B in A. Such maps are therefore 'defined' on all of B. In the case that A is an open subset of a manifold and the *closure* of B (denoted $\mathrm{cl}(B)$) is contained in A, then we shall talk about the smooth germ of a smooth map and mean by it the equivalence class of smooth maps defined as above.

Consider a smooth map $f: I \to O$ where $I \subset \mathbb{R}$ is an interval, and $\phi: O \to U \subset M$ is a frame. Call such an f a *path* in U (although it takes values on O); f gives rise to a path in $T_1(U)$ by the rule

$$\bar{s} \to \left[(\phi \circ t_{f(\bar{s})})_1, \frac{\partial f^i}{\partial s}_{|\bar{s}}\right],$$

taking $I \to T_1(U)$. We use the following shorthand: let the germ $\phi \circ f \circ t_s: (\mathbb{R}, 0) \to (M, \phi(f(s))) = f^{\#}(s)$ give a 1-sector at $\phi(f(s))$ which we denote $\bar{f}_1^{\#}(s)$. Here it is understood that s is the standard linear coordinate for $\mathbb{R}$. Then the path in $T_1(U)$ can be described as the map $s \to \bar{f}_1^{\#}(s)$.

In a similar way, let $I \times J \subset \mathbb{R} \times \mathbb{R}$ be a product of intervals. Denote the standard linear coordinates for $\mathbb{R} \times \mathbb{R}: (s, t)$. Let $F: I \times J \to O$ be smooth map with $\phi: O \to U \subset M$ a frame. Call such an F a *ribbon* in U; F gives rise to a ribbon in $T_2(U)$ by the rule

$$(\bar{s}, \bar{t}) \to \left[(\phi \circ t_{F(\bar{s}, \bar{t})})_2, \left(\frac{\partial F^i}{\partial s}_{|(\bar{s}, \bar{t})} \text{———} \frac{\partial^2 F^i}{\partial s \partial t}_{|(\bar{s}, \bar{t})} \text{———} \frac{\partial F^i}{\partial t}_{|(\bar{s}, \bar{t})}\right)\right].$$

Again, let the germ $\phi \circ F \circ t_{(s,t)}: (\mathbb{R}^2, 0) \to (M, \phi \circ F(s, t)) = F^{\#}(s, t)$ give the 2-*sector* at $\phi(F(s, t))$, namely $\bar{F}_2^{\#}(s, t)$.

Definition 2.13. Suppose that with the above notation, $f: I \to O$ is a smooth path in U with respect to frame $\phi: O \to U \subset M$. A 1-*variation* of f is an *equivalence class* of germs of ribbons $F: (I \times J, I \times t_0) \to (O, \mathrm{im}(f))$, $F_{|I \times t_0} = f$ ($I \times t_0$ identified with I) with respect to the equivalence relation $F \simeq G$ if $\bar{F}_2^{\#}(s, t_0) = \bar{G}_2^{\#}(s, t_0)$ for all s.

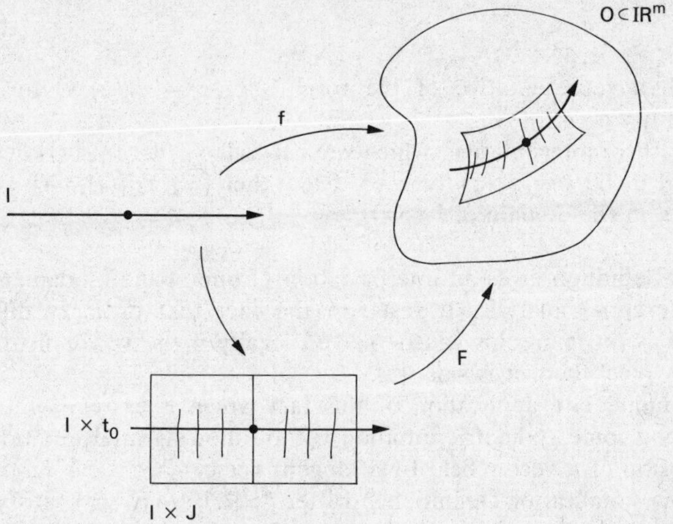

Fig. 2.2

For such ribbons it follows that $D^2(2)\bar{F}_2^{\#}(s, t_0) = \bar{f}_1^{\#}(s)$ for all s. Denote a 1-*variation* of f by the *path* $\bar{F}_2^{\#}(s, t_0): I \to T_2(U)$ given from any representative germ. ■

We think of a 1-variation of a smooth path as a smooth lifting of $\bar{f}_1^{\#}$:

$$\begin{array}{ccc} & \overset{\bar{F}_2^{\#}(s,t_0)}{\nearrow} & T_2(U) \\ & & \downarrow D^2(2) \\ I & \xrightarrow[\bar{f}_1^{\#}(s)]{} & T_1(U) \end{array}$$

subject to the additional constraint that it derive from a 'ribbon'. This last constraint is an 'integrability condition'.

In order to preceed further, it will be necessary to give a statement of the existence and uniqueness theorem for solutions of ordinary differential equations given by a smooth vector field. In particular, we define the notion of a local flow. (See Lefschetz [14] or Lang [12].)

Definition 2.14. Suppose $X = \bar{A}_1$ is a smooth vector field on a smooth manifold M, and let x_0 be a point of M. Choose a frame $\phi: O \to U \subset M$ and suppose $z_0 \in V \subset O$, V open and $\mathrm{cl}(V) \subset O$, and suppose that $\phi(z_0) = x_0$. Suppose that X has local representation $[(\phi \circ t_z)_1, X^i(z)] = \tilde{X}^i(z)$, for $z \in O$.

Under these conditions, there is an *invertible germ* γ_V called the *flow-germ for X with respect to ϕ on V* uniquely defined by the following properties:

a) $\gamma_V : (V \times \mathbb{R}, V \times O) \to (V \times \mathbb{R}, V \times O)$,
b) with a representative of the form $\gamma_V(z, s) = (\gamma_V^s(z), s)$ for γ_V^O the identity germ of V.
c) For this representative (which we also call γ_V) let $[\gamma_V(z)]$ denote the *path* in U: $s \to \gamma_V^s(z)$ for $z \in V$ fixed; then $\overline{[\gamma_V(z)]}_1^{\#}(s) = \tilde{X}^i(\gamma_V^s(z))$ for all s in the domain of $[\gamma_V(z)]$. ■

This definition gives an interpretation of one of the fundamental facts of differential analysis. It rests on the fact that ordinary differential equations on manifolds can be solved locally if the vector field is sufficiently 'nice' (here it is smooth).

An immediate application of this fact which is extremely useful for providing some geometric intuition is the following interpretation of the suspension of a vector field by a tangent vector: $\tilde{X}^i(z) \cup Y^i(z_0)$.

In the notation of Definition 2.14 let $f : (\mathbb{R}, 0) \to (V, z_0)$ satisfy

$$Y^i = \frac{\partial f^i}{\partial t}_{|0}$$

so that we may identify the tangent vector with components $Y^i(z_0)$ with the 1-jet f_1. If γ_V is the flow-germ for $\tilde{X}^i(z)$, then we may construct a ribbon at x_0 in the following way. Let $F : (\mathbb{R} \times \mathbb{R}, (0, 0)) \to (V, z_0)$, where we take (s, t) for the linear coordinates on $\mathbb{R} \times \mathbb{R}$, be the following germ:

$$F(s, t) = \gamma_V^s(f(t)).$$

F has this interpretation: it is the germ obtained by 'pushing' the tangent vector $Y^i(z_0)$ along the 'flow' given by the vector field $\tilde{X}^i(z)$.

Computing, we have

$$\frac{\partial F^i}{\partial s}_{|(0,0)} = X^i(z_0),$$

$$\frac{\partial F^i}{\partial t}_{|(0,0)} = Y^i(z_0),$$

$$\frac{\partial^2 F^k}{\partial s \partial t}_{|(0,0)} = \frac{\partial X^k}{\partial x^j}_{|z_0} Y^j(z_0).$$

Thus $\bar{F}_2^{\#}(0, 0) = \tilde{X}^i(z) \cup Y^i(z_0)$.

Returning to Definition 2.14, suppose a representative γ_V satisfying properties b) and c) of Definition 2.14 is found for a set of pairs (O_j, V_j), V_j covering M. Patch them together to give a global germ $g : (M \times \mathbb{R}, M \times$

$O) \to (M\times\mathbb{R}, M\times O)$ as a result of local uniqueness. If a representative for *this* germ is defined on a set $M\times I$ for I some open interval containing the origin, then it has a unique extension to a map $M\times\mathbb{R} \xrightarrow{G} M\times\mathbb{R}$ of the form $G(x, s)=(G^s(x), s)$ with each $G^s(x)$ a diffeomorphism of M, the correspondence $s\to G^s$ a homomorphism from $\mathbb{R}$ to the group diff(M) of diffeomorphisms of M, and finally the paths $[G(x)]: s\to G^s(x)$ for fixed $x\in M$ satisfying the condition that $\bar{A}_1(G^s(x))$ is a 1-jet of $[G(x)]: (\mathbb{R}, s)\to (M, G^s(x))$. Such a G is called a *global flow* and can be found for $\bar{A}_1 = X$, for example, when M is *compact*.

Returning to the local case, we now consider a class of differential equation which arises frequently in differential geometry and mechanics; while it is reducible to the previous case, it will be advantageous to formulate it in the setting of sectors. Suppose that M is a smooth manifold, $\phi: O\to U\subset M$ a frame and $f: I\to O$ a smooth *path* in U. Consider the diagram:

$$\begin{array}{ccc} & & T_1(U) \\ & & \downarrow q_1 \\ I & \xrightarrow[\phi\circ f]{} & U \end{array}$$

This can be extended to a *vector bundle* mapping *over* $\phi\circ f$

$$\begin{array}{ccc} I\times_U T_1(U) & \xrightarrow[(\phi\circ f)^*]{} & T_1(U) \\ q_1^*\downarrow & & \downarrow q_1 \\ I & \xrightarrow{\phi\circ f} & U \end{array}$$

for q_1^* the projection of a vector bundle with fiber isomorphic to that of q_1. This construction is entirely analogous to that given in Definition 2.2 and we will not repeat it here. The bundle with projection q_1^* is called the 'pull-back' of that with projection q_1 over $\phi\circ f$.

In a similar way, there is a *fiber bundle* projection

$$\begin{array}{c} I\times_U T_2(U) \\ q_2^*\downarrow \\ I \end{array}$$

the fiber being isomorphic as manifold to E^3.

Now consider the fiber bundle projections $\mathrm{id}\times_U D^j(2)$:

$$\begin{array}{c} I\times_U T_2(U) \\ \downarrow \\ I\times_U T_1(U) \end{array}$$

defined in the obvious way, for $j=1, 2$.

Observation 2.15. Under the above circumstances, call a (local) *linear variation equation* over the path f in U the following data:

a) a smooth section α of the bundle

$$\begin{array}{c} I \times_U T_2(U) \\ \alpha \uparrow \downarrow \mathrm{id} \times_U D^1(2) \\ I \times_U T_1(U) \end{array}$$

$\alpha(s, Y^i(f(s))) = (s, \bar{A}_2(s, Y^i(f(s))))$,

b) such that $D^2(2)\bar{A}_2(s, Y^i(f(s))) = \bar{f}_1^{\#}(s)$ for all s,

c) and such that for each s the induced map $T_1(\phi \circ f(s)) \to T_{2,2}(\bar{A}_2(s))$, $Y^i(f(s)) \xrightarrow{\pi \circ \alpha \circ \iota} (D^2(2))^{-1} f_1^{\#}(s)$ is *linear* with respect to the linear structure in $T_{2,2}(\bar{A}_2(s))$, where ι is the second coordinate inclusion, π the second coordinate projection.

These conditions are *independent of frame* and can thus be stated 'globally' for a path whose image is not contained in the image of a frame.

For a path f in U, if $\alpha : I \times_U T_1(U) \to I \times_U T_2(U)$ is a linear variation equation over f, then for s_0 *fixed* in I and $Y_0^i(f(s_0))$ *fixed* in $T_1(\phi \circ f(s))$ as 'initial conditions', there is a *unique* 1-*variation* of f represented by $F : (I \times J, I \times t_0) \to (O, \mathrm{im}(f))$ with $I \times t_0$ identified with I such that

1) $D^1(2)\bar{F}_2^{\#}(s_0, t_0) = Y_0^i(f(s_0))$,
2) $D^2(2)\bar{F}_2^{\#}(s, t_0) = \bar{f}_1^{\#}(s)$ for all s, and
3) $\bar{F}_2^{\#}(s, t_0) = \bar{A}_2(s, D^1(2)\bar{F}_2^{\#}(s, t_0))$.

Proof. Consider the vector field on $I \times E$ which associates to (s, Y^i) the tangent vector with components $(1, Z^r(s, Y^i))$ where in local components with respect to ϕ,

$$\bar{A}_2(s, Y^i(f(s))) = \left(\frac{\partial f^i}{\partial s}\bigg|_s \text{——} Z^r(s, Y^i) \text{——} Y^i \right).$$

According to the linearity condition on α, for each s, $Z^r(s, Y^i)$ is a linear function of Y^i. Thus we have defined a non-autonomous *linear* ordinary differential equation on E (see Arnold [2] or Hirsch and Smale [11]). Such differential equations have *global flows*. In particular, let $(s, W^i(s))$ be the unique solution path satisfying the initial condition $W^i(s_0) = Y_0^i$. Then we may choose representative germ F where $F(s, t) = f(s) + (t - t_0)W^i(s)$. Uniqueness is immediate. ■

Linear 1-variation equations over paths will arise in several contexts notably in the formulation of the notion of 'parallel translation' and of geodesics. They are also useful in the study of the modification in the solution of an ordinary differential equation when the 'initial conditions' suffer small perturbations (Arnold [2]) and the following formulation of the *equation of variations* is a simple application.

Proposition 2.16. Suppose M is a smooth manifold, $\phi: O \to U \subset M$ is a frame. Let $\tilde{X}^i(z)$ be a local vector field on U. Let path $f: I \to O$ for I an open interval be a solution for $\tilde{X}^i(z)$ with initial condition $f(s_0) = z_0 \in O$. Let $Y^i(z_0)$ be a tangent vector at $x_0 = \phi(z_0)$: $[(\phi \circ t_{z_0})_1, Y^i]$. Then there is a *unique* 1-variation of f with representative germ F satisfying $F: (I \times J, I \times t_0) \to (O, \text{im}(f))$

1) $D^1(2)\bar{F}_2^{\#}(s_0, t_0) = Y^i(z_0)$, and
2) $\bar{F}_2^{\#}(s, t_0) = \tilde{X}^i(f(s)) \cup [D^1(2)\bar{F}_2^{\#}(s, t_0)]$ for all s.

Proof. Define $\bar{A}_2(s, Y^i(f(s))) = \tilde{X}^i(f(s)) \cup Y^i(f(s))$. ■

Next we generalize the notion of a flow-germ for an ordinary differential equation. Suppose that $\bar{A}_2$ is a smooth section of the bundle

$$\begin{array}{c} T_2(M) \\ \downarrow q_2 \\ M \end{array}$$

and let (X, Y) denote the section of

$$\begin{array}{c} T_1^2(M) \\ \downarrow q_1(2) \\ M \end{array}$$

with X the 10-face and Y the 01-face of $\bar{A}_2$. This K-*sector field* then is the boundary of sector field $\bar{A}_2$.

Suppose further that $\phi: Z \to U \subset M$ is a frame, $V \subset Z$ is open and $\text{cl}(V) \subset Z$. Let $z \in V$, $\phi(z) = x$. Let the representation of $\bar{A}_2(x)$ with respect to frame ϕ be $[(\phi \circ t_z)_2, X^i(z) \text{——} Z^r(z) \text{——} Y^j(z)]$. Then the representation of (X, Y) will be $[(\phi \circ t_z)_1, (X^i(z), Y^j(z))]$. As usual, we represent these fields locally as $\tilde{X}^i(z)$ and $\tilde{Y}^j(z)$.

Definition 2.17. For (X, Y) a pair of vector fields as above, it is not in general true that $\tilde{X}^i(z) \cup Y^i(z_0) = X^i(z_0) \cup \tilde{Y}^i(z)$ for $z_0 \in V$. These sectors belong to the same $D(2)$ fiber over $(X, Y)(x_0)$ however, and they have a *bracket*. In particular, we may define a *vector field* over U by $[\tilde{X}^i(z) \cup Y^i(z_0); X^i(z_0) \cup \tilde{Y}^i(z)]$. This vector field over U associated to the pair (X, Y) is called the *Lie bracket* and its value at x_0 will be denoted $[X, Y](x_0)$. With respect to frame ϕ the components of $[X, Y](x_0)$ will be according to (2.6)

$$\left.\frac{\partial Y^j}{\partial x^i}\right|_{z_0} X^i(z_0) - \left.\frac{\partial X^j}{\partial x^i}\right|_{z_0} Y^i(z_0). \tag{2.8}$$

We shall say that a 2-sector field $\bar{A}_2(x)$ is *adapted to its boundary* on U if for each $z_0 \in Z$, $\phi(z_0) = x_0$

$$\tilde{X}^i(z) \cup Y^i(z_0) = \bar{A}_2(x_0) = X^i(z_0) \cup \tilde{Y}^i(z).$$

In this case, the derived pair of vector fields (X, Y) has *zero Lie bracket.* Finally, it is immediate that in general $[X, Y] = -[Y, X]$. ■

We are now in a position to state a generalization of the notion of 'flow-germ' to the case of certain elementary *partial differential equations.* The case we study here will have obvious generalization (see, for example, Spivak [21], pp. 6–11). We are concerned with the following situation; suppose $\bar{A}_2$ is a 2-sector field on M. We want to decide when it is possible to find local maps from $\mathbb{R}^2 \to M$ which 'induce' the field $\bar{A}_2$ locally.

Definition 2.18. Let M be a smooth manifold, $\phi: O \to U \subset M$ a frame, V open, $\mathrm{cl}(V) \subset O$ and let $\bar{A}_2(x)$ be a local 2-sector field on U, with $(X, Y)(x)$ the associated pair of vector fields. Then a *flow-germ* for $\bar{A}_2(x) = \tilde{X}^i_\gamma(z)$ with respect to ϕ on V is an invertible germ G_V uniquely defined by the properties:

a) $G_V: (V \times \mathbb{R}^2, V \times (0, 0)) \to (V \times \mathbb{R}^2, V \times (0, 0))$,
b) with a representative of the form $G_V(z; s, t) = (G_V^{s,t}(z); s, t)$ with $G_V^{0,0}$ the identity germ of V,
c) For this representative (which we also call G_V), let $[G_V(z)]$ denote the *ribbon* in U $(s, t) \to G_V^{s,t}(z)$ for z *fixed* in V; then $\overline{[G_V(z)]}{}^{\#}_2(s, t) = \tilde{X}^i_\gamma(G_V^{s,t}(z))$ for (s, t) in the domain of $[G_V(z)]$. ■

This is the direct generalization of the case in Definition 2.14 to the case of a 2-sector field. As it happens, a smooth 2-sector field will *not* have a flow-germ in general in this sense. This will lead eventually to such interesting notions as *curvature* on a Riemannian manifold. The following theorem, which makes an interesting application of the 1-*variation equation,* indicates the role of the Lie bracket in the study of this most elementary case.

Theorem 2.19. *Under the conditions of Definition* 2.18, $\bar{A}_2(x)$ *has a flow-germ with respect to* ϕ *on* V if and only if $\bar{A}_2(x)$ *is* adapted to its boundary *on* V. *In this case the flow-germ is unique.*

Proof. For $(X, Y)(x)$ the associated pair of vector fields, we introduce the following intermediate construction. Suppose that γ_V is the flow-germ for X with respect to ϕ on V and δ_V is the flow-germ for Y with respect to ϕ on V.

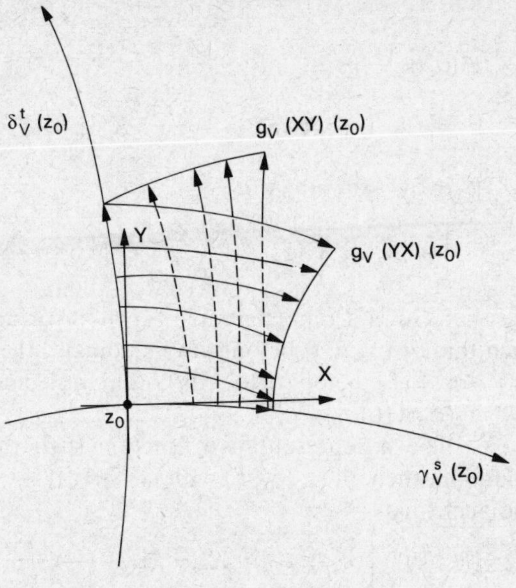

Fig. 2.3

Consider then the *germ* $g_V(XY):(V\times\mathbb{R}^2, V\times(0,0))\rightarrow(V\times\mathbb{R}^2, V\times(0,0))$ obtained by choosing *representative functions* for γ_V and δ_V which we also denote γ_V and δ_V such that for all points $(z; s, t)$ in some neighborhood of $V\times(0,0)$ the composition $\delta^t_V\circ\gamma^s_V(z)$ is defined. Then $g_V(XY)$ is the germ of the map taking $(z; s, t)\rightarrow(\delta^t_V\circ\gamma^s_V(z); s, t)$. It is easy to see that this germ is well-defined; again we let the symbol $g_V(XY)$ do double duty as the name of the germ and of the representative function given above from which it was obtained.

In a similar way, define the germ $g_V(YX):(V\times\mathbb{R}^2, V\times(0,0))\rightarrow(V\times\mathbb{R}^2, V\times(0,0))$ be choosing (perhaps different) representatives $\bar{\gamma}_V$ and $\bar{\delta}_V$ such that for all points $(z; s, t)$ in some neighborhood of $V\times(0,0)$ the composition $\bar{\gamma}^s_V\circ\bar{\delta}^t_V(z)$ is defined. Then $g_V(YX)$ is the germ of the map taking $(z; s, t)\rightarrow(\bar{\gamma}^s_V\circ\bar{\delta}^t_V(z); s, t)$. It is easy to see that $g_V(XY)$ and $g_V(YX)$ are *invertible germs.* The inverse can be constructed by reversing the two-parameter time. They are *not* flow-germs in general.

As a first step, we show that if $\bar{A}_2(x)$ has a flow-germ, that germ is equal to $g_V(XY)$ and is equal to $g_V(YX)$; further, $\bar{A}_2(x)$ is adapted to its boundary.

First, we compute: for z_0 fixed in V, let $[g_V(XY)(z_0)]$ denote the *ribbon* in U $(s, t)\rightarrow g_V(XY)^{s,t}(z_0)$; let $[g_V(YX)(z_0)]$ have similar meaning. Then the following equations give an interpretation of the Lie bracket (and of suspensions) in terms of the local flows associated to a pair of vector fields

$(X, Y)(x)$:

$$\overline{[g_V(XY)(z_0)]_2^{\#}}(0,0) = X^i(z_0) \cup \tilde{Y}^j(z)$$

$$= \left[(\phi \circ t_{z_0})_2, X^i(z_0) \text{——} \frac{\partial Y^r}{\partial x^a}\bigg|_{z_0} X^a(z_0) \text{——} Y^j(z_0)\right] \quad (2.9)$$

$$\overline{[g_V(YX)(z_0)]_2^{\#}}(0,0) = \tilde{X}^i(z) \cup Y^j(z_0)$$

$$= \left[(\phi \circ t_{z_0})_2, X^i(z_0) \text{——} \frac{\partial X^r}{\partial x^b}\bigg|_{z_0} Y^b(z_0) \text{——} Y^j(z_0)\right] \quad (2.10)$$

Now suppose that G_V is a representative function for a *flow-germ* for $\bar{A}_2(x)$ on V such that, if $(z_0; s, t)$ belongs to its domain, then all $(z_0; s', t')$ with $(s')^2 + (t')^2 \leq s^2 + t^2$ belong to its domain, and such that $\overline{[G_V(z_0)]_2^{\#}}(s, t) = \bar{A}_2(\phi \circ G_V^{s,t}(z_0))$.

Let now $g_V(XY)$ be a representative function such that, if $(z_0; s, t)$ belongs to its domain, then all $(z_0; s', t')$ with $(s')^2 + (t')^2 \leq s^2 + t^2$ belong to its domain, and such that

$$\overline{[g_V(XY)(z_0)]_2^{\#}}(s, 0) = \left[(\phi \circ t_{\gamma_V^s(z_0)})_2, X^i(\gamma_V^s(z_0)) \text{——} \frac{\partial Y^r}{\partial x^a}\bigg|_{\gamma_V^s(z_0)} \cdot X^a(\gamma_V^s(z_0)) \text{——} Y^j(\gamma_V^s(z_0))\right] \quad (2.11)$$

and

$$\overline{[g_v(XY)(z_0)]_2^{\#}}(s, \tau) = \left[(\phi \circ t_{\delta_V^\tau \circ \gamma_V^s(z_0)})_2, W^i(\delta_V^\tau \circ \gamma_V^s(z_0) \text{——} \frac{\partial Y^r}{\partial x^a}\bigg|_{\delta_V^\tau \circ \gamma_V^s(z_0)} \cdot W^a(\delta_V^\tau \circ \gamma_V^s(z_0)) \text{——} Y^j(\delta_V^\tau \circ \gamma_V^s(z_0))\right]. \quad (2.12)$$

[Note that $W^i(\gamma_V^s(z_0)) = X^i(\gamma_V^s(z_0))$.]

Then equation (2.11) implies (from the existence and uniqueness theorem for ordinary differential equations) that $g_V(XY)(z_0)(s_0, 0) = G_V(z_0; s_0, 0)$ for $|s_0| \leq |s|$. Another application of the existence and uniqueness theorem, this time with equation (2.12), gives $g_V(XY)(z_0)(s_0, t_0) = G_V(z_0; s_0, t_0)$ for s_0 as above, $|t_0| \leq |t|$. This shows that G_V and $g_V(XY)$ determine the *same germ*. A similar argument shows that G_V and $g_V(YX)$ determine the same germ also. Thus, if G_V exists, *it is unique*, and it follows from equations (2.9) and (2.10) that under this circumstance $\bar{A}_2(x)$ is *adapted to its boundary* on V.

Next we show that if $\bar{A}_2(x)$ is adapted to its boundary then it has a flow-germ on V. We let $g_V(XY)$ and $g_V(YX)$ be as above with the stated restrictions on the representative functions applying to their common domains.

Let I be an open interval containing the origin in $\mathbb{R}$ and let $z_0 \in V$. Suppose that for all $s \in I$, $(z_0; s, t_0)$ belongs to the common domain for some fixed t_0. Let $f(s) = g_V(YX)^{s,t_0}(z_0)$ for $s \in I$ define a path in U.

Clearly, this is a solution path for $\tilde{X}^i(z)$ with initial condition $f(0)=\bar{\delta}^{t_0}_{V}(z_0)$. Now let F be the 1-variation of the path f obtained from the germ

$$F(s,\tau)=g_V(YX)^{s,t_0+\tau}(z_0).$$

Following equation (2.12) we write the expression

$$\begin{aligned}\overline{[g_V(YX)(z_0)]}^{\#}_2(s,t_0)&=\bar{F}^{\#}_2(s,0)\\&=\left[(\phi\circ t_{\bar{\gamma}^s_V\circ\bar{\delta}^{t}_{V}(z_0)})_2,\ X^i(\bar{\gamma}^s_V\circ\bar{\delta}^{t_0}_{V}(z_0))\text{——}\frac{\partial X^r}{\partial x^a}_{|\bar{\gamma}^s_V\circ\bar{\delta}^{t_0}_{V}(z_0)}\right.\\&\qquad\left.\cdot\, U^a(\bar{\gamma}^s_V\circ\bar{\delta}^{t_0}_{V}(z_0))\text{——}U^i(\bar{\gamma}^s_V\circ\bar{\delta}^{t_0}_{V}(z_0))\right].\end{aligned}\tag{2.13}$$

Then for each $s\in I$ we have

1) $D^1(2)\bar{F}^{\#}_2(0,0)=U^i(\bar{\delta}^{t_0}_{V}(z_0))=Y^i(\bar{\delta}^{t_0}_{V}(z_0))$,
2) $\bar{F}^{\#}_2(s,0)=\tilde{X}^i(f(s))\cup[D^1(2)\bar{F}^{\#}_2(s,0)]$.

Thus F is the unique 1-variation of f which solves the 'equation of variations' for the initial condition 1) above guaranteed by Proposition 2.16.

Now consider the variation F' of f given by the germ $F'(s,\tau)=f(s)+\tau Y^i(f(s))$. It follows directly from the fact that $[X,Y]=0$ that F' is another solution to the equation of variations with initial condition $Y^i(f(0))$. Therefore, for each $s_0\in I$, $Y^i(f(s_0))=U^i(f(s_0))$. Referring again to (2.13) this shows that $g_V(YX)$ is a flow-germ for $\bar{A}_2(x)$ on V. Thus the representative functions $g_V(XY)$ and $g_V(YX)$ agree on their common domains. ■

The next construction gives a generalization of the bracket operation on sectors which will be useful for the study of iterated covariant derivatives and for an interpretation of the Jacobi identity.

Definition 2.20. Let γ_h be an $(h-1)$ face of $\Gamma(h+r-1)$, say $\gamma_h=(g_1,\ldots,g_h,g_{h+1},\ldots,g_{h+r})$ with $i_1<i_2<\ldots<i_h$ and $g_{i_j}=1$ for $1\leq j\leq h$, and $l_1<l_2<\ldots<l_r$ with $g_{l_k}=0$ for $1\leq k\leq r$. Define $C[\gamma_h]\subset\Gamma(h+r-1)$, the γ_h *star complex*, to be the set of faces of $\Gamma(h+r-1)$, $\gamma=(\gamma_1,\ldots,\gamma_{h+r})$, with $\gamma_{l_k}=0$ for at least one k with $1\leq k\leq r$.

$C[\gamma_h]$ has r maximal faces, each of order $(h+r-2)$ and these may be labelled $[D_{l_1},D_{l_2},\ldots,D_{l_r}]$ with $D_{l_k}=(1,1,\ldots,0,\ldots,1)$.

A fairly routine modification of Theorem 1.27 will allow an *identification* of

$$\begin{array}{c}T_{h+r}[M;C[\gamma_h]]\\ \big\downarrow{\scriptstyle \bar{q}_{h+r,h}}\\ T_h(M)\end{array}$$

with the bundle of r-boundaries

$$\begin{array}{c} B_r[T_h(m)] \\ \downarrow{\scriptstyle q_{r-1}(r)} \\ T_h(M) \end{array}$$

the trick being to define $\tilde{q}_{h+r,r}$ in this context. ■

Examples

A) Let γ_1 be a 0-face (vertex) of $\Gamma(2)$, then $C[\gamma_1]$ has one of the three forms:

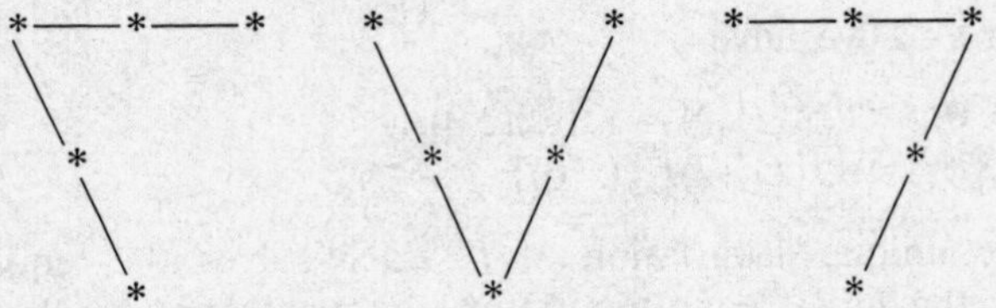

depending on the choice of vertex.

B) γ_0 the imposition of no constraints, yields the ordinary boundary complex.

C) $\gamma_{k-1} \in \Gamma(k-1)$ corresponds to the case $r=1$, $h=k-1$. In this case $T_k[M; C(\gamma_{k-1})]$ is essentially $T_{k-1}(M)$.

Exercises

8. Prove the statement in example C) via an explicit isomorphism of fiber bundles.
9. Given γ_h as in Definition 2.20, construct the map $\tilde{q}_{h+r,h}$ needed to make the identification stated there.

Now given $\gamma_h \in \Gamma(h+k-1)$ we consider the natural fiber bundle projection

$$\begin{array}{c} T_{h+r}(M) \\ \downarrow{\scriptstyle F} \\ T_{h+r}[M; C(\gamma_h)] \end{array}$$

(Definition 2.3). In light of the identification mentioned in Definition 2.20, we see that this is the projection of an affine bundle over a certain vector bundle, and it will be possible to define what we shall call a *generalized bracket* of pairs of elements in a fiber of F.

We proceed in the following way. For $\gamma_h \in \Gamma(h+r-1)$ *represent* elements of $T_{h+r}(M)$ *with respect to* γ_h following the convention established in Theorem 1.27. Letting $i_1 < i_2 < \ldots < i_h$ and $l_1 < l_2 < \ldots < l_r$ be as in Definition 2.20 with respect to γ_h write faces of $\Gamma(h+r-1)$:

1) $\gamma' = (g_1, \ldots, g_{h+r})$ with $g_{l_k} = 0$ for all $1 \leq k \leq r$ is a face of *type* 1,
2) $\gamma'' = (g_1, \ldots, g_{h+r})$ with $g_{i_j} = 0$ for all $1 \leq j \leq h$ is a face of *type* 2,
3) a face of *type* 3 has the form $\gamma' \cup \gamma''$.

An $(h+r)$ sector is *represented with respect to* γ_h when it is written in the form $\bar{A}_{h+r}(x_0) = [\phi_{h+r}, X^i_{\gamma'} — Z^k_{\gamma' \cup \gamma''} — Y^j_{\gamma''}]$ following the convention of Theorem 1.27.

Definition 2.21. Suppose that γ_h and $\bar{\gamma}_h$ are fixed $(h-1)$ faces of $\Gamma(h+r-1)$ and $\Gamma(h)$, respectively. Two elements $\bar{A}_{h+r}(x_0)$ and $\bar{B}_{h+r}(x_0)$ of $T_{h+r}(x_0)$ which belong to the same F fiber of the projection

$$\begin{array}{c} T_{h+r}(M) \\ \downarrow F \\ T_{h+r}[M; C(\gamma_h)] \end{array}$$

will be called γ_h*-coincident.*

Letting $\bar{A}_{h+r}(x_0)$ and $\bar{B}_{h+r}(x_0)$ be γ_h-coincident, represent them with respect to γ_h:

$$\bar{A}_{h+r}(x_0) = [\phi_{h+r}, X^i_{\gamma'} — Z^k_{\gamma' \cup \gamma''} — Y^j_{\gamma''}]$$

and

$$\bar{B}_{h+r}(x_0) = [\phi_{h+r}, X^i_{\gamma'} — \hat{Z}^k_{\gamma' \cup \gamma''} — \hat{Y}^j_{\gamma''}].$$

Define the $\gamma_h, \bar{\gamma}_h$ *bracket* of $\bar{A}$ and $\bar{B}$, denoted $[\bar{A}_{h+r}; \bar{B}_{h+r}]_{\gamma_h, \bar{\gamma}_h}$, to be the element of $T_{h+1}(x_0)$ whose representation *with respect to* $\bar{\gamma}_h$ is $[\phi_{h+1}, X^i_{\gamma'} — W^k_{\gamma' \cup e} — V^j_e]$, where e is the *unique* 0-face (vertex) of $\Gamma(0)$, and for E the unique $(r-1)$ face of $\Gamma(r-1)$ we define

$$V^j_e = \hat{Y}^j_E - Y^j_E,$$
$$W^k_{\gamma' \cup e} = \hat{Z}^k_{\gamma' \cup E} - Z^k_{\gamma' \cup E}.$$

Now a straightforward verification using the special form of the transformation law for sectors represented with respect to a face given in Theorem 1.27 shows that $[\bar{A}_{h+r}; \bar{B}_{h+r}]_{\gamma_h, \bar{\gamma}_h}$ is well-defined (that is, it transforms like an $(h+1)$ sector). This fact, together with the following 'naturality' property, will follow from the exercises below. This naturality property is that, if $f: M \to N$ is smooth, then

$$[T_{h+r}(f)\bar{A}_{h+r}; T_{h+r}(f)\bar{B}_{h+r}]_{\gamma_h, \bar{\gamma}_h} = T_{h+1}(f)[\bar{A}_{h+r}; \bar{B}_{h+r}]_{\gamma_h, \bar{\gamma}_h}. \qquad (2.14)$$

■

Exercise

10. Let γ_h and $\bar{\gamma}_h$ be as in the definition above, and consider the bundle

$$\begin{array}{c} T_{h+r}(M) \\ {\scriptstyle F}\downarrow \\ T_{h+r}[M;C(\gamma_h)] \end{array}$$

We show this is an affine bundle over a vector bundle in the following steps.

a) There is a bundle projection

$$\begin{array}{c} T_{h+r}[M;C(\gamma_h)] \\ \downarrow{\scriptstyle \bar{q}_{h+r,h}} \\ T_h(M) \end{array}$$

defined from γ_h using the standard ordering in $\Gamma(h+r-1)$ (exercise 9). Using $\bar{\gamma}_h$ define a *vector* bundle projection

$$\begin{array}{c} T_{h+1}(M) \\ \downarrow{\scriptstyle p} \\ T_h(M) \end{array}$$

b) Describe explicitly the vector bundle projection giving the linear structure of the fiber

$$\begin{array}{c} T_{h+r}[M;C(\gamma_h)]\times_{T_h(M)} T_{h+1}(M) \\ \downarrow{\scriptstyle p^*} \\ T_{h+r}[M;C(\gamma_h)] \end{array}$$

c) Now show that

$$\begin{array}{c} T_{h+r}(M) \\ \downarrow{\scriptstyle F} \\ T_{h+r}[M;C(\gamma_h)]. \end{array}$$

is an *affine bundle* over the *vector* bundle with projection p^* and that the *bracket* defined above reduces to translation in the fiber.

d) Argue naturality of this affine structure, and deduce the naturality formula (2.14) for the bracket.

Examples

A) $h=1$, $r=2$, let $\gamma_1=(0,0,1)$ and $\bar{\gamma}_1=(0,1)$, and let $\bar{A}_3$ and $\bar{B}_3$ be γ_1-coincident 3-sectors at $x_0\in M$. This means they have the same first two faces. Represent them as

$$\begin{array}{c}\bar{A}_3 = [\phi_3, X^i \text{——} A^r \text{——} Y^j] \\ \diagdown \qquad\qquad \diagup \\ B^s \; E^u \; C^t \\ \diagdown \quad \diagup \\ Z^k\end{array}$$

and

$$\begin{array}{c}\bar{B}_3 = [\phi_3, X^i \text{——} \bar{A}^r \text{——} Y^j] \\ \diagdown \qquad\qquad \diagup \\ B^s \; \bar{E}^u \; C^t \\ \diagdown \quad \diagup \\ Z^k\end{array}$$

Then

$$[\bar{A}_3; \bar{B}_3]_{\gamma_1, \bar{\gamma}_1} = [\phi_2, (\bar{A}^i - A^i) \text{——} (\bar{E}^r - E^r) \text{——} Z^j].$$

B) $h=2$, $r=1$; $\gamma_2=(0, 1, 1)$ and $\bar{\gamma}_2=(1, 1)$. In this case, the bracket of a pair of 3-sectors is a 3-sector also: it is their *difference* computed with respect to the additive structure of $T_{3,1}$.

Now consider the case of $\bar{A}_3(x)$ a local 3-sector *field* with respect to $\phi: Z \to U \subset M$. For notational convenience, we shall denote it

$$\begin{array}{c}[(\phi \circ t_z)_3, W_1^i(z) \text{———} A^r(z) \text{———} W_2^j(z)] \\ \diagdown \qquad\qquad\qquad \diagup \\ B^s(z) \quad E^u(z) \quad C^t(z) \\ \diagdown \qquad \diagup \\ W_3^k(z)\end{array}$$

The 0-faces (vertices) of $\bar{A}_3(x)$ give the following section of $T_1^3(M)$ over U: $(\tilde{W}_1^i(z), \tilde{W}_2^j(z), \tilde{W}_3^k(z))$ for $\phi(z)=x$. This section determines $3!=6$ distinct 3-sectors at a point $x_0 \in U$.

Definition 2.22. Let $\bar{A}_3(x)$ be as above and let $S(3)$ be the symmetric group on three letters. For $s \in S(3)$ arbitrary, we may form

$$W^i_{s^{-1}(1)}(z_0) \cup [W^i_{s^{-1}(2)}(z) \cup \tilde{W}^i_{s^{-1}(3)}(z)], \qquad \text{at } x_0 = \phi(z_0).$$

This 3-sector at x_0 has ordered sequence of 0-faces:

$$(W^i_{s^{-1}(1)}(z_0), W^i_{s^{-1}(2)}(z_0), W^i_{s^{-1}(3)}(z_0)).$$

Denote this sector $\bar{A}_3^s(x_0)$. ■

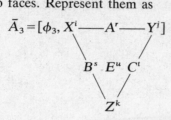

Suppose $e = \mathrm{id}$ in $S(3)$ and let $\bar{A}_3(x)$ be as above computed with respect to frame ϕ. Then we have

$$\bar{A}_3^e(x_0) = \left[\begin{array}{ccccc} (\phi \circ t_{z_0})_3,\ W_1^i(z_0) & \text{———} & \dfrac{\partial \tilde{W}_2^r}{\partial x^a}\Big|_{z_0} W_1^a(z_0) & \text{———} & W_2^j(z_0) \\ & \diagdown & & \diagup & \\ & \dfrac{\partial W_3^s}{\partial x^a}\Big|_{z_0} W_1^a(z_0) & (**) & \dfrac{\partial \tilde{W}_3^t}{\partial x^b}\Big|_{z_0} W_2^b(z_0) & \\ & & \diagdown \quad \diagup & & \\ & & W_3^k(z_0) & & \end{array}\right] \tag{2.15}$$

where

$$(**) = \frac{\partial^2 \tilde{W}_3^u}{\partial x^a \partial x^b}\Big|_{z_0} W_1^a(z_0) W_2^b(z_0) + \frac{\partial \tilde{W}_3^u}{\partial x^b}\Big|_{z_0} \frac{\partial \tilde{W}_2^b}{\partial x^a}\Big|_{z_0} W_1^a(z_0). \tag{2.15}$$

This formula is calculated following Definition 2.11 for determining the suspension by tangent vectors of arbitrary local sector fields. It is clear that in general this will *not* equal $\bar{A}_3(x_0)$. In fact, we shall say that a local sector field $\bar{A}_3(x)$ is *completely adapted* if it is equal to the *six* local fields $[\bar{A}_3^s(x)]s$ for $s \in S(3)$.

Let us denote

$$\frac{\partial \tilde{W}_2^i}{\partial x^j}\Big|_{z_0} W_1^j(z_0)\colon C_{12}^i(z_0) \qquad \text{and} \qquad \frac{\partial \tilde{W}_1^i}{\partial x^j}\Big|_{z_0} W_2^j(z_0)\colon C_{21}^i(z_0),$$

$$\frac{\partial \tilde{W}_3^i}{\partial x^j}\Big|_{z_0} W_1^j(z_0)\colon C_{13}^i(z_0) \qquad \text{and} \qquad \frac{\partial \tilde{W}_1^i}{\partial x^j}\Big|_{z_0} W_3^j(z_0)\colon C_{31}^i(z_0),$$

$$\frac{\partial \tilde{W}_3^i}{\partial x^j}\Big|_{z_0} W_2^j(z_0)\colon C_{23}^i(z_0) \qquad \text{and} \qquad \frac{\partial \tilde{W}_2^i}{\partial x^j}\Big|_{z_0} W_3^j(z_0)\colon C_{32}^i(z_0),$$

and let

$$E_{uvw}^i(z_0) = \frac{\partial \tilde{C}_{vw}^i}{\partial x^j}\Big|_{z_0} W_u^j(z_0)$$

for each triple uvw of distinct elements of $\{1, 2, 3\}$. In this notation, we may represent $[\bar{A}_3^s(x)]s$ as the sector:

$$\left[\begin{array}{ccccc} (\phi \circ t_{z_0})_3,\ W_1^i(z_0) & \text{———} & C_{\Sigma(1,2)}^r(z_0) & \text{———} & W_2^j(z_0) \\ & \diagdown & & \diagup & \\ & C_{\Sigma(1,3)}^s(z_0) & E_{\Sigma(1,2,3)}^u(z_0) & C_{\Sigma(2,3)}^t(z_0) & \\ & & \diagdown \quad \diagup & & \\ & & W_3^k(z_0) & & \end{array}\right] \tag{2.16}$$

where the action of Σ on an ordered pair (u, v) is defined by this rule: let $\{s(u), s(v)\} = \{j, k\}$ (as sets) and let $j < k$; then $\Sigma(u, v) = (s^{-1}(j), s^{-1}(k))$. Also, $\Sigma(1, 2, 3) = (s^{-1}(1), s^{-1}(2), s^{-1}(3))$.

Calculating Σ for the six permutations, we have the components for $[\bar{A}_3^s(x_0)]s$ as follows:

1) $s=(1,2,3)$:
$$\begin{array}{ccccc} W_1^i & \text{—} & C_{12}^i & \text{—} & W_2^i \\ & \diagdown & & \diagup & \\ & C_{13}^i & E_{123}^i & C_{23}^i & \\ & \diagdown & & \diagup & \\ & & W_3^i & & \end{array}$$

2) $s=(3,1,2)$:
$$\begin{array}{ccccc} W_1^i & \text{—} & C_{21}^i & \text{—} & W_2^i \\ & \diagdown & & \diagup & \\ & C_{31}^i & E_{231}^i & C_{23}^i & \\ & \diagdown & & \diagup & \\ & & W_3^i & & \end{array}$$

3) $s^2=(2,3,1)$:
$$\begin{array}{ccccc} W_1^i & \text{—} & C_{12}^i & \text{—} & W_2^i \\ & \diagdown & & \diagup & \\ & C_{31}^i & E_{312}^i & C_{32}^i & \\ & \diagdown & & \diagup & \\ & & W_3^i & & \end{array}$$

4) $t=(1,3,2)$:
$$\begin{array}{ccccc} W_1^i & \text{—} & C_{12}^i & \text{—} & W_2^i \\ & \diagdown & & \diagup & \\ & C_{13}^i & E_{132}^i & C_{32}^i & \\ & \diagdown & & \diagup & \\ & & W_3^i & & \end{array}$$

5) $st=(3,2,1)$:
$$\begin{array}{ccccc} W_1^i & \text{—} & C_{21}^i & \text{—} & W_2^i \\ & \diagdown & & \diagup & \\ & C_{31}^i & E_{321}^i & C_{32}^i & \\ & \diagdown & & \diagup & \\ & & W_3^i & & \end{array}$$

6) $ts=(2,1,3)$:
$$\begin{array}{ccccc} W_1^i & \text{—} & C_{21}^i & \text{—} & W_2^i \\ & \diagdown & & \diagup & \\ & C_{13}^i & E_{213}^i & C_{23}^i & \\ & \diagdown & & \diagup & \\ & & W_3^i & & \end{array}$$

Now for each vertex W_j^i, $j=1,2,3$, it is possible to form *two* generalized brackets for $h=1$, $r=2$. The following exercise will give a proof of the *Jacobi identity* and will set the stage for a later operation: iteration of covariant derivatives.

Exercise

11. Letting $h = 1$, $r = 2$, compute for *each* vertex W_j^i, $j = 1, 2, 3$, *two* generalized brackets of 3-sectors. These brackets are themselves 2-sectors and they have pairwise brackets. Compute these brackets also, giving an ordered triple of tangent vectors at each point.

    Geometrically these tangent vectors have the following interpretation as iterated Lie brackets:

    $$[W_1, [W_2, W_3]], \qquad [W_2, [W_3, W_1]], \qquad [W_3, [W_1, W_2]].$$

    Prove this using formula (2.15) and its counterparts.

    Show directly that at each point the *sum* of these tangent vectors is 0. This is the Jacobi identity.

This ends the series of elementary constructions and examples. In the next chapter we take up the properties of the objects *dual* to sectors. We shall call these 'sectorforms' and their relation with sectors is the same as the relation of 1-forms to tangent vectors. In fact, covariant tensors will be special cases of sectorforms. However, there will be important examples of sectorforms which are *not* tensors. These will be central to our development of the basic facts of Riemannian geometry.

3

Sectorforms

§3.1. Covariant tensor fields and sectorforms

It may appear that the cotangent bundle functor T^1 is weaker than the tangent bundle functor T_1 since it only applies to local diffeomorphisms from M to N, whereas the tangent bundle functor applies to all smooth maps. For this deficiency, there is a complementary advantage over the tangent bundle functor which makes the study of its cross sections the starting point for a rich and fruitful calculus.

The basic idea is this. Given a cross section of the cotangent bundle $s: N \to T^1(N)$, then s gives rise to a smooth map $S: T_1(N) \to \mathbb{R}$ which is *linear on each fiber.* Now if $f: M \to N$ is smooth, then, since $T_1(f): T_1(M) \to T_1(N)$ is fiber-preserving and linear on each fiber, $S \circ T_1(f): T_1(M) \to \mathbb{R}$ is linear on each fiber and can be identified with a cross section of $T_1(M)$.

Thus there is a natural way to 'pull back' smooth sections of $T^1(N)$ to smooth sections of $T^1(M)$ for arbitrary smooth maps $f: M \to N$. These sections of the cotangent bundle (global 1-forms) are well-behaved, then, under arbitrary smooth maps. Sections of the tangent bundle (vector fields), on the other hand, do not behave nicely. For $f: M \to N$ smooth, unless f is a diffeomorphism it will not be possible to use $T_1(f)$ to 'push' a vector field on M over to a vector field on N.

In this chapter, we generalize this strategy to the 'duals' of tangent sectors: objects which we call *sectorforms.* In particular, sectorform fields will correspond to mappings from $T_k(M) \to \mathbb{R}$ satisfying certain linearity conditions on the intrinsic spaces $T_{k,i}$. All covariant tensor fields are sectorform fields, but the latter form a wider class and can be used to isolate and study in a systematic fashion certain 'non-tensorial' invariants which arise notably in Riemannian geometry.

We formalize the discussion on the behavior of sections in some exercises.

Exercises

1. Let

$$\begin{array}{c} T_1^k(M) \\ {\scriptstyle q_1(k)}\downarrow \\ M \end{array}$$

denote the product of the tangent bundle with itself k times. Also let $C(M, \mathbb{R})$ denote the algebra of smooth real-valued functions on M. Let $S_k(M)$ denote, for each M, the $C(M, \mathbb{R})$ module of smooth maps $T_1^k(M) \to \mathbb{R}$ which are *k-linear on each fiber.* Associate $M \mapsto S_k(M)$ and for $f: M \to N$ smooth let $S_k(f): S_k(N) \to S_k(M)$ by the rule $S_k(f)(\phi) = \phi \circ T_1^k(f)$. Then $S_k(f)$ is a module homomorphism over the algebra homomorphism $C(N, \mathbb{R}) \to C(M, \mathbb{R})$ given by composition with f. Show that this procedure gives a *contravariant functor* from *manifolds* to *modules.*

2. Let $V = L(E_1, E_2, \ldots, E_k; \mathbb{R})$ (Definition 1.13) and let π_V be the projection of the V-tensor bundle

$$\begin{array}{c} F_1(M) \times_G V \\ \downarrow \\ M \end{array}$$

Let $\gamma[\pi_V]$ denote the $C(M, \mathbb{R})$ module of smooth cross sections of π_V. For A such a section, and $(X^1, X^2, \ldots, X^k)$ a section of $q_1(k)$, define the *contraction* $A\langle X^1, \ldots, X^k\rangle$ to be the smooth function taking $M \to \mathbb{R}$ given over a local frame $\theta: V \to U \subset M$ by

$$A\langle X^1, \ldots, X^k\rangle(x) = \tilde{A}_{i_1 i_2 \ldots i_k}(z)\tilde{X}^{1i_1}(z) \ldots \tilde{X}^{ki_k}(z) \qquad \text{for } x = \theta(z).$$

Show that this prescription is unambiguous (does not depend on frame) and defines a smooth function.

3. Use the contraction operation to establish an *isomorphism* of $C(M, \mathbb{R})$ modules for each M, $\nu_M: S_k(M) \to \gamma[\pi_V(M)]$. If, for $f: M \to N$ smooth, we let $f^* = \nu_M \circ S_k(f) \circ \nu_N^{-1}$ so that $f^*: \gamma[\pi_V(N)] \to \gamma[\pi_V(M)]$ is the *unique* homomorphism commuting the diagram

$$\begin{array}{ccc} S_k(N) & \xrightarrow{S_k(f)} & S_k(M) \\ {\scriptstyle \nu_N}\downarrow{\scriptstyle \cong} & & {\scriptstyle \cong}\downarrow{\scriptstyle \nu_M} \\ \gamma[\pi_V(N)] & \xrightarrow{f^*} & \gamma[\pi_V(M)] \end{array}$$

show that the association that takes $M \to \gamma[\pi_V(M)]$ and $f \to f^*$ gives a *contravariant functor* from manifolds to modules which is *naturally isomorphic* via ν to the one given in exercise 1.

The first step in the generalization of covectors will be the construction of a *bundle of sectorforms*. Cross sections of this bundle will be called *sectorform fields* and will be seen to have pull-back properties like those outlined in the exercises. The appropriate modules of cross sections will then be seen to be naturally isomorphic with certain modules of mappings $T_k(M) \to \mathbb{R}$ defined in terms of the behavior of those mappings on the *intrinsic spaces* $T_{k,i}$ introduced earlier. Thus, a notion of contraction can be defined (contraction of a sectorform with a sector), and we shall be concerned with the calculus of these sectorforms which develops from the existence of certain natural *differentials* taking sectorform fields of order k to sectorform fields of order $(k+1)$. In fact, if a smooth real-valued function on M is thought of (harmlessly) as an order-0 sectorform field, then the ordinary differential of that mapping is an order-1 sectorform field (a covector field). The higher-order differentials give the direct generalization of this operation. Finally, we shall be concerned with the contraction of *sectorform fields* on *K-sector fields*.

Definition 3.1. For each k, let E_k^* denote the vector space of mappings from $E_k \to \mathbb{R}$ with the property that, for each $f \in E_k^*$, the restriction of f to *each* $E_{k,i}(X)$ is linear (for $1 \leq i \leq k$, X arbitrary). Unless $k=1$, the maps $f: E_k \to \mathbb{R}$ in E_k^* are *not* linear with respect to the global linear structure in E_k. ∎

The first topic we take up in this section is the characterization of the functions in E_k^*. Once this is done, and a notation convenient for calculating the action of $G_k(m)$ on the left on E_k^* is developed, we give the general definition of *sectorforms* and *sectorform fields*.

Suppose we begin with the case $k=2$. What is the form of the functions in E_2^*? First of all, E_2 is a vector space of dimension $3m$ whose elements are represented in components by X^i——Z^k——Y^j, $1 \leq i, j, k \leq m$.

It will be convenient to develop an informal local representation for higher-rank tensors. Recall that any f in $L(E_1, E_2, \ldots, E_k; \mathbb{R})$ is uniquely determined by its effect on the k-tuples of basis vectors $(e_{i_1}, e_{i_2}, \ldots, e_{i_k})$. There are m^k such k-tuples. Denote by the symbol $e^{j_1} \otimes e^{j_2} \otimes \ldots \otimes e^{j_k}$ the element of $L(E_1, E_2, \ldots, E_k; \mathbb{R})$ which has value 1 on the k-tuple of basis vectors $(e_{j_1}, e_{j_2}, \ldots, e_{j_k})$ and which has value 0 on every other k-tuple of basis vectors. A straightforward algebraic argument shows that the set $\{e^{j_1} \otimes e^{j_2} \otimes \ldots \otimes e^{j_k}\}$ for $1 \leq j_i \leq m$ forms a *basis* for $L(E_1, E_2, \ldots, E_k; \mathbb{R})$. (In fact, in the matrix representation, these elements give matrices with zeros in every position except the $(j_1, j_2, \ldots, j_k)$ entry where there is a 1.) Classically, the basis vector $e^{j_1} \otimes e^{j_2} \otimes \ldots \otimes e^{j_k}$ is denoted $dx^{j_1} \otimes dx^{j_2} \otimes \ldots \otimes dx^{j_k}$, and we shall continue this usage.

In particular if $\mathbb{A}$ is a tensor field on m of type $\binom{0}{k}$ and if $\phi: O \to U$ is a local frame, then, if the local representation of $\mathbb{A}$ is $[(\phi \circ t_z)_1, Y_{i_1 \ldots i_k}(z)]$, we shall often represent the field $\mathbb{A}$ with respect to frame ϕ as

$$Y_{i_1 \ldots i_k}(z)\, dx^{i_1} \otimes dx^{i_2} \otimes \ldots \otimes dx^{i_k} : \tilde{Y}_{i_1 i_2 \ldots i_k}(z).$$

Similarly, any f in $L(E_1^*, E_2^*, \ldots, E_k^*; \mathbb{R})$ is uniquely determined by its effect on the m^k k-tuples of basis vectors $(e^{i_1}, e^{i_2}, \ldots, e^{i_k})$. We shall denote by the symbol $e_{j_1} \otimes e_{j_2} \otimes \ldots \otimes e_{j_k}$ the element of $L(E_1^*, E_2^*, \ldots, E_k^*; \mathbb{R})$ which has value 1 on the k-tuple $(e^{j_1}, e^{j_2}, \ldots, e^{j_k})$ and which has value 0 on all other k-tuples of basis vectors. The elements $\{e_{j_1} \otimes e_{j_2} \otimes \ldots \otimes e_{j_k}\}$ then form a basis for $L(E_1^*, E_2^*, \ldots, E_k^*; \mathbb{R})$. We often follow classical usage and represent $e_{j_1} \otimes e_{j_2} \otimes \ldots \otimes e_{j_k}$ by

$$\frac{\partial}{\partial x^{j_1}} \otimes \ldots \otimes \frac{\partial}{\partial x^{j_k}}.$$

Let $f \in E_2^*$. Then $f(X^i \text{——} 0 \text{——} Y^j) = f_{ij} X^i Y^j$ for some 2-linear function in $L(E, E; \mathbb{R}) : f_{ij} e^i \otimes e^j$ or $f_{ij}\, dx^i \otimes dx^j$. Further $f(0 \text{——} Z^k \text{——} 0) = f_k Z^k$ for some linear map $f_k e^k$ in $L(E; \mathbb{R})$. It is also easy to see that $f(0 \text{——} 0 \text{——} Y^j) = 0$, and therefore that $f(0 \text{——} Z^k \text{——} Y^j) = f(0 \text{——} Z^k \text{——} 0) = f_k Z^k$. Since $f(X^i \text{——} 0 \text{——} Y^j) = f_{ij} X^i Y^j$ and $f(0 \text{——} Z^k \text{——} Y^j) = f_k Z^k$, we have

$$f(X^i \text{——} Z^k \text{——} Y^j) = f_k Z^k + f_{ij} X^i Y^j. \tag{3.1}$$

Thus the element $f \in E_2^*$ is entirely determined by the choice of *linear* function $f_k e^k$ in $L(E; \mathbb{R})$ and *bilinear* function $f_{ij} e^i \otimes e^j$ in $L(E, E; \mathbb{R})$. Conversely, any function of the form (3.1) above is easily seen to belong to E_2^*.

Now consider the case $k = 3$. Suppose that $f \in E_3^*$ and the general element of E_3^* is represented in components:

$$\begin{array}{ccccc} X^i & \text{——} & A^r & \text{——} & Y^j = X. \\ & \diagdown & & \diagup & \\ & B^s & E^u & C^t & \\ & & \diagdown\ \diagup & & \\ & & Z^k & & \end{array}$$

First of all, it is clear that

$$\begin{array}{ccccc} f(0 & \text{——} & 0 & \text{——} & 0) = f_u E^u \\ & \diagdown & & \diagup & \\ & 0 & E^u & 0 & \\ & & \diagdown\ \diagup & & \\ & & 0 & & \end{array}$$

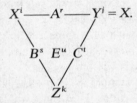

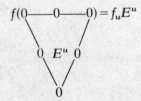

for some linear function $f_u e^u$. Similarly,

$$f\left(\begin{array}{ccccc} X^i & \text{—} & 0 & \text{—} & Y^j \\ & \diagdown & & \diagup & \\ & 0 & 0 & 0 & \\ & & \diagdown\diagup & & \\ & & Z^k & & \end{array}\right) = f_{ijk}X^iY^jZ^k$$

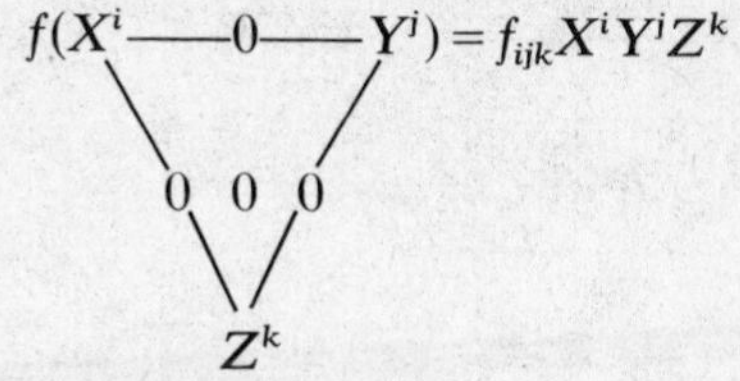

for some 3-linear function $f_{ijk}e^i \otimes e^j \otimes e^k \in L(E, E, E; \mathbb{R})$.

Now consider the effect of f on an element of the form

$$\begin{array}{ccccc} 0 & \text{—} & A^r & \text{—} & 0 \\ & \diagdown & & \diagup & \\ & 0 & 0 & 0 & \\ & & \diagdown\diagup & & \\ & & Z^k & & \end{array}$$

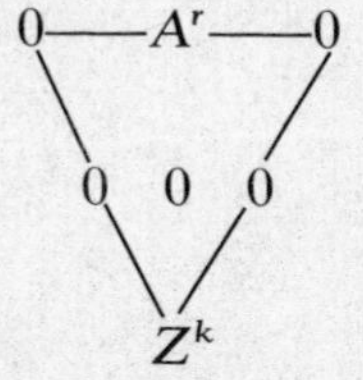

Using linearity of f on $E_{3,3}$ we see that f is linear in Z^k, and linearity of f on $E_{3,1}$ or $E_{3,2}$ shows that it is linear in A^r. Thus for $a_{rk}e^r \otimes e^k$ a bilinear function in $L(E, E; \mathbb{R})$, the value of f on this element is $a_{rk}A^rZ^k$. Similarly, there are bilinear functions $b_{sj}e^s \otimes e^j$ and $c_{ti}e^t \otimes e^i$ with

$$f\left(\begin{array}{ccccc} 0 & \text{—} & 0 & \text{—} & Y^j \\ & \diagdown & & \diagup & \\ & B^s & 0 & 0 & \\ & & \diagdown\diagup & & \\ & & 0 & & \end{array}\right) = b_{sj}B^sY^j$$

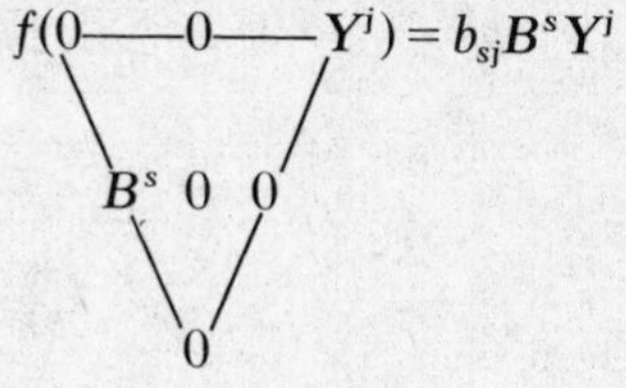

and

$$f\left(\begin{array}{ccccc} X^i & \text{—} & 0 & \text{—} & 0 \\ & \diagdown & & \diagup & \\ & 0 & 0 & C^t & \\ & & \diagdown\diagup & & \\ & & 0 & & \end{array}\right) = c_{ti}C^tX^i$$

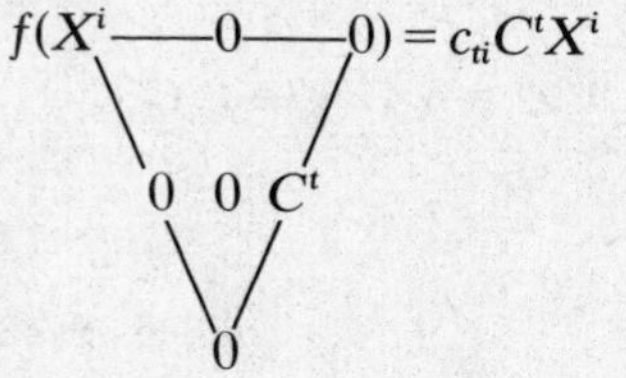

Now, for example,

$$f\left(\begin{array}{ccccc} X^i & \text{—} & 0 & \text{—} & 0 \\ & \diagdown & & \diagup & \\ & 0 & 0 & 0 & \\ & & \diagdown\diagup & & \\ & & Z^k & & \end{array}\right) = 0$$

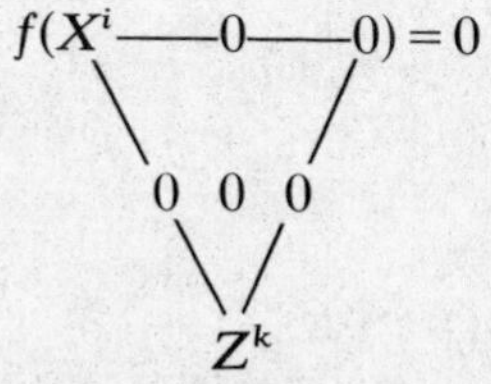

using linearity of f on $E_{3,2}$. Therefore

$$f\left(\begin{array}{ccccc} X^i & — & A^r & — & 0 \\ \setminus & 0 & 0 & 0 & / \\ & \setminus & Z^k & / & \end{array}\right) = a_{rk}A^rZ^k \cdot f\left(\begin{array}{ccccc} X^i & — & 0 & — & 0 \\ \setminus & 0 & 0 & C^t & / \\ & \setminus & Z^k & / & \end{array}\right) = c_{ti}C^tX^i.$$

Putting these last equations together with

$$f\left(\begin{array}{ccccc} X^i & — & 0 & — & Y^j \\ \setminus & 0 & 0 & 0 & / \\ & \setminus & Z^k & / & \end{array}\right) = f_{ijk}X^iY^jZ^k$$

we have

$$f\left(\begin{array}{ccccc} X^i & — & A^r & — & Y^j \\ \setminus & 0 & 0 & C^t & / \\ & \setminus & Z^k & / & \end{array}\right) = f_{ijk}X^iY^jZ^k + a_{rk}A^rZ^k + c_{ti}C^tX^i.$$

Next,

$$f\left(\begin{array}{ccccc} X^i & — & A^r & — & Y^j \\ \setminus & B^s & 0 & 0 & / \\ & \setminus & 0 & / & \end{array}\right) = b_{sj}B^sY^j$$

and so

$$f\left(\begin{array}{ccccc} X^i & — & A^r & — & Y^j \\ \setminus & B^s & 0 & C^t & / \\ & \setminus & Z^k & / & \end{array}\right) = f_{ijk}X^iY^jZ^k + a_{rk}A^rZ^k + b_{sj}B^sY^j + c_{ti}C^tX^i.$$

Finally observe that

$$f\left(\begin{array}{ccccc} 0 & — & 0 & — & Y^j \\ \setminus & 0 & E^u & 0 & / \\ & \setminus & 0 & / & \end{array}\right) = f_uE^u.$$

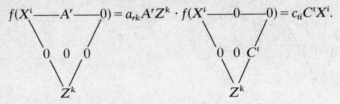

From this it follows that

$$\begin{array}{ccccc} f(X^i & \text{---} & A^r & \text{---} & Y^j) = f_u E^u. \\ & \backslash & & / & \\ & 0 & E^u & 0 & \\ & & \backslash\;/ & & \\ & & 0 & & \end{array}$$

Thus, finally, we have using linearity of f on $E_{3,3}$ that

$$\begin{array}{ccccc} f(X^i & \text{---} & A^r & \text{---} & Y^j) = f_{ijk} X^i Y^j Z^k + a_{rk} A^r Z^k + b_{sj} B^s Y^j + c_{ti} C^t X^i + f_u E^u. \\ & \backslash & & / & \\ & B^s & E^u & C^t & \\ & & \backslash\;/ & & \\ & & Z^k & & \end{array} \qquad (3.2)$$

The element $f \in E_3^*$ is uniquely determined by the multilinear maps

$$f_{ijk} e^i \otimes e^j \otimes e^k + a_{rk} e^r \otimes e^k + b_{sj} e^s \otimes e^j + c_{ti} e^t \otimes e^i + f_u e^u).$$

Again, it is easy to see that any map in the form (3.2) gives rise to an element of E_3^*.

Now in order to give an economical statement of the general form of elements in E_k^*, we must introduce a new notation. Suppose that e is the $(k-1)$ face of $\Gamma(k-1)$. Let $\gamma_{i_1} \cup \gamma_{i_2} \cup \ldots \cup \gamma_{i_r} = e$ be some join decomposition of e. Then the symbol $f_{k_1 k_2 \ldots k_r} e_{\gamma_1}^{k_1} \otimes \ldots \otimes e_{\gamma_r}^{k_r}$ is the element of E_k^* whose effect on the element of E_k whose components are given as X_γ^i (that is the element $X_\gamma^i e_i^\gamma$) is the sum $f_{k_1 k_2 \ldots k_r} X_{\gamma_{i_1}}^{k_1} X_{\gamma_{i_2}}^{k_2} \ldots X_{\gamma_{i_r}}^{k_r}$ for the $f_{k_1 \ldots k_r} \in \mathbb{R}$ and the summation taken over all $1 \leq k_j \leq m$ for all $1 \leq j \leq r$. Note that the symbol e_i^γ is meant to denote a basis vector of E_k: the one which maps γ to e_i and which vanishes on all other faces. The symbol e_γ^i represents its dual. Now since there will be several partitions with r sets in general, we shall write the *components* of the function $f_{k_1 \ldots k_r} e_{\gamma_1}^{k_1} \otimes \ldots \otimes e_{\gamma_r}^{k_r}$ *not* as $f_{k_1 k_2 \ldots k_r}$, which would lead to ambiguity concerning the partition to which it referred, but as $(f_{k_1 k_2 \ldots k_r}^{\gamma_{i_1} \gamma_{i_2} \ldots \gamma_{i_r}})$, where the *ordered* sequence $\gamma_{i_1} \gamma_{i_2} \ldots \gamma_{i_r}$ gives the data relating each γ_{i_j} with its index k_j. We may now state our next theorem.

Theorem 3.2. *A basis for E_k^* is given by the functions $e_{\gamma_{i_1}}^{k_1} \otimes e_{\gamma_{i_2}}^{k_2} \otimes \ldots \otimes e_{\gamma_{i_r}}^{k_r}$ $1 \leq k_j \leq m$, $\gamma_{i_1} \cup \ldots \cup \gamma_{i_r} = e$. The effect of this function on X_γ^i is $X_{\gamma_{i_1}}^{k_1} X_{\gamma_{i_2}}^{k_2} \ldots X_{\gamma_i}^{k_r}$.*

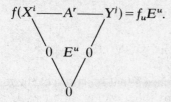

In particular, each element f of E_k^* *is uniquely described as a function*

$$\sum_{\gamma_{i_1}\cup\gamma_{i_2}\cup\ldots\cup\gamma_{i_r}=e} f^{\gamma_{i_1}\gamma_{i_2}\cdots\gamma_{i_r}}_{k_1k_2\ldots k_r} e^{k_1}_{\gamma_{i_1}} \otimes e^{k_2}_{\gamma_{i_2}} \otimes \ldots \otimes e^{k_r}_{\gamma_{i_r}}. \tag{3.3}$$

(The sum is taken over all distinct partitions $\gamma_{i_1}\cup\gamma_{i_2}\cup\ldots\cup\gamma_{i_r}=e$.*)*

Proof. The following simple argument shows that functions of the form (3.3) are indeed linear on each $E_{k,j}$. For any join decomposition $\gamma_{i_1}\cup\ldots\cup\gamma_{i_r}=e$, precisely *one* face γ_{i_s} fails to have the form $(\gamma_1, \gamma_2, \ldots, \gamma_{j-1}, 0, \gamma_{j+1}, \ldots, \gamma_k)$. Each summand then is linear on $E_{k,j}(X)$ since only the $X^{k_s}_{\gamma_{i_s}}$ factor can vary in $f^{\gamma_{i_1}\gamma_{i_2}\cdots\gamma_{i_r}}_{k_1k_2\ldots k_r} X^{k_1}_{\gamma_{i_1}}\ldots X^{k}_{\gamma_{i_r}}$ when a sum is formed.

Now the difficult part of the theorem is the demonstration that every element of E_k^* has the form (3.3).

Let γ' represent a general face of $\Gamma(k-1)$ of the form $(\gamma'_1, \gamma'_2, \ldots, \gamma'_{k-1}; 0)$ and let γ'' be the vertex of $\Gamma(k-1)$: $(0, 0, \ldots, 0; 1)$. We represent any element of E_k in the form $X^i_{\gamma'}$——$Z^k_{\gamma'\cup\gamma''}$——$Y^i_{\gamma''}$. This follows again the notation of Theorem 1.27.

Now suppose $f\in E_k^*$.

a) $f(X^i_{\gamma'}$——0——$0)=0$ by linearity of f on $E_{k,k}(X)$.

b) Now consider $f(X^i_{\gamma'}$——0——$Y^j_{\gamma''})$. For fixed $Y^j_{\gamma''}$ this gives a function taking $E_{k-1}\to\mathbb{R}$ where $\Gamma(k-2)$ is identified in the natural way with the 'front face' of $\Gamma(k-1)$. Now it is easy to see that this function belongs to E^*_{k-1}. Call it $f[Y^j_{\gamma''}]$. Inductively, we may write

$$f[Y^j_{\gamma''}](X^i_{\gamma'}) = \sum_{\gamma'_{i_1}\cup\ldots\cup\gamma'_{i_s}=e'} g^{\gamma'_{i_1}\gamma'_{i_2}\cdots\gamma'_{i_s}}_{k_1k_2\ldots k_s} X^{k_1}_{\gamma'_{i_1}} X^{k_2}_{\gamma'_{i_2}} \ldots X^{k_s}_{\gamma'_{i_s}}$$

for e' the $(k-2)$ face of $\Gamma(k-2)$ and some set of multilinear functions $g^{\gamma'_{i_1}\gamma'_{i_2}\cdots\gamma'_{i_s}}_{k_1k_2\ldots k_s} e^{k_1}_{\gamma_{i_1}} \otimes\ldots\otimes e^{k_s}_{\gamma_{i_s}}$.

Now the map $Y^j_{\gamma''}\to f[Y^j_{\gamma''}]$ is *linear* using linearity of f on $E_{k,k}$. Thus, there are multilinear functions $f^{\gamma'_{i_1}\gamma'_{i_2}\cdots\gamma'_{i_s},\gamma''}_{k_1k_2\ldots k_s,j} e^{k_1}_{\gamma'_{i_1}}\otimes\ldots\otimes e^{k_s}_{\gamma'_{i_s}}\otimes e^j_{\gamma''}$ such that

$$f(X^i_{\gamma'}\text{——}0\text{——}Y^j_{\gamma''}) = \sum_{\gamma'_{i_1}\cup\ldots\cup\gamma'_{i_s}=e'} f^{\gamma'_{i_1}\gamma'_{i_2}\cdots\gamma'_{i_s},\gamma''}_{k_1k_2\ldots k_s,j} X^{k_1}_{\gamma'_{i_1}} X^{k_2}_{\gamma'_{i_2}}\ldots X^{k_s}_{\gamma'_{i_s}} Y^j_{\gamma''}.$$

c) Suppose now that g' is a *fixed* face of the form $(\gamma'_1, \gamma'_2, \ldots, \gamma'_{k-1}; 0)$ in $\Gamma(k-1)$. Then consider $f(X^i_{\gamma'}$——$Z^k_{g'\cup\gamma''}$——$0)$. Suppose in fact that for g', $\gamma'_{w_1}=\gamma'_{w_2}=\ldots=\gamma'_{w_t}=1$ and $\gamma'_{v_1}=\gamma'_{v_2}=\ldots=\gamma'_{v_{k-t-1}}=0$ with $w_1<w_2<\ldots<w_t$ and $v_1<\ldots<v_{k-t-1}<k$.

Now let e'' be the face of $\Gamma(k-1)$ with entries 1 in positions $v_1, \ldots, v_{k-t-1}$ and entries 0 elsewhere. We also let $\bar{\gamma}'$ denote a *general* face of $\Gamma(k-1)$ with entries 0 in positions $w_1, w_2, \ldots, w_t$ and k.

Now let $\bar{X}^i_{\gamma'}$ be the element of E_{k-1} (via the obvious identification) such that $\bar{X}^i_{\bar{\gamma}'}=X^i_{\bar{\gamma}'}$ for all $\bar{\gamma}'$ as above and $\bar{X}^i_{\gamma'}=0$ if γ' is *not* a $\bar{\gamma}'$.

Then we observe first that

$$f(X^i_{\gamma'} \text{——} Z^k_{g'\cup\gamma''} \text{——} 0) = f(\bar{X}^i_{\bar{\gamma}'} \text{——} Z^k_{g'\cup\gamma''} \text{——} 0).$$

This follows from linearity of f on $E_{k,w_1}, E_{k,w_2}, \ldots, E_{k,w_t}$. We now sketch the argument. Let $\hat{\gamma}'$ denote the general face of $\Gamma(k-1)$ with entries 0 in positions w_1 and k. Let also $\hat{X}^i_{\hat{\gamma}'}$ be the element of E_{k-1} such that $\hat{X}^i_{\hat{\gamma}'} = X^i_{\hat{\gamma}'}$ for all $\hat{\gamma}'$, and $\hat{X}^i_{\gamma'} = 0$ if γ' is not of the form $\hat{\gamma}'$. Then using linearity of f on E_{k,w_1},

$$f(X^i_{\gamma'} \text{——} Z^k_{g'\cup\gamma''} \text{——} 0) = f(\hat{X}^i_{\hat{\gamma}'} \text{——} Z^k_{g'\cup\gamma''} \text{——} 0)$$

since $f(X^i_{\gamma'} \text{——} 0 \text{——} 0) = 0$.

We apply a series of reductions using linearity on $E_{k,w_2}, \ldots, E_{k,w_t}$ successively and ending with the observation above.

Next we calculate $f(\bar{X}^i_{\bar{\gamma}'} \text{——} Z^k_{g'\cup\gamma''} \text{——} 0)$. For $Z^k_{g'\cup\gamma''}$ fixed, $\bar{X}^i_{\bar{\gamma}'} \to f(\bar{X}^i_{\bar{\gamma}'} \text{——} Z^k_{g'\cup\gamma''} \text{——} 0)$ gives a map $E_{k-t-1} \to \mathbb{R}$ where $\Gamma(k-t-2)$ is identified with the collection of faces $\bar{\gamma}'$. Via this natural identification, the correspondence above is seen to belong to E^*_{k-t-1}; call it $f[Z^k_{g'\cup\gamma''}]$. Again, inductively, we may write

$$f[Z^k_{g'\cup\gamma''}](\bar{X}^i_{\bar{\gamma}'}) = \sum_{\bar{\gamma}'_{i_1}\cup\bar{\gamma}'_{i_2}\cup\ldots\cup\bar{\gamma}'_{i_r}=e''} g^{\bar{\gamma}'_{i_1}\bar{\gamma}'_{i_2}\ldots\bar{\gamma}'_{i_r}}_{k_1k_2\ldots k_r} \cdot \bar{X}^{k_1}_{\bar{\gamma}'_{i_1}} \bar{X}^{k_2}_{\bar{\gamma}'_{i_2}} \ldots \bar{X}^{k_r}_{\bar{\gamma}'_{i_r}}.$$

Again applying linearity of f on $E_{k,k}$ we see that the map $Z^k_{g'\cup\gamma''} \to f[Z^k_{g'\cup\gamma''}]$ is itself linear, and therefore there are multilinear functions

$$f^{\bar{\gamma}'_{i_1}\bar{\gamma}'_{i_2}\ldots\bar{\gamma}'_{i_r},(g'\cup\gamma'')}_{k_1k_2,\ldots,k_r,k} e^{k_1}_{\bar{\gamma}'_{i_1}} \otimes e^{k_2}_{\bar{\gamma}'_{i_2}} \otimes \ldots \otimes e^{k_r}_{\bar{\gamma}'_{i_r}} \otimes e^k_{g'\cup\gamma''}, \qquad \bar{\gamma}'_{i_1}\cup\ldots\cup\bar{\gamma}'_{i_r} = e''$$

with the property that

$$f(X^i_{\gamma'} \text{——} Z^k_{g'\cup\gamma''} \text{——} 0)$$
$$= \sum_{\bar{\gamma}'_{i_1}\cup\bar{\gamma}'_{i_2}\cup\ldots\cup\bar{\gamma}'_{i_r}=e''} f^{\bar{\gamma}'_{i_1}\bar{\gamma}'_{i_2}\ldots\bar{\gamma}'_{i_r},(g'\cup\gamma'')}_{k_1k_2\ldots k_r k} \bar{X}^{k_1}_{\bar{\gamma}'_{i_1}} \bar{X}^{k_2}_{\bar{\gamma}'_{i_2}} \ldots \bar{X}^{k_r}_{\bar{\gamma}'_{i_r}} Z^k_{g'\cup\gamma''}$$

the sum being taken over all join decompositions of e''.

We do this for *each* face of the form $g'\cup\gamma''$, recall case b) and apply linearity of f on $E_{k,k}$ to reach the conclusion of the theorem. ■

Just as we do with elements of E_k, we shall often represent the elements of E^*_k in *components*. Thus let P_r represent an *ordered partition* of $\{1, 2, \ldots, k\}$ into r non-empty disjoint sets; P_r gives a join decomposition of the $(k-1)$ face of $\Gamma(k-1)$: $\gamma_{i_1}\cup\gamma_{i_2}\cup\ldots\cup\gamma_{i_r} = e$. Then if $f \in E^*_k$ has the form given in formula (3.3), we represent it simply $(f^{P_r}_{k_1k_2\ldots k_r})$, this being understood to be a set of components of multilinear functions indexed by the partitions P_r (of various lengths $1 \le r \le k$). Notice that there is a unique partition of length 1 (giving a linear function) and a unique partition of length k (giving a k-linear function).

Now there is a natural action of $G_k(m)$ on the left on E^*_k. This action generalizes the one which gives rise to the covariant tensors of type $\binom{0}{k}$. In fact, it is already obvious that it will be possible to interpret such tensors (when a frame is fixed) as elements of E_k^*. This idea will be formalized shortly. In order to motivate the definition of this action, let $g_k \in G_k(m)$ and consider the diagram analogous to diagram (1.8):

$$\begin{array}{ccccc} & E_k & \cdots\cdots & (E,0) & \xleftarrow{\phi_k^{-1}} & (M,x) \\ \swarrow^{g_k f} & \downarrow g_k^{-1} & & \downarrow & \swarrow^{g^{-1}(\phi g)_k^{-1}} & \\ \mathbb{R} \xleftarrow{f} & E_k & \cdots\cdots & (E,0) & & \end{array}$$

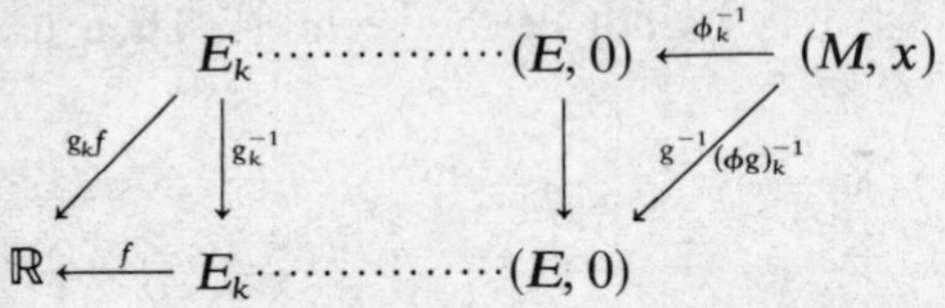

The left action of $G_k(m)$ on E_k^* will be defined by the rule

$$g_k f = f \circ g_k^{-1} \tag{3.4}$$

where the left action of g_k on E_k is given by formula (1.26).

Definition 3.3. Suppose $f = (f^{Q_s}_{k_1 k_2 \ldots k_s}) \in E_k^*$, and g_k (the k-jet of a germ from $(E, 0) \to (E, 0)$) belongs to $G_k(m)$. Suppose in fact that g_k has the form

$$\sum_{|\alpha| \leqslant k} \left(\frac{1}{\alpha!}\right) \frac{\partial^{|\alpha|} g^i}{\partial x^\alpha} x^\alpha.$$

This is the 'polynomial representative' of the jet given in standard coordinates $x^1, x^2, \ldots, x^m$ for E (formula (1.14)).

Introduce the following notation. Say that a partition Q_s: $\gamma'_{j_1} \cup \gamma'_{j_2} \cup \ldots \cup \gamma'_{j_s} = e$ is *weaker than* a partition P_r: $\gamma_{i_1} \cup \gamma_{i_2} \cup \ldots \cup \gamma_{i_r} = e$ (denoted $Q_s \leqslant P_r$) if each set of P_r is contained in some set of Q_s. Then we may say that

$$\gamma'_{j_1} = \gamma_{i_{a_1(1)}} \cup \ldots \cup \gamma_{i_{a_1(t_1)}}$$

$$\vdots$$

$$\gamma'_{j_s} = \gamma_{i_{a_s(1)}} \cup \ldots \cup \gamma_{i_{a_s(t_s)}}$$

(for P_r fixed, the q_n determined by Q_s) with $t_1 + t_2 + \ldots + t_s = r$.

Then $g_k f = (\bar{f}^{P_r}_{h'_1 h_2 \ldots h_r})$ with (for $\hat{g}$ written for the germ g^{-1}):

$$\bar{f}^{P_r}_{h'_1 h_2 \ldots h_r} = \sum_{Q_s \leqslant P_r} f^{Q_s}_{k'_1 \ldots k_s} \left(\frac{\partial^{t_1} \hat{g}^{k_1}}{\partial x^{h_{a_1(1)}} \ldots \partial x^{h_{a_1(t_1)}}}\right) \cdots \left(\frac{\partial^{t_s} \hat{g}^{k_s}}{\partial x^{h_{a_s(1)}} \ldots \partial x^{h_{a_s(t_s)}}}\right). \tag{3.5}$$

■

Formula (3.5) follows directly from (3.4) and (2.14). Fortunately, we shall seldom have need to apply this formula in the most general form.

Now suppose that M is a smooth manifold. For a point $x \in M$, let $\phi_k : (E, 0) \to (M, x)$ be a frame-jet at x. A *k sectorform at x* (element of $T^k(x)$) will be the data of a *pair* $(\phi_k, f^{P_r}_{k'_1 k_2 \ldots k_r})$ denoted generally $A^k(x)$ or $B^k(x)$. We now elaborate upon this.

Definition 3.4. Let M be a smooth manifold. Recall the k-frame bundle

$$\begin{array}{c} F_k(M) \\ \downarrow^{\pi_k} \\ M \end{array}$$

Define the following *left* action of $G_k(m)$ on $F_k(M)\times E_k^*$:

$$g_k(\phi_k, f^{P_r}_{k'_1\ldots k_r}) = (\phi_k g_k^{-1}, g_k, \cdot f^{P_r}_{k'_1\ldots k_r}).$$

Denote the set of *orbits* with respect to this action $T^k(M) = F_k(M)\times_{G_k(m)} E_k^*$. With the quotient topology, this has a smooth manifold structure and there is a 'factored' projection

$$\begin{array}{c} T^k(M) \\ \downarrow^{q^k} \\ M \end{array}$$

taking each orbit to the source of its jet. Then q^k is the projection of a *vector* bundle with fiber the Euclidean space E_k^*. $T^k(M)$ is the total space of the *bundle of k sectorforms* over M. Each element of $T^k(M)$ will be denoted by the orbit (equivalence class) of its pair with square brackets: $[\phi_k, f^{P_r}_{k'_1\ldots k_r}]$. These classes are subject to the relation:

$$[\phi_k g_k, f^{P_r}_{k'_1\ldots k_r}] = [\phi_k, g_k f^{P_r}_{k'_1\ldots k_r}]. \tag{3.6}$$

Finally, the symbol $f^{P_r}_{k'_1\ldots k_r}$ represents a set of 'matrices' of dimension m^r, $1\leq r\leq k$, indexed by the partitions P_r of $\{1, 2, \ldots, k\}$. P_r (and therefore r) are allowed to take all possible values. ■

Suppose now that $\phi: 0\to U\subset M$ is a *frame*. A *local sectorform field* over U with respect to ϕ will be denoted $[(\phi\circ t_z)_k, f^{P_r}_{k'_1\ldots k_r}(z)]$ for $z\in O$. When mention of ϕ can be suppressed, we denote the field informally as $\tilde{f}^{P_r}_{k'_1\ldots k_r}(z) = A^k(x)$. (The 'tilde' indicates the fact that it is a 'field' and not simply defined at a point; the variable z still belongs to O.) Such a field may also be given simply as $\mathbb{A}$.

We treat these notions rather briskly because we intend to make use of a *different interpretation* of sectorforms and sectorform fields. In the case of covariant tensors, we passed from the 'bundle' interpretation to their interpretation as *mappings* from Whitney products of $T_1(M)$ with itself to $\mathbb{R}$ which are multilinear on each fiber of the Whitney product. A similar viewpoint will be developed for sectorform fields. Certain naturality properties of those fields will be transparent from this point of view and, in particular, a notion of 'contraction' of k sectorform fields with k-sector fields will be formulated. It will be shown how the classical covariant tensor fields give special cases of sectorform fields.

Observation 3.5. Let M be a smooth manifold. The set of smooth maps $T_k(M) \to \mathbb{R}$ with the property that *on each fiber* of the projection

$$\begin{array}{c} T_k(M) \\ \downarrow{\scriptstyle q_k} \\ M \end{array}$$

the restriction to the vector spaces $T_{k,1}(X), T_{k,2}(X), \ldots, T_{k,k}(X)$ *for each X in the fiber* is *linear* will be denoted $\mu_k(M)$.

It is clear (following exercise 1) that $\mu_k(M)$ has $C(M, \mathbb{R})$ module structure; define multiplication pointwise over M. There is a *module isomorphism* $\nu : \mu_k(M) \to \gamma[q^k]$ where $\gamma[q^k]$ is the $C(M, \mathbb{R})$ module of smooth sections of the k sectorform bundle

$$\begin{array}{c} T^k(M) \\ \downarrow{\scriptstyle q_k} \\ M \end{array}$$

Proof. Define the $C(M, \mathbb{R})$ module homomorphism $\nu : \mu_k(M) \to \gamma[q^k]$ in the following way. Let $\mathbb{B} \in \mu_k(M)$ and let $\phi : O \to U \subset M$ be a frame. There is defined a map $b : O \times E_k \to \mathbb{R}$ given by $(z, X^i_\gamma) \to \mathbb{B}([((\phi \circ t_z)_k, X^i_\gamma)])$. Now for each $z \in O$, the map $X^i_\gamma \to b(z, X^i_\gamma)$ belongs to E^*_k. Call it $f^{P_r}_{k'_1 \ldots k_r}(z) e^{k_1}_{\gamma_{i_1}} \otimes e^{k_2}_{\gamma_{i_2}} \otimes \ldots \otimes e^{k_r}_{\gamma_{i_r}}$ (or simply in components $f^{P_r}_{k'_1 \ldots k_r}(z)$). The map $z \to f^{P_r}_{k'_1 \ldots k_r}(z)$ is smooth with respect to the manifold structure on E^*_k, and this gives a local sectorform field represented with respect to $\phi : \tilde{f}^{P_r}_{k'_1 k_2 \ldots k_r}(z)$ *over* U.

Now the argument in the proof of exercise 2 can be extended painlessly to this situation to show that the construction of the local section is *independent of frame* ϕ. This shows well-definedness, and shows that these local sections can be 'patched together' to give a global section in $\gamma[q^k]$. In a similar manner, following the proof of that exercise, we show that ν is *bijective*, a $C(M, \mathbb{R})$ module isomorphism. ■

Definition 3.6. Suppose $f : M \to N$ is a smooth map of manifolds. There is a *unique* module homomorphism $f^* : \gamma[q^k(N)] \to \gamma[q^k(M)]$ over ring homomorphism $\hat{f} : C(N, \mathbb{R}) \to C(M, \mathbb{R})$ induced by composition with f which commutes the diagram:

$$\begin{array}{ccc} \mu_k(M) & \xleftarrow{\mu_k(f)} & \mu_k(N) \\ {\scriptstyle \nu_M} \Big\downarrow {\scriptstyle \cong} & & {\scriptstyle \cong} \Big\downarrow {\scriptstyle \nu_N} \\ \gamma[q^k(M)] & \xleftarrow{f^*} & \gamma[q^k(N)] \end{array} \tag{3.7}$$

The association $M \mapsto \gamma[q^k(M)]$ and $f \mapsto f^*$ gives a contravariant functor from the category of smooth manifolds to the category of modules. Then ν is a natural isomorphism of functors.

This parallels the construction given in the exercises (and extends it as we shall see). ■

This shows that just as in the case of covariant tensor fields, sectorform fields can be *pulled back* by arbitrary smooth maps. This is an important fact. We see, for example, that the metric tensor has this property (under certain regularity conditions giving rise to a new metric tensor). In the next chapter, we shall show how a metric tensor can be used to construct various sectorform fields, some of these having an interpretation as metric tensors on certain sector bundles. The naturality result above can then be applied to give a simple formulation of D'Alembert's principle, essentially as a 'gradient' principle.

Now it is clear from the definitions that a covariant tensor field of type $\binom{0}{k}$ gives rise to a k sectorform field in the following way: suppose $\phi : O \to U \subset M$ is a frame. Let $\tilde{Y}_{j_1 j_2 \ldots j_k}(z)$ be a local representation of the covariant tensor field (with respect to ϕ). Then associate to this local field the local k sectorform field (represented with respect to ϕ) $\tilde{f}^{P_r}_{h_1 \ldots h_r}(z)$ with all terms zero except that corresponding to the partition $(1)(2)(3)\ldots(k) = Q_k$. And $\tilde{f}^{Q_k}_{h_1 \ldots h_k}(z) = \tilde{Y}_{h_1 \ldots h_k}(z)$. It follows that the k sectorform corresponding to $\tilde{Y}_{j_1 j_2 \ldots j_k}(z)$ transforms under transformation law (3.5) to the image of the tensor transform of $\tilde{Y}_{j_1 j_2 \ldots j_k}(z)$ via formula (1.9) under a change of frame. In fact, this construction gives a smooth map $F_1(M) \times_G V \to T^k(M)$ (where the former is the V-tensor bundle of type $\binom{0}{k}$) which will allow us to *identify* each covariant tensor of type $\binom{0}{k}$ with a k sectorform.

There is a natural bundle projection taking $T_k(M) \to T_1^k(M)$ associating to $[\phi_k, X^i_\gamma]$, $[\phi_1, (X^i_{\gamma_1}, X^i_{\gamma_2}, \ldots, X^i_{\gamma_k})]$ for $\gamma_1, \ldots, \gamma_k$ some enumeration of the vertices (0-faces) of $\Gamma(k-1)$. We shall define below a 'contraction' of k-sectors with k sectorforms, then show its relation to the earlier notion of contraction.

Definition 3.7. Suppose that M is a smooth manifold and that $\phi : O \to U \subset M$ is a frame. Suppose also that $\mathbb{A}$ is a k sectorform field with local representation $\tilde{f}^{\gamma_{i_1} \gamma_{i_2} \cdots \gamma_{i_r}}_{k_1 k_2 \ldots k_r}(z)$ with respect to ϕ and that $\mathbb{F}$ is a k-sector field with local representation $\tilde{X}^i_\gamma(z)$ with respect to ϕ. Then the *contraction of* $\mathbb{A}$ *with* $\mathbb{F}$ over U, denoted $\mathbb{A}\langle\mathbb{F}\rangle$, is a function from $U \to \mathbb{R}$ defined by

$$x \to \sum_{P_r = \gamma_{i_1} \cup \ldots \cup \gamma_{i_r}} \tilde{f}^{\gamma_{i_1} \gamma_{i_2} \cdots \gamma_{i_r}}_{k_1 \ldots k_r}(\phi^{-1}(x)) \tilde{X}^{k_1}_{\gamma_{i_1}}(\phi^{-1}(x)) \ldots \tilde{X}^{k_r}_{\gamma_{i_r}}(\phi^{-1}(x))$$

this summation being over *all* partitions P_r of $\{1, 2, \ldots, k\}$ and for each partition over all repeated upper and lower indices k_j. An argument practically identical to the one assigned in exercise 2 shows that the smooth function $\mathbb{A}\langle\mathbb{F}\rangle : \mathrm{U} \to \mathbb{R}$ is independent of the frame ϕ. Thus, for example, if $\mathbb{A}$ and $\mathbb{F}$ are global sectorform and sector fields, they define a smooth map $\mathbb{A}\langle\mathbb{F}\rangle : M \to \mathbb{R}$. ■

The next topic of this section will be the determination of the effect of $f^*(\mathbb{A})$ for $\mathbb{A}$ a k sectorform field on N and $f: M \to N$ a smooth map of manifolds. Referring to diagram (3.7), we see that $\nu_M^{-1} f^*(\mathbb{A}) = \mu_k(f)\nu_N^{-1}(\mathbb{A})$. Now in general if $\mathbb{B}$ is a k sectorform field on M and $\mathbb{F}$ is a k-sector field on M, then $\nu_M^{-1}(\mathbb{B})(\mathbb{F}) = \mathbb{B}\langle\mathbb{F}\rangle$ as functions mapping $M \to \mathbb{R}$. This follows directly from the definition of contraction. Let $\mathbb{B} = f^*(\mathbb{A})$. Then for $\mathbb{F}$ a k-sector field on M,

$$\nu_M^{-1} f^*(\mathbb{A})(\mathbb{F}) = f^*(\mathbb{A})\langle\mathbb{F}\rangle = \nu_N^{-1}(\mathbb{A})(T_k(f)(\mathbb{F})) = \mathbb{A}\langle T_k(f)(\mathbb{F})\rangle. \tag{3.8}$$

According to this equation, $f^*(\mathbb{A})\langle\mathbb{F}\rangle = \mathbb{A}\langle T_k(f)(\mathbb{F})\rangle$ and this is the naturality property for contraction. Eventually, we shall have use for an explicit form of equation (3.8) which will be derived easily using the transformation law (3.5) as model. It should be observed here that $T_k(f)(\mathbb{F})$ is not a sector field on N, but rather is a smooth map from $M \to T_k(N)$. The 'contraction' in this case is defined by a simple extension of the one defined in Definition 3.7.

Now recalling the projection

$$\begin{array}{c} T_k(M) \\ \downarrow \\ T_1^k(M) \end{array}$$

suppose that $\mathbb{F}$ is a k-sector field and that its projection is $\bar{\mathbb{F}}$ (a k-tuple of vector fields on M). Let $\mathbb{A}$ be a covariant tensor field on M of type $\binom{0}{k}$. Then $\mathbb{A}$ is identified with a k sectorform field which we shall also denote $\mathbb{A}$. A simple check shows that the *tensor* contraction $\nu_M^{-1}(\mathbb{A})(\bar{\mathbb{F}})$ (the contraction of $\mathbb{A}$ with the contravariant field of type $\binom{k}{0}$ derived from $\bar{\mathbb{F}}$) is equal to the contraction $\mathbb{A}\langle\mathbb{F}\rangle$.

Thus, covariant tensor fields of type $\binom{0}{k}$ can be seen to act in a natural way on k-sector fields. That action 'passes to quotients' to give the more familiar one on type $\binom{k}{0}$ tensors via restriction to vertices.

Exercise

4. Let $C \subset \Gamma(k-1)$ be a subcomplex, and $F: T_k(M) \to T_k[M; C]$ as in Definition 2.3. A k sectorform field on M, say B, will be called a *C-sectorform field* if $\nu_M^{-1}(B)$ *factors through* F. In this case, there is a well-defined contraction of B with C-sector fields on M. Show that it satisfies the naturality property (3.8) and state explicitly what conditions B should satisfy in terms of subcomplex C in order for it to be a C-sectorform field.

[The notion of K-sectorform field will be important in the sequel and the previous exercise is strongly recommended.]

We end this section with a description of the *symmetric group action* on sectorforms. There is a *right* action of $S(k)$ defined on $T_k(M)$ by the rule $[\phi_k, X^i_\gamma]s = [\phi_k, X^i_{s(\gamma)}]$. This takes $D^i(k)$ fibers to $D^i(k)$ fibers and is linear on each fiber. Therefore it gives rise to a *left* action of $S(k)$ on $T^k(M)$. We make this explicit in the following definition.

Definition 3.8. If $A^k(x_0)$ is a k sectorform at x_0 and $s \in S(k)$, define the sectorform $s \cdot A^k(x_0)$ to be the k sectorform at x_0 determined by the contraction formula

$$(s \cdot A^k(x_0))\langle \bar{A}_k(x_0)\rangle = A^k(x_0)\langle \bar{A}_k(x_0) \cdot s\rangle \tag{3.9}$$

for each k-sector $\bar{A}_k(x_0)$ at x_0. This action is easily computed in components. If

$$A^k(x_0) = [\phi_k, a^{\gamma_{i_1}\gamma_{i_2}\cdots\gamma_{i_t}}_{h_1 h_2 \ldots h_t}]$$

then

$$sA^k(x_0) = [\phi_k, a^{s(\gamma_{i_1})\ldots s(\gamma_{i_t})}_{h_1 h_2 \ldots h_t}]. \tag{3.10}$$

■

[Now to illustrate this definition and to facilitate calculation in the low dimensions, we adopt the following convention which will be in force throughout the remainder of this book.]

Convention 3.9. Elements of E_1, E_2, and E_3 will be represented in components as they were in formulas (1.27), (1.28), and (1.29) with *characteristic superscripts* (i, j, k, r, s, t, u) corresponding in each order to particular faces:

E_1: X^i

E_2: X^i—A^r—Y^j

E_3:

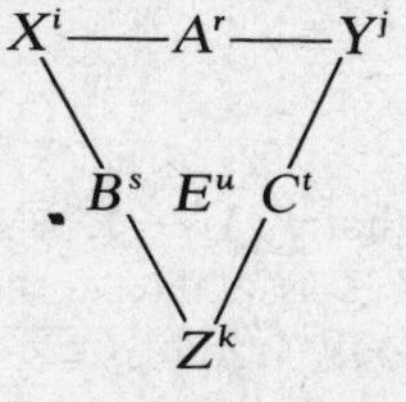

Thus, with rare exceptions where this usage will be impossible because of prior choices, the superscript can be used to designate the face. In fact, formulas (3.1) and (3.2) represent 'sectorforms' without explicit reference to the partition using the corresponding subscript labels.

We shall represent order-1, order-2, and order-3 sectorforms without explicit reference to partitions but using these labels (i, j, k, r, s, t, u) to designate the face. Thus if A^2 has components $a_{h_1h_2}^{(1)(2)} - b_{k_1}^{(12)}$ we represent it simply $a_{ij} - b_r$. The action of the transposition σ of $S(2)$ yields: σA^2 has components $a_{ji} - b_r$. Now according to the summation convention we have $a_{ji}X^iY^j = a_{ij}X^jY^i$ and so these are *different* sectorforms, in general.

[Whatever the order or the names of the subscripts, a given letter will designate, in the course of a calculation, a fixed matrix.]

Thus if A^3 has components $a_{h_1h_2}^{(12)(3)} + b_{k_1k_2k_3}^{(1)(2)(3)}$, then its contraction with the 3-sector with the components given above is $a_{h_1h_2}A^{h_1}Z^{h_2} + b_{k_1k_2k_3}X^{k_1}Y^{k_2}Z^{k_3}$. According to this convention, we may represent the components simply as $a_{rk} + b_{ijk}$ and the contraction above is $a_{rk}A^rZ^k + b_{ijk}X^iY^jZ^k$.

Now let s be the 'shift' of $S(3)$: $s(1)=2$, $s(2)=3$, $s(3)=1$. Then, according to Definition 3.8, sA^3 has components $a_{h_1h_2}^{(23)(1)} + b_{k_1k_2k_3}^{(2)(3)(1)}$ and this is represented simply as $a_{ti} + b_{jki}$. The contraction of this with the 3-sector with components above is $a_{ti}C^tX^i + b_{jki}X^iY^jZ^k$. Notice that $b_{jki}X^iY^jZ^k = b_{ijk}X^kY^iZ^j$. ■

Exercises

5. Show that, with respect to Convention 3.9, the symmetric group action on sectorforms can be represented by letting $S(k)$ act on the subscript labels $(i, j, k, \ldots)$. How should it act?

6. Show that the symmetric group action takes the intrinsic spaces to intrinsic spaces and that it is a linear isomorphism on them, as was asserted.

7. Define a sectorform to be *antisymmetric* if $sA^k = (\operatorname{sgn} s)A^k$ for $s \in S(k)$. First of all, show that this definition makes sense at a point $x_0 \in M$: the fiber $T^k(x_0)$ has a *linear* structure. Next show that an antisymmetric k-sectorform is a covariant tensor of type $\binom{0}{k}$, a differential form.

§3.2. Differentials of sectorform fields

Let M be a smooth manifold and suppose that $f: M \to \mathbb{R}$ is a smooth real-valued function. For each $x \in M$ associate the 1-jet $(t_{f(x)}^{-1} \circ f)_1 : (M, x) \to (\mathbb{R}, 0)$. With respect to a local frame $\phi: V \to U \subset M$ we have local section of the cotangent bundle with local representative

$$\left[(\phi \circ t_z)_1, \frac{\partial (f \circ \phi)}{\partial x^i}_{|z}\right].$$

This cotangent vector field is called the *differential* of f and may be denoted

$$df_{|x} = \frac{\partial(f \circ \phi)}{\partial x^i}_{|z}\, dx^i$$

with respect to local frame ϕ (where $\phi(z) = x$).

The goal of this section is to extend the definition of 'differential' in such a way that its most useful properties and interpretations are preserved in the extension. In particular, we may view the real-valued function f as an order-0 sectorform field, and its differential as an order-1 sectorform field. Each order-k sectorform field will be shown to have $(k+1)$ differentials, each of which is a $(k+1)$ sectorform field. By means of this extension, it will be possible to define a natural notion of *differentiation* of a vector-valued function on a smooth manifold M by a k-sector.

We begin with *another* operation on sectorforms related to the *promotion* operator (Definition 2.12). Instead of beginning with the data of sector field as we did in the afore-mentioned definition, let us use the natural isomorphism $T_{r+1}(M) \to T_r[T_1(M)]$ introduced in Theorem 1.27 to give a slightly more general notion of promotion.

Definition 3.10. Let $\gamma_1 \in \Gamma(r)$ be the $(r+1)$ vertex $(0, 0, \ldots, 0; 1)$. Following the discussion immediately preceding Definition 2.21, we may represent an $(r+1)$ sector $\bar{A}_{r+1}(x_0)$ *with respect to* γ_1 by the following convention:

1) $\gamma' = (0, 0, \ldots, 0; 1) = \gamma_1$ is the unique *type* 1 face of $\Gamma(r)$,
2) $\gamma'' = (g_1, g_2, \ldots, g_r; 0)$ represents a general *type* 2 face of $\Gamma(r)$,
3) $\gamma' \cup \gamma''$ represents a general *type* 3 face of $\Gamma(r)$.

Then we shall represent $\bar{A}_{r+1}(x_0)$ in the following form

$$[\phi_{r+1}, X^i_{\gamma'} \text{---} Z^i_{\gamma' \cup \gamma''} \text{---} Y^i_{\gamma''}],$$

and call this the representation *with respect to the last vertex.* The transformation law for this representation is given in the proof of Theorem 1.27, with minor modifications. In particular the 1-sector $[\phi_1, X^i_{\gamma'}]$ is simply the $(r+1)$ vertex of $\bar{A}_{r+1}(x_0)$.

Now for M a smooth manifold modelled on Euclidean space E, for $x_0 \in M$ and ϕ a frame-germ at x_0, let $\tilde{\phi}^1 \circ t_{(0, X^i_{\gamma'})}$ be the frame-germ at $[\phi_1, X^i_{\gamma'}]$ defined in Theorem 1.27 for $T_1(M)$. We give a name for the modified natural isomorphism ν in this case.

The mapping $T_{r+1}(M) \underset{p}{\overset{\cong}{\Rightarrow}} T_r[T_1(M)]$ represented in components

$$[\phi_{r+1}, X^i_{\gamma'} \text{---} Z^i_{\gamma' \cup \gamma''} \text{---} Y^i_{\gamma''}] \to [(\tilde{\phi}^1 \circ t_{(0, X^i_{\gamma'})})_r, W^I_{\gamma''}]$$

will be called the *pointing* operator. According to the discussion in Theorem 1.27, the components $W^I_{\gamma''}$ have the following description in terms of the components of $\bar{A}_{r+1}(x_0)$. I has the form $(i, 0)$ or (i, γ') and takes $2m$ values for each γ'' of type 2, $W^{(i,0)}_{\gamma''} = Y^i_{\gamma''}$ and $W^{(i,\gamma')}_{\gamma''} = Z^i_{\gamma'\cup\gamma''}$. Here $\dim(M) = m$. ■

Example 1. Letting $r = 1$, $\bar{A}_2 = [\phi_2, X^i — Z^r — Y^j]$ and $p\bar{A}_2$ is the 1-sector at $[\phi_1, Y^i]$ with component representation

$$\left[(\tilde{\phi}^1 \circ t_{(0,Y^i)})_1, \begin{bmatrix} X^i \\ Z^r \end{bmatrix}\right]. \tag{3.11}$$

Example 2. Letting $r = 2$, $\bar{A}_3$ has component representation

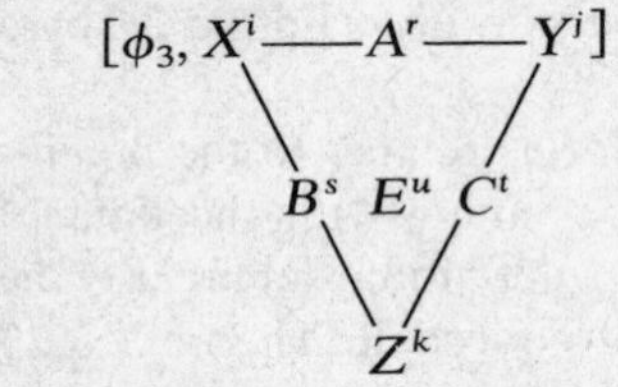

then $p\bar{A}_3$ is a 2-sector at $[\phi_1, Z^k]$ with component representation

$$\left[(\tilde{\phi}^1 \circ t_{(0,Z^k)})_2, \begin{bmatrix} X^i \\ B^s \end{bmatrix} — \begin{bmatrix} A^r \\ E^u \end{bmatrix} — \begin{bmatrix} Y^j \\ C^t \end{bmatrix}\right]. \tag{3.12}$$

Exercises

8. Use the transformation law for sectors represented in components such as (3.12) above (Theorem 1.27) to show that, if $g \in G_3(m)$ is a coordinate change, then the components of (3.12) transform under g to

$$\begin{bmatrix} \bar{X}^i \\ \bar{B}^s \end{bmatrix} — \begin{bmatrix} \bar{A}^r \\ \bar{E}^u \end{bmatrix} — \begin{bmatrix} \bar{Y}^j \\ \bar{C}^t \end{bmatrix}$$

with

$$\begin{bmatrix} \bar{X}^i \\ \bar{B}^s \end{bmatrix} = \begin{bmatrix} \dfrac{\partial g^i}{\partial x^a} X^a \\ \dfrac{\partial g^s}{\partial x^b} B^b + \dfrac{\partial^2 g^s}{\partial x^a \partial x^k} X^a Z^k \end{bmatrix}$$

$$\begin{bmatrix} \bar{Y}^j \\ \bar{C}^t \end{bmatrix} = \begin{bmatrix} \dfrac{\partial g^j}{\partial x^c} Y^c \\ \dfrac{\partial g^t}{\partial x^d} C^d + \dfrac{\partial^2 g^t}{\partial x^c \partial x^k} Y^c Z^k \end{bmatrix}$$

and

$$\begin{bmatrix}\bar{A}^r\\ \bar{E}^u\end{bmatrix}=\begin{bmatrix}\dfrac{\partial g^r}{\partial x^e}A^e\\ \dfrac{\partial g^u}{\partial x^f}E^f+\dfrac{\partial^2 g^u}{\partial x^e\partial x^k}A^eZ^k\end{bmatrix}+\begin{bmatrix}\dfrac{\partial^2 g^r}{\partial x^i\partial x^j}X^iY^j\\ \dfrac{\partial^2 g^u}{\partial x^a\partial x^d}X^aC^d+\dfrac{\partial^2 g^u}{\partial x^c\partial x^b}Y^cB^b\\ +\dfrac{\partial^3 g^u}{\partial x^i\partial x^j\partial x^k}X^iY^jZ^k\end{bmatrix}.$$

Argue that these components satisfy the transformation law for 2-sectors.

9. Show that the pointing map $p: T_{r+1}(M)\to T_r[T_1(M)]$ takes $T_{r+1,i}(X)\to T_{r,i}[pX]$ for $1\leq i\leq r$ and that it is *linear* on these intrinsic spaces.

10. Let $A^{r+1}(x_0)$ be an $(r+1)$ sectorform at $x_0\in M$. Show that the association $\bar{X}_1(x_0)\to A^{r+1}(x_0)_{|\bar{X}_1(x_0)}$ [where the last symbol has the following meaning: $A^{r+1}(x_0)_{|\bar{X}_1(x_0)}$ is the mapping $T_r[\bar{X}_1(x_0)]\to\mathbb{R}$ (the fiber over tangent vector $\bar{X}_1(x_0)\to\mathbb{R}$) given by $A^{r+1}(x_0)\circ p^{-1}$] gives a *smooth* mapping $T_1(x_0)\to T^r[T_1(M)]$ associating to each tangent vector at $x_0\in M$ an *r sectorform* on $T_1(M)$ *at* that tangent vector (a 'cross section' of $T^r[T_1(m)$ over the fiber $T_1(x_0)$).

11. The symbol $A^{r+1}(x_0)_{|\bar{X}_1(x_0)}$ will be read: the sectorform $A^{r+1}(x_0)$ pointed by the tangent vector $\bar{X}_1(x_0)$. Using naturality of the transformation p (formulated in Theorem 1.27), show that this operation has the following naturality property. If $f:(M,x_0)\to(N,y_0)$ is smooth, $\bar{X}_1(x_0)$ is a tangent vector at x_0, $B^{r+1}(y_0)$ an $(r+1)$ sectorform at y_0, then

$$f^*[B^{r+1}(y_0)]_{|\bar{X}_1(x_0)}=T_1(f)^*[B^{r+1}(y_0)_{|T_1(f)(X_1(x_0))}]. \tag{3.13}$$

12. Let $e''=(1,1,\ldots,1;0)\in\Gamma(r)$. Suppose that ϕ_{r+1} is a fixed frame-jet at $x_0\in M$ and that $A^{r+1}(x_0)$ is an $(r+1)$ sectorform at x_0. Compute $A^{r+1}(x_0)_{|\bar{A}_1(x_0)}$ for $A^{r+1}(x_0)$ with ϕ components $a^{P_t}_{h_1h_2\ldots h_t}$, and $\bar{A}_1(x_0)$ with ϕ components $X^i_{\gamma'}$. Show that the ϕ components of $A^{r+1}(x_0)_{|\bar{A}_1(x_0)}$ are $\bar{a}^{Q_s}_{\alpha_1\alpha_2\ldots\alpha_s}$ $(Q_s:\gamma''_{i_1}\cup\ldots\cup\gamma''_{i_s}=e'')$ with:

1) $\bar{a}^{Q_s}_{\alpha_1\alpha_2\ldots\alpha_s}=0$ if two or more of the α_i have γ' for second component.

2) $\bar{a}^{\gamma''_{i_1}\gamma''_{i_2}\ldots\gamma''_{i_s}}_{(w_1,0),(w_2,0),\ldots,(w_s,0)}=a^{\gamma''_{i_1}\gamma''_{i_2}\ldots\gamma''_{i_s},\gamma'}_{w_1w_2\ldots w_s,i}X^i_{\gamma'}$ (sum over i).

3) $\bar{a}^{\gamma''_{i_1}\ldots\gamma''_{i_t}\ldots\gamma''_{i_s}}_{(w_1,0),\ldots,(w_t,\gamma'),\ldots,(w_s,0)}=a^{\gamma''_{i_1}\ldots(\gamma'\cup\gamma''_{i_t})\ldots\gamma''_{i_s}}_{w_1,\ldots,w_t,\ldots,w_s}.$ (3.14)

The general definition of the $(k+1)$ differential of a local k sectorform field is fairly easy to give in components. What will require work is the demonstration that this definition is frame-independent.

Thus, suppose that for frame $\phi: O \to U \subset M$, $A^k(x)$ is a sectorform field with local representation $\tilde{a}^{P_t}_{h_1 \dots h_t}(z)$. The representation of $d_{k+1}A^k(x)$ is the $(k+1)$ sectorform field with local representation $\tilde{b}^{Q_s}_{k_1 k_2 \dots k_s}(z)$ for Q_s a partition of $\{1, 2, \dots, k, k+1\}$ and b determined by the following rules:

1) if Q_s is of the form $\gamma_{i_1} \cup \dots \cup \gamma_{i_{s-1}} \cup (k+1)$ (where $(k+1)$ is the $(k+1)$ vertex), then

$$\tilde{b}^{Q_s}_{k_1 k_2 \dots k_s}(z) = \frac{\partial(\tilde{a}^{\gamma_{i_1} \gamma_{i_2} \dots \gamma_{i_{s-1}}}_{k_1 k_2 \dots k_{s-1}})}{\partial x^{k_s}}_{|z} ;$$

2) if Q_s is of the form $\gamma_{i_1} \cup \dots \cup (\gamma_{i_t} \cup (k+1)) \cup \dots \cup \gamma_{i_s}$ for $\gamma_{i_1} \cup \dots \cup \gamma_{i_t} \cup \dots \cup \gamma_{i_s}$ a join decomposition of the front face of $\Gamma(k)$, $(1, 1, \dots, 1; 0)$, then

$$\tilde{b}^{Q_s}_{k_1 k_2 \dots k_s}(z) = \tilde{a}^{\gamma_{i_1} \gamma_{i_2} \dots \gamma_{i_t} \dots \gamma_{i_s}}_{k_1 \dots \hat{k}_t \dots k_s}(z). \tag{3.15}$$

Proposition 3.11. The definition of $\tilde{b}^{Q_s}_{k_1 \dots k_s}(z)$ given in formula (3.15) is *independent of frame* in the following sense: if $\theta: O \to U$ is another frame, and $f = \theta^{-1} \circ \phi$, then for $w = f(z)$, $t_w^{-1} \circ f \circ t_z = \bar{f}$, $\bar{f}_{k+1} \cdot (b^{Q_s}_{k_1 \dots k_s})(w)$ corresponds under (3.15) to $\bar{f}_k \cdot (a^{P_t}_{h_1 \dots h_t})(z)$.

Proof. We make use of the transformation formula (3.5) for sectorforms. Suppose that we calculate $\bar{f}_{k+1} \cdot (b^{Q_s}_{k_1 \dots k_s}) = c^{R_s}_{j_1 j_2 \dots j_s}$. There are two cases:

1) R_s has the form $\gamma_{i_1} \cup \dots \cup (\gamma_{i_t} \cup (k+1)) \cup \dots \cup \gamma_{i_s}$. Let $R'_s = \gamma_{i_1} \cup \dots \cup \gamma_{i_t} \cup \dots \cup \gamma_{i_s}$. Then according to (3.5),

$$c^{R_s}_{j_1 \dots j_s} = \sum_{P_r \leq R'_s} a^{P_r}_{h_1 h_2 \dots h_r} \left(\frac{\partial^{t_1} \hat{\bar{f}}^{h_1}}{\partial x^{j_{a_1(1)}} \dots \partial x^{j_{a_1(t_1)}}} \right) \dots \left(\frac{\partial^{t_r} \hat{\bar{f}}^{h_r}}{\partial x^{j_{a_r(1)}} \dots \partial x^{j_{a_r(t_r)}}} \right) \tag{3.16}$$

for $\hat{\bar{f}}$ the *inverse* germ $\bar{f}^{-1}$, $P_r = \gamma'_{i_1} \cup \dots \cup \gamma'_{i_r}$ with $\gamma'_{i_u} = \gamma_{i_{a_u}(1)} \cup \dots \cup \gamma_{i_{a_u(t_u)}}$, $s = t_1 + \dots + t_r$.

2) R_s has the form $\gamma_{i_1} \cup \dots \cup \gamma_{i_{s-1}} \cup (k+1)$. Let $R''_{s-1} = \gamma_{i_1} \cup \dots \cup \gamma_{i_{s-1}}$. In

this case, there are *two* types of partitions weaker than R_s. We have

$$\begin{aligned} c^{R_s}_{j_1 \dots j_s} = \sum_{P_r \leq R''_{s-1}} & \frac{\partial(\tilde{a}^{P_r}_{h'_1 \dots h_r})}{\partial x^v}\Big|_z \frac{\partial^{t_1} \hat{\bar{f}}^{h_1}}{\partial x^{j_{q_1(1)}} \dots \partial x^{j_{q_1(t_1)}}} \\ & \dots \frac{\partial^{t_r} \hat{\bar{f}}^{h_r}}{\partial x^{j_{q_r(1)}} \dots \partial x^{j_{q_r(t_r)}}} \left(\frac{\partial \hat{\bar{f}}^v}{\partial x^{j_s}}\right) + \sum_{\substack{P_r \leq R''_{s-1} \\ 1 \leq \tau \leq r}} a^{P_r}_{h'_1 \dots h_r} \\ & \times \frac{\partial^{t_1} \hat{\bar{f}}^{h_1}}{\partial x^{j_{q_1(1)}} \dots \partial x^{j_{q_1(t_1)}}} \dots \frac{\partial^{t_\tau + 1} \hat{\bar{f}}^{h_\tau}}{\partial x^{j_{q_\tau}(1)} \dots \partial x^{j_{q_\tau(t_\tau)}} \delta x^{j_s}} \\ & \dots \frac{\partial^{t_r} \hat{\bar{f}}^{h_r}}{\partial x^{j_{q_r(1)}} \dots \partial x^{j_{q_r(t_r)}}}. \end{aligned} \tag{3.17}$$

Considering the expression for $\bar{f}_k \cdot (a^{P_t}_{h'_1 \dots h_t})$ given by (3.5), its transform under rule (3.15) gives $\bar{c}^{R_s}_{j_1 \dots j_s}$ with the two types of partitions R_s. For R_s of type 1) above, checking (3.16) we have immediately that $c^{R_s}_{j_1 \dots j_s} = \bar{c}^{R_s}_{j_1 \dots j_s}$. For R_s of type 2) above, (3.17) results from the 'product formula' for the partial differentiation operator $\partial/\partial x^{j_s}$. So $c^{R_s}_{j_1 \dots j_s} = \bar{c}^{R_s}_{j_1 \dots j_s}$ in this case also. ■

Definition 3.12. In light of the previous proposition, we are entitled to refer to the $(k+1)$ *differential* $d_{k+1}A^k(x)$ of a (local) k sectorform field. It is calculated in components unambiguously according to formula (3.15).

For $1 \leq i \leq k+1$, define the *i-differential* $d_i A^k(x)$ of the (local) k sectorform field by the rule

$$d_i A^k(x) = \sigma^i(d_{k+1} A^k)(x)$$

for σ the 'shift' in $C(k+1) \subseteq S(k+1)$. Thus a k sectorform field has $(k+1)$ differentials $d_1 A^k, d_2 A^k, \dots, d_{k+1} A^k$. ■

Now for $k \geq 1$, the differentials of a tensor are never tensors. They are, however, sectorforms and can be manipulated algebraically as such. Further, *pointing* each differential of A^k yields an order-k sectorform on the tangent space (see exercise 10). After some examples, we discuss this operation.

The differential operators d_i are *natural* in the following sense. Let $f: M \to N$ be a smooth map and let $B^k(y)$ be a local sectorform field on $V \subset N$. Suppose $U \subset M$ and $f(U) \subset V$. We let $A^k(x) = f^*[B^k(y)]$ be the pull-back sectorform field on U. Say $f(x) = y$.

Then for each i, $1 \leq i \leq k+1$, $d_i A^k(x) = d_i f^*[B^k(y)] = f^*[d_i B^k(y)]$. This naturality relation is the basis for most of the applications which sectorform fields have for differential geometry and mechanics. It will lead in the next section, for example, to the naturality of the coboundary operator on differential forms, a property of fundamental importance for the calculus of differential forms.

In view of Proposition 3.11, there is little to prove to establish the naturality relation. Returning to the argument given there, we replace the map $\theta^{-1}\circ\phi$, which represents a change of frame, with the map $\theta^{-1}\circ f\circ\phi$ with f as above. Next, the transformation law (3.5) is replaced with the component formula which describes the effect of f^* on components. We state that formula below.

Suppose that in the above notation $A^k(x)=f^*[B^k(y)]$ and that frames $\phi: O\to U\subset M$ and $\theta: O'\to V\subset N$ are chosen. Suppose that $f(x_0)=y_0$ and that, in frame θ, $B^k(y_0)$ has components $b^{Q_s}_{k_1k_2\ldots k_s}(w_0)$, $\theta(w_0)=y_0$. Then $A^k(x_0)=f^*[B^k(y_0)]$ has the components $a^{P_r}_{h_1h_2\ldots h_r}(z_0)$ in frame ϕ where $\phi(z_0)=x_0$, where for each partition P_r of $\Gamma(k-1)$ $(P_r\colon \gamma_{i_1}\cup\gamma_{i_2}\cup\ldots\cup\gamma_{i_r}=e)$ we say that a partition Q_s of $\Gamma(k-1)$ $(Q_s\colon \gamma'_{j_1}\cup\gamma'_{j_2}\cup\ldots\cup\gamma'_{j_s}=e)$ is weaker than P_r, which is denoted $Q_s\leq P_r$, if the criterion in Definition 3.3 is satisfied. Let g be the germ $g:(\mathbb{R}^m,0)\to(\mathbb{R}^n,0)$ $t^{-1}_{w_0}\circ\theta^{-1}\circ f\circ\phi\circ t_{z_0}=g$. Then with $\gamma'_{j_u}=\gamma_{i_{a_u(1)}}\cup\ldots\cup\gamma_{i_{a_u(t_u)}}$

$$a^{P_r}_{h_1h_2\ldots h_r}=\sum_{Q_s\leq P_r} b^{Q_s}_{k_1k_2\ldots k_s}\left(\frac{\partial^{t_1}g^{k_1}}{\partial x^{h_{a_1(1)}}\ldots\partial x^{h_{a_1(t_1)}}}\right)\cdots\left(\frac{\partial^{t_s}g^{k_s}}{\partial x^{h_{a_s(1)}}\ldots\partial x^{h_{a_s(t_s)}}}\right). \quad (3.18)$$

The argument of Proposition 3.11 easily applies now to give naturality.

In order to interpret the operation d_{k+1} geometrically, we describe its contraction with a $(k+1)$ sector in a slightly less abstract way than we have done. This can be done by generalizing the notion of 'ribbon' to describe the F-related classes of maps represented by sectors.

Definition 3.13. Let $\phi: V\to U\subset M$ be a frame on a smooth manifold M. Also let $f:(\mathbb{R}^k,(0,0,\ldots,0))\to(V,z_0)$ be a smooth germ. Equip $\mathbb{R}^k$ with 'standard' linear coordinates $(s^1,s^2,\ldots,s^k)$. Denote the *F-related class* (with $F=\mathbb{R}^k$ with linear coordinates (s^i)) of the k-jet $\phi\circ f:(F,(0,0,\ldots,0))\to(M,\phi(z_0))$ by $\bar{f}^{\#}_k(0,0,\ldots,0)$. In general, if $f:F\to V$ is a local smooth mapping, then for $(s_0^1,s_0^2,\ldots,s_0^k)\in F=\mathbb{R}^k$, denote the F-related class of the k-jet of $\phi\circ f\circ t_{(s_0^1,\ldots,s_0^k)}:(F,(0,0,\ldots,0))\to(M,\phi\circ f(s_0^1,s_0^2,\ldots,s_0^k))$ simply as $\bar{f}^{\#}_k(s_0^1,\ldots,s_0^k)$. This latter is the k-sector $[(\phi\circ t_{f(s_0^1,\ldots,s_0^k)})_k, X^i_\gamma(f(s_0^1,\ldots,s_0^k))]$ with $X^i_\gamma(f(s_0^1,\ldots,s_0^k))$ equal to

$$\frac{\partial^r f^i}{\partial s^{m_1}\ldots\partial s^{m_r}}\bigg|_{(s_0^1,\ldots,s_0^k)}$$

for $\gamma=(\gamma_1,\gamma_2,\ldots,\gamma_k)$, $\gamma_{m_j}=1$ for $j=1,\ldots,r$, otherwise $\gamma_i=0$.

Thus $f:F\to V$ gives rise to a smooth map $\bar{f}^{\#}_k: F\to T_k(V)$. When $k=1$, this is the tangent vector map, and when $k=2$, this is the 'ribbon' map defined just before Definition 2.13. ■

Observation 3.14. Under the conditions of Definition 3.13, suppose $\phi(z_0) = x_0 \in M$ and $A^k(x)$ is a local sectorform field with local coordinates $\tilde{a}^{P_t}_{h_1 \ldots h_t}(z)$ and let $\bar{f}^{\#}_{k+1}(0, \ldots, 0)$ be a $(k+1)$ sector at $x_0 \in M$. For $\tau \in \mathbb{R}$, let $g(\tau)$ denote the germ $(s_0^1, \ldots, s_0^k) \to f(s_0^1, \ldots, s_0^k; \tau)$. This defines the germ of a path in $T_k(U)$ by passage to F-related classes of maps by the rule $\tau \to \overline{g(\tau)}^{\#}_k(0, 0, \ldots, 0)$. Then the *contraction*

$$d_{k+1}A^k(x_0)\langle \bar{f}^{\#}_{k+1}(0, 0, \ldots, 0)\rangle = \frac{\partial}{\partial \tau_{|0}} [A^k(f(0, \ldots, 0; \tau))\langle \overline{g(\tau)}^{\#}_k(0, \ldots, 0)\rangle].$$

Proof. Compute $A^k(f(0, \ldots, 0; \tau))\langle \overline{g(\tau)}^{\#}_k(0, \ldots, 0)\rangle$ using the expression for components given in Definition 3.13. We have, for τ fixed

$$\sum_{P_t = \gamma_{i_1} \cup \ldots \cup \gamma_{i_t}} \tilde{a}^{P_t}_{h_1 h_2 \ldots h_t}(f(0, \ldots, 0; \tau)) \frac{\partial^{k_1} f^{h_1}}{\partial s^{\gamma_{i_1}}}_{|(0, \ldots, 0; \tau)} \cdots \frac{\partial^{k_t} f^{h_t}}{\partial s^{\gamma_{i_t}}}_{|(0, \ldots, 0; \tau)}$$

with $k_j = |\gamma_{i_j}|$ the number of vertices in γ_{i_j}. Differentiating at $\tau = 0$ gives the desired result. ∎

This gives an interpretation of the differential $d_{k+1}A^k(x)$ of a local k sectorform field in terms of its contraction with a germ of a path in $T_k(U)$. Such a path defines a tangent vector at $T_k(x_0)$ in $T_k(U)$ and therefore a $(k+1)$ sector at x_0.

The next interpretation generalizes the relation between differentials of functions and integrals along paths. It will be useful later for the discussion of the *Poincaré lemma*. First we need a generalization of the notion of 1-variation of a path (Definition 2.13).

Suppose $\phi: V \to U \subset M$ is a frame, and $f: I \to V$ is a smooth path in U. An *r-variation* of f is an *equivalence class* of germs $F: (I \times \mathbb{R}^r, I \times (0, \ldots, 0)) \to (V, \text{im}(f))$ with $F_{|I \times (0, \ldots, 0)} = f$ ($I \times (0, \ldots, 0)$ identified with I) with respect to the equivalence relation $F \simeq G$ if $\bar{F}^{\#}_{r+1}(s; 0, \ldots, 0) = \bar{G}^{\#}_{r+1}(s; 0, \ldots, 0)$ for all $s \in I$. Denote the r-variation of f by the path $\bar{F}^{\#}_{r+1}(s, (0, \ldots, 0)): I \to T_{r+1}(U)$, or simply $\bar{F}^{\#}_{r+1}(s)$. Now suppose that $A^r(x)$ is an r sectorform field on U with local components $\tilde{a}^{P_t}_{h_1 h_2 \ldots h_t}(z)$, and modifying the preceding notation slightly let $\bar{F}^{\#}_{r+1}(s)$ be an r-variation of path $f: I \to V$. For each $s \in I$ let $\bar{g}^{\#}_r(s) = D^1(r+1)\bar{F}^{\#}_{r+1}(s)$. This is the F-related class of the germ $(t^1, t^2, \ldots, t^r) \to F(s; t^1, \ldots, t^r)$. In this way, the r-variation of f defines a path of r-sectors.

For $I = [0, 1]$, the previous observation allows us to conclude that the *contraction* $d_1 A^r(f(s_0))\langle \bar{F}^{\#}_{r+1}(s_0)\rangle$ for $s_0 \in (0, 1)$ is equal to

$$\frac{\partial}{\partial s}\Big|_{s_0} [A^r(f(s))\langle \bar{g}^{\#}_r(s)\rangle].$$

Computing integrals, we may say that

$$\int_0^1 d_1 A^r(f(\sigma))\bar{F}^{\#}_{r+1}(\sigma)\, d\sigma = A^r(f(1))\langle \bar{g}^{\#}_r(1)\rangle - A^r(f(0))\langle \bar{g}^{\#}_r(0)\rangle. \quad (3.19)$$

In this sense, the differential operators play the role for r sectorform fields and r-variations of paths that the differential of a smooth real-valued function plays with respect to a path.

Exercises

13. Prove component change formula (3.18):
    a) in the case that $B^k(y_0)$ is a covariant *tensor* of type $\binom{0}{k}$,
    b) in the general case.

14. Suppose that $A^1(x)$ is a local covariant tensor field of type $\binom{0}{1}$ (a local 1-form) and that it has the component representation $\tilde{a}_i(z)$. Using Convention 3.9 show that $d_2A^1(x)$ has components

$$\tilde{a}_r + \frac{\partial \tilde{a}_i}{\partial x^j}_{|z}$$

and $d_1A^1(x)$ has

$$\tilde{a}_r + \frac{\partial \tilde{a}_j}{\partial x^i}_{|z}$$

for components. Compute the contraction at x_0 with a sector whose components are X^i——A^r——Y^j.

15. Suppose that $A^2(x)$ is a covariant tensor field of type $\binom{0}{2}$ in local components $A^2(x) = \tilde{a}_{ij}(z)$. Show that in local components (following Convention 3.9):

    a) $d_3A^2(x) = \tilde{a}_{sj}(z) + \tilde{a}_{it}(z) + \frac{\partial \tilde{a}_{ij}}{\partial x^k}_{|z}$,

    b) $d_1A^2(x) = \tilde{a}_{rk}(z) + \tilde{a}_{js}(z) + \frac{\partial \tilde{a}_{jk}}{\partial x^i}_{|z}$,

    c) $d_2A^2(x) = \tilde{a}_{ti}(z) + \tilde{a}_{kr}(z) + \frac{\partial \tilde{a}_{ki}}{\partial x^j}_{|z}$.

16. Compute $d_3 \circ d_2A^1(x)$ for $A^1(x)$ as in exercise 14.

17. Show that if $\tilde{a}_{ij}(z)$ is antisymmetric (or skew-symmetric) for all z for $A^2(x)$ as in exercise 15, then the *sum* $d_1A^2 + d_2A^2 + d_3A^2$ is a *tensor* of type $\binom{0}{3}$. This will be the classical *coboundary* of the 2-form $A^2(x)$.

18. A *metric tensor* on a smooth manifold M is a covariant tensor field of type $\binom{0}{2}$ with the property that if it has representation in local frame

$A^2(x) = \tilde{a}_{ij}(z)$ then for each z the matrix $\tilde{a}_{ij}(z)$ is *symmetric* and *invertible*. Show that for each x, if these conditions hold for a single frame they hold for all frames at x_0. Does any linear combination of differentials of A^2 yield a *tensor*?

We end this section with a definition which will lead to some noteworthy constructions in Chapter 4 on Riemannian geometry. Pointing a sectorform field of order $(k+1)$ yields a 'natural' sectorform field on the tangent bundle and this field has order k. We shall be interested in the following situation: starting with $A^k(x)$, a sectorform field of order k on M, construct its $(k+1)$ differential $d_{k+1}A^k(x)$, an order $(k+1)$ sectorform field on the manifold M. Next, *point* the field $d_{k+1}A^k(x)$ in order to obtain a k sectorform field *on the tangent bundle*. This operation will be especially significant when the original field $A^2(x)$ is a metric tensor; in this case, the operation yields a *metric tensor* on the tangent bundle which has naturality properties to be spelled out below.

Definition 3.15. Suppose $\phi: V \to U \subset M$ is a frame. Let $A^k(x)$ be a local k sectorform field on U. Then $d_{k+1}A^k(x)$ is a local $(k+1)$ sectorform field on U, and it gives rise, via pointing, to a local k sectorform field on $T_1(U)$. We do this by the association

$$\bar{A}_1(x) \to [d_{k+1}A^k(x)]_{|\bar{A}_1(x)} = \delta A^k[\bar{A}_1(x)].$$

Call the latter k sectorform field the *promoted field* and the operation on sectorform fields $A^k(x) \to \delta A^k[\bar{A}_1(x)]$ *promotion* (not to be confused with promotion of sector fields) of sectorform fields.

This operation has the following naturality property: if $f: M \to N$ is smooth, and $B^k(y)$ is a k sectorform field on N and if $f(x) = y$, then

$$\delta[f^*B^k(y)][\bar{A}_1(x)] = [T_1(f)^*(\delta B^k)][\bar{A}_1(x)]. \tag{3.20}$$

That is $\delta \circ f^* = T_1(f)^* \circ \delta$. This follows directly from the naturality of the differential operator d_{k+1}. ■

Exercises

19. Show that a promoted metric is a metric on the tangent bundle.

20. Prove relation (3.20).

21. We often denote the promoted field simply as δA^k. Show that for the natural isomorphism in Theorem 1.27

$$\delta A^k \langle \nu \circ \sigma^{j-1}[\bar{B}_{k+1}(x)] \rangle = d_j A^k \langle \bar{B}_{k+1}(x) \rangle. \tag{3.21}$$

§3.3. Differential forms

In this section, we extend some of the observations in §1.2 on covariant tensor fields to an important special case: differential forms. Suppose $\phi: O \to U \subset M$ is a frame in smooth manifold M. Let us agree to represent a local covariant tensor field of type $\binom{0}{r}$ with respect to ϕ as $\tilde{a}_{i_1 i_2 \ldots i_r}(z)$ and to think of such a field as a local r sectorform field via the identification given in §3.1. (Here the partition is $(1)(2)\ldots(r)$.) As usual, we think of these components $\tilde{a}_{i_1 i_2 \ldots i_r}(z)$ of the sectorform $A^r(x)$ as a *variable* r-dimensional matrix (m entries in each dimension, so m^r entries in all), and its contraction with r-sector field $\bar{A}_r(x)$ with components $\tilde{X}^i_\gamma(z)$ is the local function of $z: \tilde{a}_{i_1 i_2 \ldots i_r}(z) \times \tilde{X}^{i_1}_{\gamma_{i_1}}(z)\tilde{X}^{i_2}_{\gamma_{i_2}}(z)\ldots\tilde{X}^{i_r}_{\gamma_{i_r}}(z)$ with $\gamma_{i_1}, \ldots, \gamma_{i_r}$ the *vertices* of $\Gamma(r-1)$. In writing the contraction above, we shall *drop* mention of the γ_{i_j} in the future.

Differential forms are a class of covariant tensor with the property that certain operators derived in a simple way from the 'differential operators' take them to higher-order covariant tensors, in fact to differential forms. These operations will clearly be *natural.* While we shall not pursue the theory in much detail here, it should be observed that the differential forms can be manipulated algebraically in an extremely simple 'calculus' and they can, via an integration theory on manifolds (see, for example, Spivak [20]), give important *geometric* and *global* properties of smooth manifolds.

We begin with the simplest non-trivial case, the differential 1-forms. For example, if $f: M \to \mathbb{R}$ is smooth, we observed that the *differential* of f, $df: T_1(M) \to \mathbb{R}$ by the rule (in local coordinates)

$$df\langle \tilde{X}^i(z)\rangle = \frac{\partial(f_o\phi)}{\partial x^i}\bigg|_z X^i(z),$$

gives a map which is smooth and *linear on each fiber,* and hence which can be identified via ν_M with a global covariant tensor field of type $\binom{0}{1}$. [In general, a local covariant tensor field of type $\binom{0}{1}$ is the data of a local 1-form.] And while it is not true in general that every local 1-form is of the form df for local map f, it will be possible to state easily checked conditions for this to be so for a 1-form. In the presence of a *metric form* on a smooth manifold, a 1-form will be associated with a *vector field* on the manifold: the gradient field. Often the local evolution of a dynamical system will be interpreted in terms of such a field.

Now suppose that $A^1(x) = \tilde{a}_{i_1}(z)$ is a local 1-form represented with respect to frame $\phi: O \to U \subset M$. Following Convention 3.9, we form the differentials

$$d_1A^1(x)=\frac{\partial a_j}{\partial x^i}_{|z}+a_r(z)=\sigma^1 d_2A^1(x),$$
$$d_2A^1(x)=\frac{\partial a_i}{\partial x^j}_{|z}+a_r(z)=\sigma^2 d_2A^1(x). \tag{3.22}$$

Of course, these 2 sectorforms are *not* tensors. Suppose we form the sum

$$d_1A^1(x)-d_2A^1(x)=\sum_{j=1}^{2}\operatorname{sgn}(\sigma^{j-1})\sigma^j d_2A^1(x)=dA^1(x).$$

We denote this 2 sectorform $dA^1(x)$ (unsubscripted d) and it will be called the *coboundary* of $A^1(x)$. It is obvious that $dA^1(x)$ is a *covariant* 2-*tensor*. In particular, we have

$$dA^1(x)=\sum_{j=1}^{2}\operatorname{sgn}(\sigma^{j-1})\sigma^j d_2A^1(x)=\frac{\partial a_j}{\partial x^i}_{|z}-\frac{\partial a_i}{\partial x^j}_{|z}. \tag{3.23}$$

The important point here is that the a_r terms cancel. This will give a clue for the general definition. Also observe that, owing to the equality of mixed partial derivatives, we have for a smooth real-valued function $f:U\to\mathbb{R}$ that $d(df)=0$ on U.

Finally, observe that for σ^1 the transposition of $S(2)$,

$$\sigma^1\, dA^1(x)=-dA^1(x) \tag{3.24}$$

which expresses the fact that $dA^1(x)$ is an antisymmetric 2 sectorform, or an antisymmetric covariant 2-tensor. We now formalize this last property.

Definition 3.16. Suppose that with respect to frame $\phi:O\to U\subset M$, $\tilde{a}_{i_1i_2\ldots i_r}(z)=A^r(x)$ is a local covariant tensor field of type $\binom{0}{r}$. For $s\in S(r)$, we have defined the *left action* $s\cdot A^r(x)=\tilde{a}^{(s(1))\ldots(s(r))}_{i_{s(1)}\cdots i_{(s(r))}}(z)$ where $(s(j))$ denotes the $s(j)$th vertex. We shall sometimes write the tensor $s\cdot A^r(x)$ in components simply: $\tilde{a}_{i_{s(1)},\ldots,i_{s(r)}}(z)$.

a) $A^r(x)$ is *symmetric* if $s\cdot A^r(x)=A^r(x)$ for all $s\in S(r)$, all $x\in U$.
b) $A^r(x)$ is *antisymmetric* or *skew-symmetric* if $s\cdot A^r(x)=\operatorname{sgn}(s)A^r(x)$ for all $s\in S(r)$, all $x\in U$.

Call such an asymmetric tensor an *r-form.* It is easy to see from the transformation laws for covariant tensors that properties a) and b) are independent of frame. ■

Differential forms, say r-forms, are covariant tensor fields of type $\binom{0}{r}$ which are *antisymmetric*. All 1-forms are trivially both symmetric and antisymmetric, but antisymmetry (which can be thought of as a property of the representative matrices in a local frame) obviously imposes constraints. As mentioned earlier, differential forms give rise in a natural way to differential forms of the next higher type. We shall show how to do this when we give a general definition for the 'coboundary' operator. Before stating it, however, it will be worthwhile to interpret the differential operator geometrically. This will lend some intuition to the definition.

Letting $A^1(x)=\tilde{a}_{i_1}(z)$ in local coordinates, we have from (3.22) that

$$d_2A^1(x)=\frac{da_i}{\partial x^j}\Big|_z+a_r(z)$$

following our convention for representing 2 sectorforms. Let a 2-sector be given at $x_0\in U$, say $\bar{A}_2(x_0)=[(\phi\circ t_{z_0})_2, X^i\text{——}Z^r\text{——}Y^j]$ with $\phi(z_0)=x_0$. $\bar{A}_2(x_0)$ can be thought of as a tangent vector *at* $X^i(x_0)\in T_1(M)$ via the identification given in Theorem 1.27. In particular, we may choose a *ribbon* (discussion preceding Definition 2.13) $F:(\mathbb{R}^2,(0,0))\to(0,z_0)$ such that $\bar{F}_2^{\#}(0,0)=\bar{A}_2(x_0)$. Then we have the components

$$X^i=\frac{\partial F^i}{\partial s},\qquad Y^j=\frac{\partial F^j}{\partial t},\qquad\text{and}\qquad Z^r=\frac{\partial^2F^r}{\partial s\delta t}$$

in this frame. Essentially, $\bar{A}_2(x_0)$ is the F-related class of the ribbon. For geometric insight, we shall often represent sectors this way. This allows us to write

$$\bar{A}_2(x_0)=\left[(\phi\circ t_{z_0})_2,\frac{\partial F^i}{\partial s}\text{——}\frac{\partial^2F^r}{\partial s\partial t}\text{——}\frac{\partial F^j}{\partial t}\right]=\bar{F}_2^{\#}(0,0).$$

Now compute the *contraction*

$$d_2A^1(x_0)\langle\bar{F}_2^{\#}\rangle=\frac{\partial a_i}{\partial x^j}\Big|_{z_0}\cdot\frac{\partial F^i}{\partial s}\cdot\frac{\partial F^j}{\partial t}+a_r(z_0)\frac{\partial^2F^r}{\partial s\partial t}.$$

Thus it is easy to see that

$$d_2A^1(x_0)\langle\bar{F}_2^{\#}\rangle=\frac{\partial}{\partial t}\Big|_{(0,0)}\left[a_i(F(0,t))\left\langle\frac{\partial F^i}{\partial s}\Big|_{(0,t)}\right\rangle\right]$$

and similarly that

$$d_1A^1(x_0)\langle\bar{F}_2^{\#}\rangle=\frac{\partial}{\partial s}\Big|_{(0,0)}\left[a_j(F(s,0))\left\langle\frac{\partial F^j}{\partial t}\Big|_{(s,0)}\right\rangle\right].$$

Therefore the contraction $dA^1(x_0)\langle\bar{F}_2^{\#}\rangle$ can be thought of as the limit as $\tau\to 0$ of

$$(1/\tau)\left[a_i(F(0,0))\left\langle\frac{\partial F^i}{\partial s}_{|(0,0)}\right\rangle + a_i(F(\tau,0))\left\langle\frac{\partial F^i}{\partial t}_{|(\tau,0)}\right\rangle - a_i(F(0,\tau))\left\langle\frac{\partial F^i}{\partial s}_{|(0,\tau)}\right\rangle - a_i(F(0,0))\left\langle\frac{\partial F^i}{\partial t}_{|(0,0)}\right\rangle\right]$$

the 'integral' of the 1-form around an 'infinitesimal rectangle' in the counterclockwise sense. Of course, the use of the word 'integral' is merely a heuristic to highlight this differential version of Green's theorem.

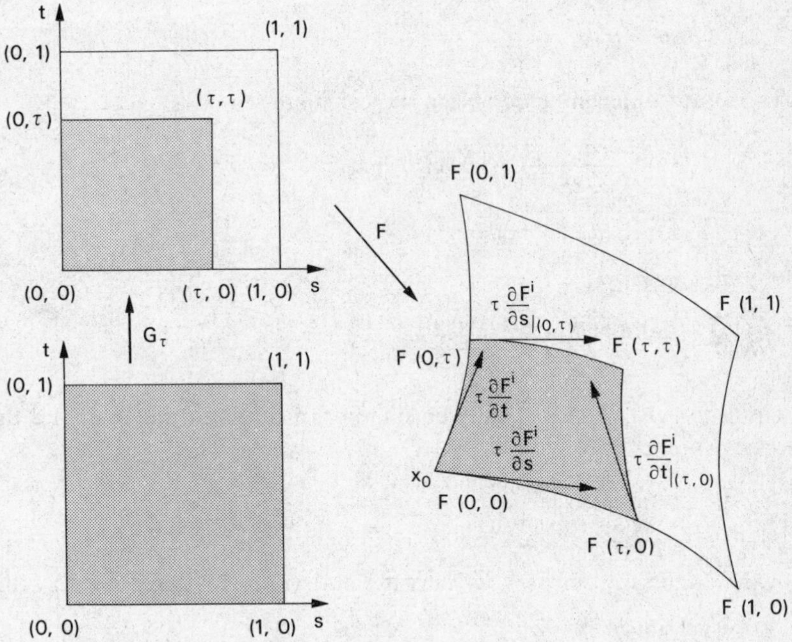

Fig. 3.1. Infinitesimal Green's theorem

$$\tau^2 dA^1(x_0)\langle(\overline{F\circ G_\tau})_2^{\#}\rangle \approx a_i(F\circ G_\tau(0,0))\left\langle\frac{\partial(F\circ G_\tau)^i}{\partial s}_{|(0,0)}\right\rangle + a_i(F\circ G_\tau(1,0))\left\langle\frac{\partial(F\circ G_\tau)^i}{\partial t}_{|(1,0)}\right\rangle - a_i(F\circ G_\tau(0,1))\left\langle\frac{\partial(F\circ G_\tau)^i}{\partial s}_{|(0,1)}\right\rangle - a_i(F\circ G_\tau(0,0))\left\langle\frac{\partial(F\circ G_\tau)^i}{\partial t}_{|(0,0)}\right\rangle$$

Indeed, 2-sectors can be likened to infinitesimal rectangles, and higher-order sectors to infinitesimal higher-order cubes, to give a theory of integration on manifolds, a theory which will not be developed here. We give now the general definition of the coboundary of a differential form.

Definition 3.17. With respect to a local frame $\phi : 0 \to U \subset M$, let an r-form $A^r(x) = \tilde{a}_{i_1 i_2 \ldots i_r}(z)$ be given. Define the *coboundary* of $A^r(x)$, denoted $dA^r(x)$, to be the $(r+1)$ form (for σ the shift of $S(r+1)$.

$$\sum_{j=1}^{r+1} \operatorname{sgn}(\sigma^{j-1}) d_j A^r(x). \tag{3.25}$$

In order to see that the $(r+1)$ sectorform

$$\sum_{j=1}^{r+1} \operatorname{sgn}(\sigma^{j-1}) d_j A^r(x)$$

is in fact an *antisymmetric tensor*, we calculate

$$d_{r+1} A^r(x) = \frac{\partial (a_{i_1 i_2 \ldots i_r})}{\partial x^{i_{r+1}}}\bigg|_z + \sum_{k=1}^{r} a_{i_1 i_2 \ldots i_k \ldots i_r}^{(1)(2)\ldots(k,r+1)\ldots(r)}(z)$$

$$d_j A^r(x) = \sigma^j d_{r+1} A^r(x).$$

First, we show that

$$\sum_{\substack{j=1,\ldots,r+1 \\ k=1,\ldots,r}} \operatorname{sgn}(\sigma^{j-1}) a_{i_1 i_2 \ldots i_k \ldots i_r}^{(\sigma^j(1))\ldots(\sigma^j(k),\sigma^j(r+1))\ldots(\sigma^j(r))}$$

is equal to zero. There is an even number of terms in the sum, and they can be *paired*:

a) $\operatorname{sgn}(\sigma^{j-1}) a_{i_1 i_2 \ldots i_k \ldots i_r}^{(\sigma^j(1))\ldots(\sigma^j(k),\sigma^j(r+1))\ldots(\sigma^j(r))}$,

b) $\operatorname{sgn}(\sigma^{j+k-1}) a_{i_1 i_2 \ldots i_{r+1-k} \ldots i_r}^{(\sigma^{j+k}(1))\ldots(\sigma^{j+k}(r+1-k),\sigma^{j+k}(r+1))\ldots(\sigma^{j+k}(r))}$.

Now if s is the shift of $S(r)$, we may use antisymmetry to replace b) with

$$\operatorname{sgn}(s^k) \operatorname{sgn}(\sigma^{j+k-1}) a_{i_1 i_2 \ldots i_r}^{(j,j+k),(j+1),(j+2),\ldots,\widehat{(j+k)},\ldots,(j+r)}$$

where superscripts are computed mod$(r+1)$ with $0 \equiv r+1$, and ^ means *omit*. Meanwhile, applying $(k-1)$ transpositions in a) we can write a) as

$$(-1)^{k-1} \operatorname{sgn}(\sigma^{j-1}) a_{i_1 i_2 \ldots i_r}^{(j+k,j),(j+1)\ldots\widehat{(j+k)},\ldots,(j+r)}$$

and since $\operatorname{sgn}(s) = -\operatorname{sgn}(\sigma)$, it is clear that these terms add to zero. Thus we have

$$dA^r(x) = \sum_{j=1}^{r+1} \operatorname{sgn}(\sigma^{j-1}) \sigma^j \left[\frac{\partial a_{i_1 i_2 \ldots i_r}}{\partial x^{i_{r+1}}}\bigg|_z \right] \tag{3.26}$$

and this is easily seen to be an $(r+1)$ form.

Another useful formula for computation is the following. If $A^r(x)$ is a local r-form on U, and $dA^r(x)$ is its coboundary $(r+1)$ form on U, then let $\bar{f}^{\#}_{r+1}(0, 0, \ldots, 0)$ be an $(r+1)$ sector at $x_0 = \phi(z_0)$ for frame $\phi: O \to U$ in the notation of Definition 3.13. Then the contraction

$$\begin{aligned} dA^r(x)\langle \bar{f}^{\#}_{r+1}(0, 0, \ldots, 0)\rangle &= \sum_{j=1}^{r+1} \operatorname{sgn}(\sigma^{j-1}) \frac{\partial a_{i_1 i_2 \ldots i_r}}{\partial x^{i_{r+1}}}\Big|_{z_0} \langle \sigma^j \bar{f}^{\#}_{r+1}(0, \ldots, 0)\rangle \\ &= \sum_{j=1}^{r+1} \operatorname{sgn}(\sigma^{j-1}) \frac{\partial a_{i_1 i_2 \ldots i_r}}{\partial x^{i_{r+1}}}\Big|_{z_0} \frac{\partial f^{i_1}}{\partial s^{\sigma^j(1)}} \cdot \frac{\partial f^{i_2}}{\partial s^{\sigma^j(2)}} \\ &\qquad \cdot \ldots \cdot \frac{\partial f^{i_{r+1}}}{\partial s^{\sigma^j(r+1)}}. \end{aligned} \tag{3.27}$$

If $f: M \to N$ is smooth and $f^*[B^r(y)] = A^r(x)$, then in view of the naturality of d_i, we have $f^*[dB^r(y)] = d[A^r(x)] = d[f^*B^r(y)]$. ■

The geometric interpretation of the contraction $dA^r(x)\langle \bar{f}^{\#}_{r+1}(0, \ldots, 0)\rangle$ above is given essentially in Definition 3.13. $\bar{f}^{\#}_{r+1}(0, \ldots, 0)$ has an interpretation as $(r+1)$ tangent vectors at points in $T_r(U)$ by each of the natural isomorphisms derivable from ν by 'shifting', or geometrically as a path-jet of sectors following the discussion in Definition 3.13. Then this contraction is a signed sum of derivatives along these tangent vectors of the contraction of $A^r(x)$ with the r-sectors along the path.

In order to develop some of the geometric applications of the coboundary operator, it will be necessary to define a new type of 'suspension'. Recall that the promotion operator takes a local vector field on M to a local vector field on $T_1(M)$. This operation can clearly be iterated to give local vector fields on $T_r(M)$ for all $r \geq 1$. The following discussion formalizes this idea.

As a preliminary, recall that sectors give a generalization of tangent vectors. We have explored many of the algebraic properties of these generalized tangent vectors. But the essential feature of a tangent vector is that it gives an operation via the derivative on smooth real-valued functions. That operation (which is a *derivation* on the algebra of real-valued smooth functions on M) is calculated by contracting the differential of the function with the tangent vector. Now that higher-order differentials have been defined, we may develop the action of an r-sector on a smooth real-valued function.

Definition 3.18. Let $f: M \to \mathbb{R}$ be a smooth function. Suppose for $x_0 \in M$, $\bar{A}_r(x_0)$ is an r-sector, and suppose that, with respect to frame $\phi: O \to U \subset M$, $\bar{A}_r(x_0)$ has components X^i_γ, $\gamma \in \Gamma(r-1)$. As usual, $\mathbb{R}$ is equipped with global frame given by standard linear coordinate function (s). The *derivative* of f at x_0 with respect to $\bar{A}_r(x_0)$ is defined in the following way. Let $T_r(f)\bar{A}_r(x_0) = [(\theta \circ t_{f(x_0)})_r, d_\gamma]$, where θ is the frame on

$\mathbb{R}$ (fixed once for all), and for each $\gamma \in \Gamma(r-1)$, $d_\gamma \in \mathbb{R}$ is the γ-component of the sector on $\mathbb{R}$ $T_r(f)\bar{A}_r(x_0)$. Now let e be the $(r-1)$ face of $\Gamma(r-1)$. Then the derivative of f at x_0 with respect to $\bar{A}_r(x_0)$ is the real number d_e.

Since $T_r(f)$ preserves the algebraic structure in the fiber $T_r(x_0)$, it is easy to see that for f fixed the derivative is linear on each $T_{r,k}(\bar{A}_r(x_0)) \subset T_r(x_0)$. It defines an r sectorform at x_0. ■

In the terminology above, it is easy to calculate the derivative of f at x_0 with respect to $\bar{A}_r(x_0)$ directly using formula (1.26). Letting $\bar{f} = \theta^{-1} \circ f \circ \phi$, then we have

$$d_e = \sum_{\gamma_{i_1} \cup \ldots \cup \gamma_{i_t} = e} \frac{\partial^t \bar{f}}{\partial x^{k_1} \ldots \partial x^{k_t}}\bigg|_{z_0} X^{k_1}_{\gamma_{i_1}} X^{k_2}_{\gamma_{i_2}} \ldots X^{k_t}_{\gamma_{i_t}}.$$

Thus the r sectorform mentioned in Definition 3.18 is $[(\phi \circ t_{z_0})_r, a^{\gamma_{i_1} \ldots \gamma_{i_t}}_{k_1 k_2 \ldots k_t}]$ with

$$a^{\gamma_{i_1} \gamma_{i_2} \ldots \gamma_{i_t}}_{k_1 k_2 \ldots k_t} = \frac{\partial^t \bar{f}}{\partial x^{k_1} \ldots \partial x^{k_t}}\bigg|_{z_0}. \tag{3.28}$$

Observation 3.19. The r sectorform given in formula (3.28) is the iteration of differentials $d_r \circ d_{r-1} \circ \ldots d_2 \circ d_1(f)$ at x_0. This has the following sense: df is a local 1 sectorform field defined near x_0, $d_2(df)$ is a local 2 sectorform field near x_0, and so forth. We will denote this sectorform $d_{r,r-1,\ldots,2,1}f(x_0)$ and observe that the derivative of f with respect to the r-sector $\bar{A}_r(x_0)$ at x_0 is the *contraction* $d_{r,\ldots,2,1}f(x_0)\langle \bar{A}_r(x_0) \rangle$. ■

Exercises

[These exercises develop some analytic applications of sector differentiation.]

22. Define an *n-block* in a manifold M as a smooth map $\phi: I^n \to M$ with $I = [0, 1]$. Let $\bar{\phi}^\#_n: I^n \to T_n(M)$ be the associated n-block in $T_n(M)$ over ϕ defined as in Definition 3.13. Here let I^n have the standard coordinates $(s^1, s^2, \ldots, s^n)$, $0 \leq s^i \leq 1$.

    The mean value theorem says that if $f: M \to \mathbb{R}$ is smooth and if $\phi: I \to M$ is a 1-block (path) then

    $$f[\phi(1)] - f[\phi(0)] = df\langle \bar{\phi}^\#_1(\sigma) \rangle \qquad \text{for some } \sigma \in I.$$

    Suppose now that $\phi: I^2 \to M$ is a 2-block (ribbon) in M and suppose $f: M \to \mathbb{R}$ is smooth satisfying

    $$f[\phi(0, 0)] = f[\phi(1, 0)] = f[\phi(0, 1)] = f[\phi(1, 1)].$$

Letting $g(s^1)=f[\phi(s^1,1)]-f[\phi(s^1,0)]$ we have from an application of Rolle's theorem that

$$\frac{dg}{ds^1}_{|\sigma}=0=\frac{\partial(f\circ\phi)}{\partial s^1}_{|(\sigma,1)}-\frac{\partial(f\circ\phi)}{\partial s^1}_{|(\sigma,0)}\qquad\text{for some } 0\leqslant\sigma\leqslant 1.$$

Then for some τ, $0\leqslant\tau\leqslant 1$,

$$\frac{\partial^2(f\circ\phi)}{\partial s^1\partial s^2}_{|(\sigma,\tau)}=0$$

by another application of the mean value theorem.

Another way to write this two-dimensional version of Rolle's theorem is:

$$d_{2,1}f\langle\bar{\phi}_2^{\#}(\sigma,\tau)\rangle=0\qquad\text{for some } (\sigma,\tau) \text{ in } I^2.$$

Next, freeing the values of f on the corner points of the 2-block, let $y_0=f[\phi(0,0)]$, $y_1=f[\phi(1,0)]$, $y_2=f[\phi(0,1)]$, and $y_3=f[\phi(1,1)]$. Define $h:I^2\to\mathbb{R}$ by

$$h(s^1,s^2)=y_0+s^1(y_1-y_0)+s^2(y_2-y_0)+s^1s^2(y_3-y_2-y_1+y_0).$$

Then $h(0,0)=y_0$, $h(1,0)=y_1$, $h(0,1)=y_2$, and $h(1,1)=y_3$. Applying the previous observation to $(f\circ\phi-h):I^2\to\mathbb{R}$ we conclude that for some $(\sigma,\tau)\in I^2$,

$$d_{2,1}f\langle\bar{\phi}_2^{\#}(\sigma,\tau)\rangle=f[\phi(1,1)]-f[\phi(1,0)]-f[\phi(0,1)]+f[\phi(0,0)].$$

This is a two-dimensional version of the mean value theorem.

Now suppose that $\phi:I^n\to M$ is an n-block in M. Let $f:M\to\mathbb{R}$ be smooth and let $\Delta_\phi(f)$ be the number

$$\Delta_\phi(f)=\sum_{\gamma\subseteq\{1,2,\ldots,n\}}(-1)^{n-|\gamma|}f[\phi(v_\gamma)]$$

with $v_\gamma=(\gamma_1,\gamma_2,\ldots,\gamma_n)$, $\gamma_i=1$ if and only if $i\in\gamma$, and $\gamma_i=0$ otherwise, $|\gamma|=\sum\gamma_i$.

Prove the following *generalized mean value theorem*: For some $(\sigma^1,\sigma^2,\ldots,\sigma^n)\in I^n$,

$$d_{n,n-1,\ldots,1}f\langle\bar{\phi}_n^{\#}(\sigma^1,\sigma^2,\ldots,\sigma^n)\rangle=\Delta_\phi(f).$$

23. What is the integral form of this theorem?

24. Suppose that $\phi:I^n\to M$ is an n-block and $f:M\to\mathbb{R}^k$ is smooth, $k\geqslant 1$. Then define $\Delta_\phi(f)$ in a manner analogous to the definition from exercise 22. Now if $\gamma\in\Gamma(n-1)$ (that is, γ is a non-empty subset of $\{1,2,\ldots,n\}$), say γ is an r-face of the $(n-1)$ simplex, then define

$i_\gamma : I^{r+1} \to I^n$ by $i_\gamma(\tau^1, \tau^2, \ldots, \tau^{r+1}) = (s^1, s^2, \ldots, s^n)$ with $s^j = 0$ if $j \notin \gamma$, and $s^j = \tau^{n(j)}$ for $j \in \gamma$ and j the $n(j)$th element of γ. Let ϕ_γ be the $(r+1)$ block $\phi \circ i_\gamma$. Show that

$$f[\phi(1, 1, \ldots, 1)] - f[\phi(0, 0, \ldots, 0)] = \sum_{\gamma \in \Gamma(n-1)} \Delta_{\phi_\gamma}(f).$$

25. For $\phi : I^n \to M$ and $f : M \to \mathbb{R}$ smooth, show that there is a set of $2^n - 1$ points in I^n $\{\sigma_\gamma \in \operatorname{im}(i_\gamma) \mid \gamma \in \Gamma(n-1)\}$ such that

$$f[\phi(1, 1, \ldots, 1)] - f[\phi(0, 0, \ldots, 0)] = \sum_{\gamma \in \Gamma(n-1)} d_{|\gamma|, \ldots, 2, 1} f\langle (\bar{\phi}_\gamma)^{\#}_{|\gamma|}(\sigma_\gamma)\rangle.$$

[This is another (perhaps more versatile) form of the generalized mean value theorem.]

26. Let $\phi : I^n \to \mathbb{R}$ by

$$\phi(s^1, s^2, \ldots, s^n) = x_0 + \left(\sum_{i=1}^{n} \frac{s^i}{n}\right) h$$

for h real (an 'equilateral' affine n-block) and let $f : \mathbb{R} \to \mathbb{R}$ be smooth. Apply the result of exercise 25 to show that

$$f(x_0 + h) - f(x_0) = \sum_{i=1}^{n} \left[\left(\frac{1}{i!}\right) \frac{n(n-1)\ldots(n-i+1)}{n^i} \frac{d^i f}{dx^i}\bigg|_{\xi_i} \right] \cdot h^i$$

with ξ_i between x_0 and $x_0 + (hi/n)$ for each i.

[Notice that for $\gamma \in \Gamma(n-1)$, $|\gamma| = i$, then $\Delta_{\phi_\gamma}(f)$ is the 'ith difference' of f on the partition $\{x_0, x_0 + (h/n), x_0 + (2h/n), \ldots, x_0 + (ih/n)\}$ in the sense of numerical analysis.]

27. Let $\bar{A}_n(0)$ be an n-sector at $0 \in \mathbb{R}^m$. Suppose $f : (\mathbb{R}^m, 0) \to (\mathbb{R}, 0)$ is a smooth map. For $\gamma \in \Gamma(n-1)$ let $\bar{B}_{|\gamma|}(0)$ be the $|\gamma|$-face of $\bar{A}_n(0)$ defined in the obvious way by the monotone inclusion $\{1, 2, \ldots, |\gamma|\} \to \{1, 2, \ldots, n\}$ given by γ. Then choosing 'small' n-blocks $\phi : I^n \to \mathbb{R}^m$ with $\phi(0, \ldots, 0) = 0$ and $\phi(1, 1, \ldots, 1) = (x^1, x^2, \ldots, x^m)$ and $\bar{\phi}^{\#}_n(0) = \bar{A}_n(0)$ gives the following 'polynomial approximation' to f near 0:

$$f(x^1, x^2, \ldots, x^m) \approx \sum_{\gamma \in \Gamma(n-1)} d_{|\gamma|, \ldots, 2, 1}(f)\langle \bar{B}_{|\gamma|}(0)\rangle.$$

For $m = 1$, choose equilateral affine n-blocks (exercise 26) and compare with the Taylor polynomial approximations. For $m > 1$, investigate the approximation at points where df vanishes for sectors 'adapted' to the eigendirections of $d^2 f$.

We are now in a position to give the definition of the suspension of a vector field by a sector. The geometric idea is this. Suppose that $\phi : O \to U \subset M$ is a frame and $\bar{A}_1(x) = \tilde{X}^i(z)$ is a local vector field on U. Let

$z_0 \in 0$ with $\phi(z_0) = x_0$ and suppose that $\bar{B}_r(x_0)$ is an r-sector at x_0. Suppose that the *flow-germ* for $\tilde{X}^i(z)$ on V is γ_V. Just as we did in defining the suspension of a tangent vector by a local vector field, we may 'push' the sector $\bar{B}_r(x_0)$ along the flow γ_V^s. First we give the algebraic definition of this 'suspension', then we establish invariance by interpreting it in terms of γ_V. Finally, we should observe that what is involved here is the differentiation of a local vector field with respect to a sector; this differs from the differentiation of a sector field by a tangent vector.

Definition 3.20. Suppose that $\bar{B}_r(x_0)$ is an r-sector at $x_0 \in M$ and that $\bar{A}_1(x)$ is a local vector field on U, $x_0 \in U$. Let $\phi : O \to U \subset M$, $\phi(z_0) = x_0$ be a frame, and suppose that with respect to ϕ, $\bar{B}_r(x_0)$ has components $W^i_\gamma(z_0) = W^i_\gamma$ for $\gamma \in \Gamma(r-1)$, and $\bar{A}_1(x) = \tilde{X}^i(z)$. Define *suspension of* $\tilde{X}^i(z)$ *by the sector* W^i_γ, written as $W^i_\gamma(z_0) \cup \tilde{X}^i(z)$, to be the following $(r+1)$ sector at x_0.

Identify $\gamma = (\gamma_1, \gamma_2, \ldots, \gamma_r)$ with $\gamma' = (\gamma_1, \gamma_2, \ldots, \gamma_r; 0) \in \Gamma(r)$ and let γ'' be the $(r+1)$ vertex $(0, 0, \ldots, 0; 1)$ of $\Gamma(r)$. Then in the notation of Theorem 1.27

$$W^i_\gamma(z_0) \cup \tilde{X}^i(z) = [(\phi \circ t_{z_0})_{r+1}, U^i_{\gamma'} \text{——} Z^i_{\gamma' \cup \gamma''} \text{——} V^i_{\gamma''}]$$

with $U^i_{\gamma'} = W^i_\gamma$ (via the identification above), $V^i_{\gamma''} = X^i(z_0)$ and $Z^i_{\gamma' \cup \gamma''}$ defined in the following way:

$$Z^i_{\gamma' \cup \gamma''} = \sum_{\gamma_{i_1} \cup \ldots \cup \gamma_{i_s} = \gamma'} \frac{\partial^s X^i}{\partial x^{k_1} \ldots \partial x^{k_s}}\bigg|_{z_0} W^{k_1}_{\gamma_{i_1}} W^{k_2}_{\gamma_{i_2}} \ldots W^{k_s}_{\gamma_{i_s}}.$$

Thus, for fixed i, $Z^i_{\gamma' \cup \gamma''}$ is just the derivative of $\tilde{X}^i(z)$ (thought of as a smooth function from $O \to \mathbb{R}$) with respect to the *sector* with ϕ components W^i_γ. ■

Of course, it is necessary to show that $[(\phi \circ t_{z_0})_{r+1}, U^i_{\gamma'} \text{——} Z^i_{\gamma' \cup \gamma''} \text{——} V^i_{\gamma''}]$ is 'independent of frame'. This requires a calculation like that given in Definition 2.10. Thus, suppose $\theta : O' \to U \subset M$ is another frame, $\theta(u_0) = \phi(z_0) = x_0$ and $\theta(u) = \phi(z) = x$. Letting $g(z) = \theta^{-1} \circ \phi(z)$ we may write:

$$[(\phi \circ t_{z_0})_r, W^i_\gamma(z_0)] = [(\theta \circ t_{u_0})_r, \bar{g}_r W^i_\gamma(z_0)]$$

with $\bar{g}$ the germ $t^{-1}_{u_0} g t_{z_0} : (O, \text{origin}) \to (O', \text{origin})$. The vector field may be written in θ components:

$$\frac{\partial g^j}{\partial x^i}\bigg|_{g^{-1}(u)} \cdot X^i(g^{-1}(u)),$$

as function of $u \in O'$. Thus we must show that the sector $[(\theta \circ t_{u_0})_{r+1}, \bar{U}^i_{\gamma'} \text{——} \bar{Z}^i_{\gamma' \cup \gamma''} \text{——} \bar{V}^i_{\gamma''}]$ with $\bar{U}^i_{\gamma'} = [\bar{g}_r W^i_\gamma(z_0)]^i$ (with the earlier identification of γ and γ'),

$$\bar{V}^i_{\gamma''} = \frac{\partial g^i}{\partial x^j}_{|g^{-1}(u_0)} \cdot X^j(g^{-1}(u_0)),$$

$$\bar{Z}^i_{\gamma' \cup \gamma''} = \sum_{\gamma'_{i_1} \cup \ldots \cup \gamma'_{i_s} = \gamma'} \frac{\partial^s}{\partial u^{k_1} \ldots \delta u^{k_s}}_{|u_0} \left[\frac{\partial g^i}{\partial x^j}_{|g^{-1}(u)} \cdot X^j(g^{-1}(u)) \right] \times \bar{U}^{k_1}_{\gamma'_{i_1}} \ldots \bar{U}^{k_s}_{\gamma'_{i_s}}$$

is equal to the sector $[(\phi \circ t_{z_0})_{r+1}, U^i_{\gamma'} \text{---} Z^i_{\gamma' \cup \gamma''} \text{---} V^i_{\gamma''}]$.

Checking the transformation law given in Theorem 1.27 for sectors represented in this form, it is clear that the components $\bar{U}^i_{\gamma'}$ and $\bar{V}^i_{\gamma''}$ are correct. The $\bar{Z}^i_{\gamma' \cup \gamma''}$ components must be shown to agree with the transforms of the $Z^i_{\gamma' \cup \gamma''}$ components. In particular, the components for the $\gamma' \cup \gamma''$ term after transformation are

$$\sum_{\gamma'_{i_1} \cup \ldots \cup \gamma'_{i_s} = \gamma'} \frac{\partial^{s+1}(\bar{g}^i)}{\partial x^{k_1} \ldots \partial x^{k_s} \delta x^h} U^{k_1}_{\gamma'_{i_1}} \ldots U^{k_s}_{\gamma'_{i_s}} X^h(z_0)$$

$$+ \sum_{\gamma'_{i_1} \cup \ldots \cup \gamma'_{i_t} \cup \ldots \cup \gamma'_{i_s} = \gamma'} \frac{\partial^s(\bar{g}^i)}{\partial x^{k_1} \ldots \partial x^{k_s}} U^{k_1}_{\gamma'_{i_1}} \ldots Z^{k_t}_{\gamma'_{i_t} \cup \gamma''} \ldots U^{k_s}_{\gamma'_{i_s}}.$$

The 'chain rule' gives the result as will be shown in Lemma 4.51.

Observation 3.21. Under the above conditions, let γ_V be the *flow-germ* for $\tilde{X}^i(z)$ on $V \subset U$, $x_0 \in V$. Suppose $\bar{B}_r(x_0) = \bar{f}^{\#}_r(0, 0, \ldots, 0)$ for germ $f: (\mathbb{R}^r, (0, 0, \ldots, 0)) \to (V, z_0)$. Here $\mathbb{R}^r$ has linear coordinates $(t^1, \ldots, t^r)$ and $\mathbb{R}^{r+1}$ has coordinates $(t^1, \ldots, t^r; s)$. And let γ^s_V be defined as in Definition 2.14: for small s the germ of a diffeomorphism from $V \to V$. Now let $F: (\mathbb{R}^{r+1}, (0, 0, \ldots, 0; 0)) \to (V, z_0)$ be defined by $F(t^1, t^2, \ldots, t^r; s) = \gamma^s_V(f(t^1, t^2, \ldots, t^r))$. Then $\bar{F}^{\#}_{r+1}(0, 0, \ldots, 0; 0) = W^i_\gamma(z_0) \cup \tilde{X}^i(z)$.

Proof. The proof of this is an exercise in the chain rule. The elementary induction is left to the reader. ■

We are now in a position to define a new operation on sectorforms, one which is dual in a sense to the suspension operation on sectors. This operation will be used to develop some of the geometric applications of the coboundary operator. Suppose that $A^{r+1}(x_0)$ is an $(r+1)$ sectorform at $x_0 \in M$, and suppose that $\bar{A}_1(x)$ is a local vector field on U, $x_0 \in U$. For $\phi: O \to U \subset M$ a frame, suppose that $A^{r+1}(x_0)$ has components $a^{P_k}_{i_1 i_2 \ldots i_k}(z_0)$ and that $\bar{A}_1(x)$ has components $\tilde{X}^i(z)$. Now suppose that $\bar{B}_r(x_0)$ is an r-sector at x_0. Iterating the promotion operator $\tilde{X}^i(z)$ gives rise to a *vector field* on $T_r(U)$. In particular, it determines an element of $T_1(T_r(x_0))$ at $\bar{B}_r(x_0)$: essentially the suspension $\bar{B}_r(x_0) \cup \bar{A}_1(x)$. Now contraction of

$A^{r+1}(x_0)$ with this $(r+1)$ sector yields an element of $\mathbb{R}$. Then the correspondence

$$\bar{B}_r(x_0) \to A^{r+1}(x_0)\langle \bar{B}_r(x_0) \cup \bar{A}_1(x)\rangle$$

will be shown to be an r sectorform at x_0: the *reduction* of the sectorform $A^{r+1}(x_0)$ *by the local vector field* $\bar{A}_1(x)$. When $A^{r+1}(x_0)$ is a certain *differential* of an r-form, reduction by a vector field will yield the *Lie derivative* of the r-form *along the vector field*, an operation with useful and colorful geometric interpretations. First, we describe the reduction operation.

Proposition 3.22. Let $\phi: O \to U \subset M$ be a frame and suppose that $\phi(z_0) = x_0$. Let $\bar{A}_1(x)$ be a local vector field on U with components $\tilde{X}^i(z)$, and let $A^{r+1}(x_0)$ be an $(r+1)$ sectorform with components $a^{P_t}_{i_1 i_2 \ldots i_t}$, then the correspondence

$$\bar{B}_r(x_0) \to A^{r+1}(x_0)\langle \bar{B}_r(x_0) \cup \bar{A}_1(x)\rangle$$

is an r sectorform at x_0. Call this sectorform the *reduction* of $A^{r+1}(x_0)$ *by the local vector field* $\bar{A}_1(x)$ and denote it $A^{r+1}(x_0)[\bar{A}_1(x)]$. It is defined by the identity:

$$A^{r+1}(x_0)[\bar{A}_1(x)]\langle \bar{B}_r(x_0)\rangle = A^{r+1}(x_0)\langle \bar{B}_r(x_0) \cup \bar{A}_1(x)\rangle. \tag{3.29}$$

Proof. The suspension mapping taking $\bar{B}_r(x_0) \to \bar{B}_r(x_0) \cup \bar{A}_1(x)$ can be described in components (following Definition 3.20)

$$W^i_\gamma \to \left(W^i_{\gamma'} \text{——} \sum_{\gamma'_{i_1} \cup \ldots \cup \gamma'_{i_s} = \gamma'} \frac{\partial^s X^i}{\partial x^{k_1} \ldots \partial x^{k_s}}\bigg|_{z_0} W^{k_1}_{\gamma'_{i_1}} \ldots W^{k_s}_{\gamma'_{i_s}} \text{——} X^i_{\gamma''}(z_0) \right),$$

the middle term giving the components of the face $\gamma' \cup \gamma''$.

From this, it is fairly easy to see that it takes $T_{r,k}(\bar{B}_r(x_0))$ to $T_{r+1,k}(\bar{B}_r(x_0) \cup \bar{A}_1(x))$ for $1 \leq k \leq r$. For each k, $1 \leq k \leq r$, this mapping is *linear*, as is checked directly using the component form above. It then follows easily from the *definition* of sectorform at x_0 that $A^{r+1}(x_0)[\bar{A}_1(x)]$ is a sectorform at x_0. ■

Now suppose that $A^r(x)$ is a *differential r-form* defined on $U \subset M$. As usual, $d_{r+1}A^r(x)$ is an $(r+1)$ sectorform (not a form in general). Letting $\bar{A}_1(x)$ be a local vector field on U we shall show that the *r sectorform* $(d_{r+1}A^r(x_0))[\bar{A}_1(x)]$ is in fact an *r-form*: that is, it is an antisymmetric tensor of type $\binom{0}{r}$.

Recall that Observation 3.14 gives a geometric interpretation for the contraction

$$(d_{r+1}A^r(x_0))[\bar{A}_1(x)]\langle \bar{B}_r(x_0)\rangle = d_{r+1}A^r(x_0)\langle \bar{B}_r(x_0) \cup \bar{A}_1(x)\rangle.$$

The flow-germ for $\bar{A}_1(x)$ 'pushes' $\bar{B}_r(x_0)$ along a path-germ of r-sectors. The derivative along this path-germ (at 0) of the contraction of $A^r(x)$ with the 'pushed' sectors gives this contraction.

Before studying this operation on r-forms more closely, some remarks concerning the *naturality* of the suspension and reduction operations are in order.

Suppose that $f: M \to N$ is smooth, $U \subset M$ and $V \subset N$ with frames $\phi: Z \to U$ and $\theta: Z' \to V$, and $f(U) \subset V$. Further, suppose $\bar{A}_1(x)$ is a local vector field on U and $\bar{B}_1(y)$ is a local vector field on V. Say that f *carries* $\bar{A}_1(x)$ *to* $\bar{B}_1(y)$ if for each $x_0 \in U$ with $y_0 = f(x_0) \in V$, $T_1(f)[\bar{A}_1(x_0)] = \bar{B}_1(y_0)$.

If in frame ϕ, $\bar{A}_1(x)$ has components $\tilde{X}^i(z)$, and in frame θ, $\bar{B}_1(y)$ has components $\tilde{Y}^i(w)$, $\theta(w) = y$, let $\gamma_Z: (Z \times \mathbb{R}, Z \times O) \to (Z \times \mathbb{R}, Z \times O)$ be the *flow-germ* for $\tilde{X}^i(z)$. Also let $\gamma_{Z'}: (Z' \times \mathbb{R}, Z \times O) \to (Z' \times \mathbb{R}, Z' \times O)$ be the flow-germ for $\tilde{Y}^i(w)$. Then if f carries $\bar{A}_1(x)$ to $\bar{B}_1(y)$, it is fairly clear that the diagram

$$\begin{array}{ccc} (Z \times \mathbb{R}, Z \times O) & \xrightarrow{\gamma_Z} & (Z \times \mathbb{R}, Z \times O) \\ {\scriptstyle f \times \mathrm{id}} \downarrow & & \downarrow {\scriptstyle f \times \mathrm{id}} \\ (Z' \times \mathbb{R}, Z' \times O) & \xrightarrow[\gamma_{Z'}]{} & (Z' \times \mathbb{R}, Z' \times O) \end{array}$$

commutes, as a diagram of germs.

Now suppose that $T_r\bar{A}_r(x_0) = \bar{B}_r(y_0)$ for r-sectors $\bar{A}_r(x_0)$ and $\bar{B}_r(y_0)$. If in frame ϕ, $\bar{A}_r(x_0)$ has components W^i_γ, and in frame θ, $\bar{B}_r(y_0)$ has components U^i_γ, then we can conclude from Observation 3.21 and the commutativity of the diagram above that

$$T_{r+1}(f)[W^i_\gamma(z_0) \cup \tilde{X}^i(z)] = U^i_\gamma(w_0) \cup \tilde{Y}^i(w). \tag{3.30}$$

Formula (3.30) expresses the naturality of the suspension operation.

The reduction operation satisfies a similar naturality condition. In particular, if, in the above situation, $B^{r+1}(y)$ is a local sectorform field on V, and if $A^{r+1}(x) = f^*[B^{r+1}(y)]$ is a local sectorform field on U, and if f carries $\bar{A}_1(x)$ to $\bar{B}_1(y)$, then it is easy to conclude that

$$A^{r+1}(x_0)[\bar{A}_1(x)] = f^*(B^{r+1}(y_0)[\bar{B}_1(x)]). \tag{3.31}$$

These naturality properties will play an important role in the sequel.

Returning now to tensors, let $A^r(x)$ be a local covariant *tensor* field on U. And suppose that $\bar{A}_1(x)$ is a local vector field on U. Then as we observed, $d_{r+1}A^r(x)$ is not a tensor in general but is an $(r+1)$ sectorform field on U. We may form the *reduction* $(d_{r+1}A^r(x))[\bar{A}_1(x)]$ to get an r sectorform field on U.

Suppose now that $\phi: O \to U \subset M$ is a frame. Suppose that $\bar{A}_1(x)$ has components $\tilde{X}^i(z)$ in this frame and that $A^r(x)$ (a *tensor* field, remember) has components $\tilde{a}_{h_1h_2\ldots h_r}(z)$. Then by definition we may write that $(d_{r+1}A^r(x))[\bar{A}_1(x)]$ has components at $x_0: b^{P_s}_{i_1i_2\ldots i_s}(z_0)$ with $b^{P_s}_{i_1i_2\ldots i_s}(z_0) = 0$ if

$P_s \neq (1)(2)\dots(r)$ and for $P_s = (1)(2)\dots(r)$

$$b^{P_s}_{i_1 i_2 \dots i_s}(z_0) = \left.\frac{\partial \tilde{a}_{i_1 i_2 \dots i_r}}{\partial x^{i_{r+1}}}\right|_{z_0} X^{i_{r+1}}(z_0) + \sum_{j=1}^{r} a_{i_1 i_2 \dots h_j \dots i_r}(z_0) \left.\frac{\partial \tilde{X}^{h_j}}{\partial x^{i_j}}\right|_{z_0}. \tag{3.32}$$

This follows directly from the definitions of d_{r+1} and the suspension. Since the only non-zero terms in this sectorform correspond to the partition $(1)(2)\dots(r)$, it is clear that this is a *tensor* of type $\binom{0}{r}$. If $A^r(x)$ is antisymmetric (symmetric), then it is easy to see from the component formula that $(d_{r+1}A^r(x))[\bar{A}_1(x)]$ is also antisymmetric (symmetric). Putting this together with our observations on naturality we have the following definition.

Definition 3.23. Let $A^r(x)$ be a local covariant tensor field on U, and let $\bar{A}_1(x)$ be a local vector field on U. Then $(d_{r+1}A^r(x))[\bar{A}_1(x)]$ is a covariant tensor field on U of type $\binom{0}{r}$, antisymmetric (symmetric) if $A^r(x)$ is antisymmetric (symmetric). It is called the *Lie derivative* of $A^r(x)$ *along* $\bar{A}_1(x)$ *on* U, denoted $L_{\bar{A}_1(x)}(A^r(x))$. If $f: M \to N$ carries $\bar{A}_1(x)$ to $\bar{B}_1(y)$ and if $B^r(y)$ is a local covariant tensor field on N, then

$$L_{\bar{A}_1(x)} f^*(B^r(y)) = f^*[L_{\bar{B}_1(y)} B^r(y)]. \tag{3.33}$$

■

This is the place for a few examples.

Example 1. $A^1(x)$ is a local 1-form with components $\tilde{a}_i(z)$ (following Convention 3.9). $\bar{A}_1(x)$ has components $\tilde{X}^j(z)$. Then $L_{\bar{A}_1(x)}(A^1(x))$ has components at x_0 given by

$$a_r(z_0) \left.\frac{\partial X^r}{\partial x^i}\right|_{z_0} + \left.\frac{\partial a_i}{\partial x^j}\right|_{z_0} X^j(z_0)$$

a 1-form at x_0.

Example 2. $A^2(x)$ is a *metric tensor* with components $\tilde{a}_{ij}(z)$ (Convention 3.9). $\bar{A}_1(x)$ has components $\tilde{X}^k(z)$. $L_{\bar{A}_1(x)}(A^2(x))$ has components at x_0 given by

$$\left.\frac{\partial a_{ij}}{\partial x^k}\right|_{z_0} X^k(z_0) + a_{sj}(z_0) \left.\frac{\partial X^s}{\partial x^i}\right|_{z_0} + a_{it}(z_0) \left.\frac{\partial X^t}{\partial x^j}\right|_{z_0}$$

This is a symmetric 2-tensor at x_0.

Example 3. $w^m(x)$ is the 'determinant form' on $O \subset E = \mathbb{R}^m$ defined with components:

$$\tilde{a}_{i_1 i_2 \dots i_m}(z) = 0 \qquad \text{if } \{i_1, i_2, \dots, i_m\} \text{ not all distinct,}$$

$$\tilde{a}_{i_{s(1)} i_{s(2)} \dots i_{s(r)}}(z) = \operatorname{sgn}(s) = \operatorname{sgn}(s^{-1}) \qquad \text{for } s \in S(m).$$

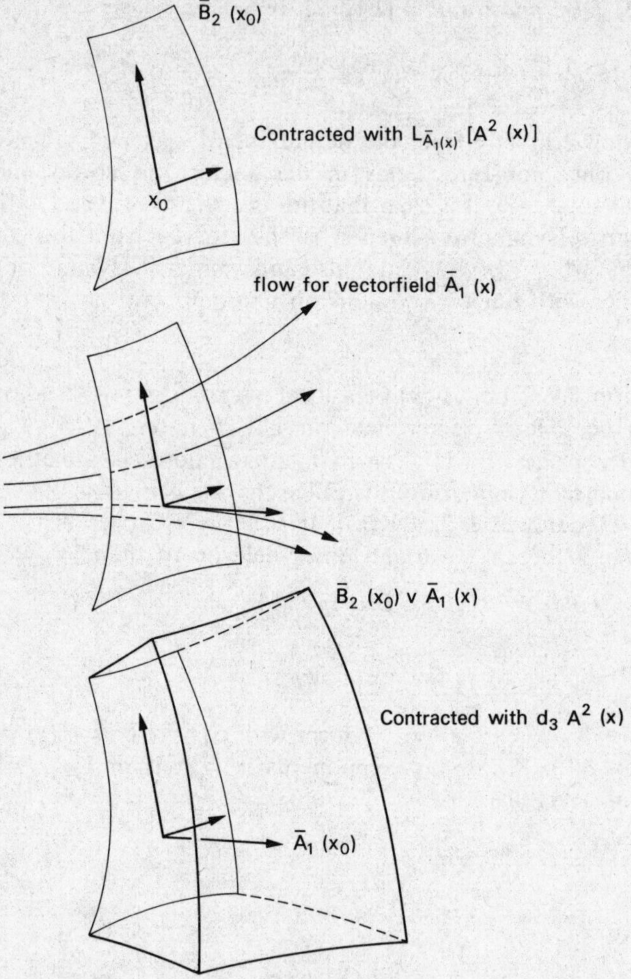

Fig. 3.2

This form can be described succinctly in the following way: let $B^m(x)$ be

$$\tilde{b}_{i_1 i_2 \ldots i_m}(z) = \begin{cases} 1 & \text{if } i_j = j \text{ for } j = 1, 2, \ldots, m, \\ 0 & \text{otherwise.} \end{cases}$$

Then

$$w^m(x) = \sum_{s \in S(m)} \operatorname{sgn}(s) s \cdot B^m(x).$$

It is clear that this is antisymmetric: it associates a 'signed volume' to each ordered m-tuple of tangent vectors. We compute $L_{\bar{A}_1(x)}(w^m(x))$ for local vector field $\bar{A}_1(x)$ with components $\tilde{X}^i(z)$.

According to formula (3.32), this Lie derivative at x_0 is

$$\sum_{j=1}^{m} \tilde{a}_{i_1 i_2 \ldots h_j \ldots i_m}(z_0) \frac{\partial X^{h_j}}{\partial x^{i_j}}_{|z_0}.$$

It is not difficult to see that this sum is

$$\left[\sum_{k=1}^{m} \frac{\partial X^k}{\partial x^k}_{|z_0}\right] w^m(x_0).$$

The interpretation of this form (Liouville's theorem) is that the instantaneous rate of change of 'signed volume' along a solution curve for $\bar{A}_1(x)$ is the *trace* of $(\partial X^k/\partial x^i)_{|z_0}$ multiplied by the form $w^m(x_0)$ for $z_0 \in O$ a point along the solution path. This trace is the *divergence* of $\bar{A}_1(x)$

Now the reduction of a *tensor* by a vector field is a *tensor*. We shall investigate somewhat more closely the case of the reduction of a local r-form by a local vector field. In this case, it will be possible to derive an important relation between the Lie derivative operation and the coboundary operation on local forms. In order to simplify the appearance of 'signs' in some of the calculations to follow, we make the following notational choice. If $A^r(x)$ is a local r-form on $U \subset M$, and $\bar{A}_1(x)$ a local vector field on U such that with respect to frame $\phi: O \to U \subset M$, $\bar{A}_1(x)$ has components $\tilde{X}^i(z)$ and $A^r(x)$ has components $\tilde{a}_{i_1 i_2 \ldots i_r}(z)$, then the $(r-1)$ form $A^r(x)[\bar{A}_1(x)]$ has local components $\tilde{a}_{i_1 i_2 \ldots i_{r-1} i_r}(z)\tilde{X}^{i_r}(z)$. Now denote the $(r-1)$ form

$$(-1)^{r-1} A^r(x)[A_1(x)]\colon A^r(x) \rfloor \bar{A}_1(x). \tag{3.34}$$

This form is called, at times, the 'inner product' of $A^r(x)$ with $\bar{A}_1(x)$.

Of particular interest is the following equation. Assuming that $A^r(x)$ and $\bar{A}_1(x)$ are as above, and W^i_γ the components of an r-sector $\bar{A}_r(x_0)$ at x_0:

$$(dA^r(x) \rfloor \bar{A}_1(x))\langle \bar{A}_r(x_0)\rangle = L_{\bar{A}_1(x)}(\bar{A}^r(x))\langle \bar{A}_r(x_0)\rangle + \sum_{j=1}^{r} (-1)^{rj} d_j A^r(x)[A_1(x)]\langle A_r(x_0)\rangle. \tag{3.35}$$

This formula follows immediately from the definitions of the differential operator and the 'inner product'. We now inspect the sum:

$$\sum_{j=1}^{r} (-1)^{rj} d_j A^r(x)[\bar{A}_1(x)]\langle \bar{A}_r(x_0)\rangle. \tag{3.36}$$

Checking the calculation made in Definition 3.17 we see that this is equal to:

$$\sum_{j=1}^{r} (-1)^{rj}\left[\sigma^j\left(\frac{\partial a_{i_1 i_2 \ldots i_r}}{\partial x^{i_{r+1}}}_{|z_0}\right) + \sum_{k=1}^{r} a_{i_1 i_2 \ldots i_k \ldots i_r}^{\sigma^j(1)\ldots(\sigma^j(k),\sigma^j(r+1))\ldots\sigma^j(r)}(z_0)\right]\langle W^i_\gamma \cup \tilde{X}^i(z)\rangle.$$

After the cancellations of paired terms made in Definition 3.17, (3.36) becomes

$$\left[\sum_{j=1}^{r} (-1)^{rj}\left(\sigma^j\left(\frac{\partial a_{i_1 i_2 \ldots i_r}}{\partial x^{i_{r+1}}}_{|z_0}\right) + a_{i_1 \ldots i_{r+1-j} \ldots i_r}^{(j+1),\ldots,r,(r+1,j),1,\ldots,(j-1)}(z_0)\right)\right]\langle W^i_\gamma \cup \tilde{X}^i(z)\rangle.$$

Next, we compute

$$d(A^r(x) \rfloor \bar{A}_1(x))\langle \bar{A}_r(x_0)\rangle. \tag{3.37}$$

This is equal to $(-1)^{r-1} d(A^r(x)[\bar{A}_1(x)])\langle \bar{A}_r(x_0)$ according to (3.34), and in components this is equal to

$$(-1)^{r-1} d(\tilde{a}_{i_1 i_2 \ldots i_{r-1}, i_s}(z)\tilde{X}^{i_s}(z))\langle W^i_\gamma\rangle$$

which becomes

$$\left[\sum_{j=1}^{r} (-1)^{j(r-1)}\left(\frac{\partial [a_{i_1 i_2 \ldots i_{r-1} i_s} X^{i_s}]}{\partial x^{i_r}}_{|z_0} + \sum_{h=1}^{r-1} a_{i_1 \ldots i_h \ldots i_{r-1} i_s}^{(1)\ldots(h,r)\ldots(r-1)}(z_0) X^{i_s}(z_0)\right)\right]\langle W^i_\gamma \tau^j\rangle$$

with τ the shift of $S(r)$.

Now since $A^r(x)[\bar{A}_1(x)]$ is an $(r-1)$ form, its coboundary is an r-form. Therefore the terms in the right-hand summand cancel pairwise, and (3.37) reduces to

$$\left[\sum_{j=1}^{r} (-1)^{j(r-1)}\left(\frac{\partial a_{i_1 i_2 \ldots i_{r-1} i_s}}{\partial x^{i_r}}_{|z_0} X^{i_s}(z_0) + a_{i_1 i_2 \ldots i_{r-1} i_s}(z_0)\frac{\partial X^{i_s}}{\partial x^{i_r}}_{|z_0}\right)\right]\langle W^i_\gamma \tau^j\rangle.$$

Now using antisymmetry, it follows with a little work that (3.36) + (3.37) = 0. We summarize the result of this calculation in the following proposition.

Proposition 3.24. If $A^r(x)$ is a local r-form on U and $\bar{A}_1(x)$ is a local vector field on U, there is the following relation connecting the Lie derivative, the coboundary operator, and the inner product:

$$L_{\bar{A}_1(x)}(A^r(x)) = dA^r(x) \rfloor \bar{A}_1(x) + d(A^r(x) \rfloor \bar{A}_1(x)). \tag{3.38}$$

■

Now suppose that J is the interval $(-\varepsilon, 1+\varepsilon)$ for $\varepsilon>0$, and I is the interval $[0, 1]$. For $\phi: O \to U \subset M$ a frame, suppose that $A^r(t, x)$ is a local r-form on the cylinder $J\times U$ where we choose coordinates $(t; x^1, x^2, \ldots, x^m)$ for $J\times O$.

If $\bar{A}_r(x_0)$ is an r-sector at $x_0 \in U$, let $\bar{F}^{\#}_{r+1}(\bar{A}_r(x_0))$ denote the r-variation of the path $t \mapsto (t, z_0)$ taking I to $I\times z_0$ defined in the following way. If $A_r(z_0)$ is the germ representing $\bar{A}_r(x_0)$, then the germ representing $\bar{F}^{\#}_{r+1}(\bar{A}_r(x_0))$ is the map $\mathrm{id}\times A_r(x_0): (I\times\mathbb{R}^r, I\times(0, \ldots, 0)) \to (I\times O, I\times z_0)$. When the sector $\bar{A}_r(x_0)$ is fixed and mention of it can be dropped, we refer to the r-variation simply as $\bar{F}^{\#}_{r+1}(t)$, determining for each value of t an $(r+1)$ sector at (t, z_0) in $I\times O$.

Referring to equation (3.19), we may say:

$$i_1^*(A^r(1, x_0))\langle\bar{A}_r(x_0)\rangle - i_0^*(A^r(0, x_0))\langle\bar{A}_r(x_0)\rangle = \int_0^1 d_1 A^r(t, x_0)\langle\bar{F}^{\#}_{r+1}(t)\rangle\, dt \tag{3.39}$$

where $i_1: U \to I\times U$ by $i_1(x) = (1, x)$ and $i_0: U \to I\times U$ by $i_0(x) = (0, x)$.

Now let $\bar{A}_1(t, x)$ be the 'vertical' vector field with components $(1; 0, \ldots, 0)$ on $J\times O$. If $i_t: U \to I\times U$ by $i_t(x) = (t, x)$, then it is clear that

$$[T_r(i_{t_0})(\bar{A}_r(x_0)) \cup \bar{A}_1(t, x)] = \bar{F}^{\#}_{r+1}(t_0)\sigma$$

where σ is the shift of $\Gamma(r)$.

Therefore for t_0 fixed,

$$d_1 A^r(t_0, x_0)\langle\bar{F}^{\#}_{r+1}(t_0)\rangle = L_{\bar{A}_1(t,x)} A^r(t, x)\langle T_r(i_{t_0})\bar{A}_r(x_0)\rangle. \tag{3.40}$$

Putting (3.39) and (3.40) together we have for each r-sector $\bar{A}_r(x_0)$ at $x_0 \in U$,

$$\begin{aligned}&[i_1^*(A^r(1, x_0)) - i_0^*(A^r(0, x_0))]\langle\bar{A}_r(x_0)\rangle\\ &= \int_0^1 [dA^r(t, x_0) \rfloor \bar{A}_1(t, x_0) + d(A^r(t, x_0) \rfloor \bar{A}_1(t, x_0))]\\ &\qquad\qquad \times\langle T_r(i_t)\bar{A}_r(x_0)\rangle\, dt.\end{aligned} \tag{3.41}$$

We now introduce the key construction. Suppose given an $(r+1)$ form $B^{r+1}(t, x)$ on $J\times U$. Suppose in local coordinates with respect to frame $\mathrm{id}\times\phi: J\times O \to J\times U$ that

$$B^{r+1}(t, x) = \tilde{b}_{h_1 h_2 \ldots h_{r+1}}(t, z).$$

Suppose also that $\bar{A}_r(x_0)$ is an r-sector at x_0 with ϕ components X^i_γ, let $\gamma_1, \gamma_2, \ldots, \gamma_r$ be the respective vertices of $\Gamma(r-1)$, and let X^i (unsubscripted) denote the $(m+1)$ tuple $(1; 0, 0, \ldots, 0)$.

Associate to X^i_γ the following number:

$$\int_0^1 b_{h_1 h_2 \ldots h_{r+1}}(t, z_0) X^{h_1} \bar{X}^{h_2}_{\gamma_1} \ldots \bar{X}^{h_{r+1}}_{\gamma_r}\, dt \tag{3.42}$$

with the convention that

$$\bar{X}^{h_i}_{\gamma_i} = X^{h_i-1}_{\gamma_i} \qquad \text{if } 2 \leq h_i \leq r+1,$$

and

$$\bar{X}^{h_i}_{\gamma_i} = 0 \qquad \text{if } h_i = 1.$$

This association is clearly r-linear and antisymmetric as a function $(X^i_{\gamma_1}, \ldots, X^i_{\gamma_r}) \to \mathbb{R}$ thus gives an r-form at x_0. Differentiating 'under the integral', we see that we have in fact a local r-form on U letting x_0 vary. Now this integral has the description:

$$\int_0^1 [B^{r+1}(t, x_0) \rfloor \bar{A}_1(t, x_0)]\langle T_r(i_t)\bar{A}_r(x_0)\rangle \, dt.$$

Suppose given a frame change considered as a diffeomorphism $U \xrightarrow{\alpha} U$. Then $\mathrm{id} \times \alpha$ carries $\bar{A}_1(t, x)$ to $\bar{A}_1(t, x)$ and the naturality property for inner products given in formula (3.31) guarantees that (3.42) does not depend on frame.

Denote the r-form defined by (3.42) as follows: on U, $C(B^{r+1})(x_0)$ so that

$$C(B^{r+1})(x_0)\langle \bar{A}_r(x_0)\rangle = \int_0^1 [B^{r+1}(t, x_0) \rfloor \bar{A}_1(t, x_0)]\langle T_r(i_t)\bar{A}_r(x_0)\rangle \, dt. \tag{3.43}$$

Now (3.41) translates to

$$[i_1^*(A^r(1, x_0)) - i_0^*(A^r(0, x_0))]\langle \bar{A}_r(x_0)\rangle = (C\,d + d\,C)A^r(t, x_0)\langle \bar{A}_r(x_0)\rangle \tag{3.44}$$

with a little work. We now summarize this.

Theorem 3.25. If $A^r(t, x)$ is an r-form on the cylinder $J \times U$, J an interval $(-\varepsilon, 1+\varepsilon)$ for $\varepsilon > 0$, then the r-forms on U are equal:

$$i_1^*(A^r(1, x)) - i_0^*(A^r(0, x)) = (C\,d + d\,C)A^r(t, x). \tag{3.45}$$

■

This theorem (essentially the Poincaré lemma, although we shall cite one of its consequences under that title) illustrates with simple clarity the fundamental utility of differential forms. Aside from providing the foundation for integration theory on manifolds, forms can be manipulated in a calculus of amazing applicability.

Now it is an easy consequence of formula (3.26) that $d \circ d = 0$ as an operator on local forms. Theorem 3.25 gives a partial converse.

Proposition 3.26 (the Poincaré lemma). Suppose $A^r(x)$ is a local r-form on $U \subset M$ with $\phi : O \to U$ a frame where O is *contractible*, that is, there is a smooth map $f: I\times O\to O$ such that $f(0, x)=x$ for all $x\in O$ and for some $x_0\in O$, $f(1, x)=x_0$ for all $x\in O$. Then $dA^r(x)=0$ on U *if and only if* $A^r(x)=dB^{r-1}(x)$ for some local $(r-1)$ form $B^{r-1}(x)$ on U.

Proof. Apply Theorem 3.25 to the form $f^*(A^r(x))$. ■

As a consequence, any local form with zero coboundary is *locally* (although perhaps not globally on its domain of definition) the coboundary of a lower-dimensional form.

The forms $A^r(x)$ on a manifold of dimension $m<r$ are necessarily identically zero; this follows from elementary and general reasoning in multilinear (exterior) algebra. Therefore, the coboundary of an m-form on an m-dimensional manifold is zero. Now suppose $O\subset E=\mathbb{R}^m$ and $w^m(x)$ is the determinant form on O defined in example 3 above (recall that E has a fixed ordered basis dual to its standard system of linear coordinates). Then $dw^m=0$. The following consequence of Proposition 3.24, known as the *divergence theorem* will close this section and Chapter 3.

Observation 3.27. Suppose O is as above, and $\bar{A}_1(x)$ is a vector field on O. Also suppose that $D\subset O$ is a compact smooth m-dimensional *manifold with boundary* (the boundary which is an orientable compact $(m-1)$ dimensional manifold we shall call $\delta(D)$). To say that $\delta(D)$ is orientable is to say that there is a global *non-vanishing* $(m-1)$ form defined on $\delta(D)$. Such a form can be constructed in this case by taking the *inner product* of the outward pointing normal with the determinant form on O.

Now let γ_0 be the *flow-germ* for $\bar{A}_1(x)$ on O. For small s, the maps γ_0^s give *diffeomorphisms* from $D\to O$ (they cannot be defined globally on O, however). Now the Lie derivative of w^m with respect to $\bar{A}_1(x)$ has the interpretation (Observation 3.14)

$$L_{\bar{A}_1(x)}w^m(x)\langle \bar{A}_m(x_0)\rangle=\frac{\partial}{\partial s_{|0}}(\gamma_0^s)^*[w^m(\gamma_0^s(x_0)]\langle \bar{A}_m(x_0)\rangle.$$

Now $\int_D w^m(x)$ has the interpretation as ±volume of D (depending on sign conventions for computing integrals which we shall not develop here). Therefore, differentiating under the integral gives

$$\int_D L_{\bar{A}_1(x)}w^m(x)=\pm\frac{\partial}{\partial s_{|0}}[\text{volume } \gamma_0^s(D)]. \tag{3.46}$$

Next, applying Proposition 3.24 and the fact that $dw^m(x)=0$, we have

$$\int_D L_{\bar{A}_1(x)}w^m(x)=\int_D d[w^m(x)\rfloor \bar{A}_1(x)]=\int_{\delta(D)} w^m(x)\rfloor \bar{A}_1(x)$$

where the last equality follows from *Stokes's theorem* and utilizes an implicit *sign convention* for orienting $\delta(D)$ and, of course, requires an interpretation for computing an integral of an $(m-1)$ form over $\delta(D)$. When these conventions are made, the right-hand integral is interpreted as the integral of 'flux across the boundary of D'. The left-hand side (in light of example 3) is interpreted as the integral of the divergence of $\bar{A}_1(x)$ over D; properly speaking, it is the integral of div $\bar{A}_1(x) \cdot w^m(x)$ over D. One integrates *forms and only forms* on manifolds. Both integrals have the interpretation of being $\pm(\partial/\partial s)_{|0}[\text{volume } \gamma_0^s(D)]$. ■

With these few examples, we end the discussion of differential forms, and the *general* treatment of sectorform fields. Riemannian geometry and Lagrangian dynamics require the study of certain sectorform fields derived from the data of a metric (or of a 'kinetic energy' function), and these fields are far from being antisymmetric; in a few cases, they are not even tensors. The final part of this book addresses itself to this situation: when the manifold has the *additional* structure of a (possibly indefinite) metric, certain characteristic sectorform fields arise and satisfy enough naturality conditions to lead to a rich and profound local geometry.

4

Riemannian geometry

§4.1. Metric structure and Gaussian curvature

When a smooth manifold is endowed with a *metric tensor* (exercise 3.18), it inherits a rich and subtle *local* geometry which it will be the intention of the remaining chapter of this book to investigate. The study of the *global* geometry of Riemannian manifolds, requiring as it does a more sophisticated use of integration theory than we can make here, will be treated only in cursory ways.

We shall *not* in general assume that the metric $A^2(x)$ on M is positive (or negative) *definite*, although this restriction will be necessary in a few special cases. In any event, the manifold will be assumed connected.

Intuitively, a metric tensor gives the 'infinitesimal line element', the 'line element', or the 'length of a tangent vector' in the positive definite case; and in the indefinite case (for example, the Lorentz metric in special relativity) it must have a less concrete interpretation. With an eye to the applications in classical mechanics, we observe here that the association $X \to \frac{1}{2}A^2(x)\langle X, X\rangle$ taking $T_1(M) \to \mathbb{R}$ can be thought of as 'kinetic energy' in the positive definite case. As is well-known, a Riemannian manifold (with a positive definite metric) can be assimilated to a conservative mechanical system on which there is no potential (no real forces), and for which the law of motion is given by the Euler–Lagrange equations. We shall in §4.2 place more general conservative systems into this geometric framework.

Suppose now that M is a smooth manifold and that $A^2(x)$ is a metric tensor on M. With respect to frame $\phi : O \to U \subset M$ say that $A^2(x)$ has components $\tilde{a}_{ij}(z)$. Now compute the differential d_3A^2 with components

$$\tilde{a}_{sj}(z) + \tilde{a}_{it}(z) + \frac{\partial \tilde{a}_{ij}}{\partial x^k}_{|z}$$

according to Convention 3.9 (exercise 3.15).

The 3 sectorform field d_3A^2 on M satisfies certain naturality properties. Suppose for example that $f: N \rightarrow M$ is a smooth *immersion.* Then if $A^2(x)$ is a *definite* metric on M, it follows that $f*A^2(w)$ is a definite metric on N; if $A^2(x)$ is *indefinite* then it is possible that $f*A^2(w)$ is degenerate, and hence not a metric. Let us assume that $f*A^2(w)$ is a metric on N. Then since the differential operators are natural, we have

$$d_3[f*A^2(w)] = f*[d_3A^2(x)].$$

This requires no non-degeneracy assumption, of course, but we intend to make the following interpretation of d_3A^2.

Theorem 4.1. *Suppose that* $A^2(x)$ *is a metric tensor on* M. *Then according to Definition* 3.15 *the* 3 *sectorform field* $d_3A^2(x)$ pointed at the last vertex (*exercise* 3.11) *gives rise to a* 2 *sectorform field on* $T_1(M)$, *namely* $(d_3A^2(x))_{|\bar{A}_1(x)}$, *which we denote simply* $\delta A^2[\bar{A}_1(x)]$ or δA^2 *and have called the* promotion *of the metric tensor* $A^2(x)$.

Now δA^2 *is a metric tensor on* $T_1(M)$, *in general indefinite even if* A^2 *is definite. It satisfies the following naturality property. If* $f: N \rightarrow M$ *is a smooth immersion,* $A^2(x)$ *a metric tensor on* M and *if* $f*A^2(w)$ *is a metric tensor on* N, *then for* $T_1(f): T_1(N) \rightarrow T_1(M)$

$$T_1(f)*[\delta A^2] = \delta[f*A^2]. \tag{4.1}$$

This naturality formula says that the induced metric on the tangent bundle $T_1(N)$ *can be computed as the promotion of* $f*A^2$ or *as the pull-back of the metric* δA^2 *on* $T_1(M)$.

Proof. First we show that δA^2 is a metric tensor on $T_1(M)$ giving its local components with respect to frame-germ $\tilde{\phi}^1$ derived from frame-germ ϕ (see Theorem 1.27 or Definition 3.10 for notation). Next we consider the naturality condition.

Start with a frame $\phi: O \rightarrow U \subset M$. Then $\tilde{\phi}^1: (O \times E_1) \rightarrow q_1^{-1}(U) \subset T_1(U)$ is that special frame defined by the rule $\tilde{\phi}^1(z, Z^k) = [(\phi \circ t_z)_1, Z^k]$.

Now the mapping introduced in Definition 3.10, $p: T_3(x_0) \rightarrow T_2[T_1(x_0)]$, which was represented in components:

$$[\phi_3, X^i_{\gamma'} \text{——} Z^i_{\gamma' \cup \gamma''} \text{——} Y^i_{\gamma''}] \rightarrow [(\tilde{\phi}^1 \circ t_{(0, X^i_{\gamma'})})_2, W^I_{\gamma''}] \qquad \text{for } \gamma' = (0, 0, 1),$$

was called the *pointing* operator. By means of it, the 3-sector

$$\begin{array}{c} [\phi_3, X^i \text{——} A^r \text{——} Y^j] \\ \diagdown \qquad\qquad \diagup \\ B^s \ E^u \ C^t \\ \diagdown \quad \diagup \\ Z^k \end{array}$$

is identified with the 2-sector at $[\phi_1, Z^k]$:

$$\left[(\tilde{\phi}^1 \circ t_{(z_0,\, Z^k)})_2, \begin{bmatrix} X^i \\ B^s \end{bmatrix} \!-\!\!-\! \begin{bmatrix} A^r \\ E^u \end{bmatrix} \!-\!\!-\! \begin{bmatrix} Y^j \\ C^t \end{bmatrix}\right]. \tag{4.2}$$

Now for γ' as above, recall the γ'-*star complex* (Definition 2.20)

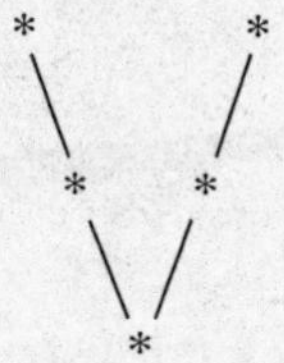

which we call $C[\gamma'] \subseteq \Gamma(2)$. There is a natural identification (Definition 2.20)

$$\begin{array}{ccc} T3[M; C[\gamma']] & \underset{\cong}{\longleftrightarrow} & B_2[T_1(M)] \\ \downarrow \tilde{q}_{3.1} & & \downarrow q_1(2) \\ T_1(M) & & T_1(M) \end{array}$$

If in local components $A^2(x) = \tilde{a}_{ij}(z)$, then

$$d_3 A^2(x) = \tilde{a}_{sj}(z) + \tilde{a}_{it}(z) + \frac{\partial \tilde{a}_{ij}}{\partial x^k}_{|z},$$

and it is fairly easy to see that $d_3 A^2(x)$ is a $C[\gamma']$ sectorform field (exercise 3.4). Suppose the 3 K-sector

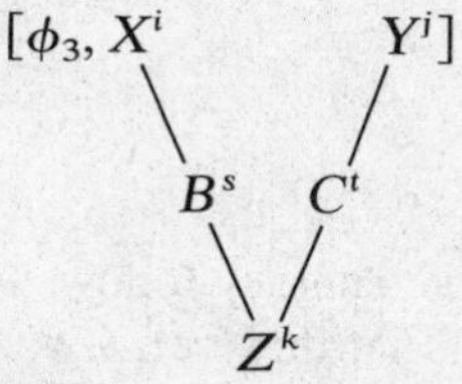

is called $\bar{K}_3(x_0)$, and let $[\phi_1, Z^k] = \bar{A}_1(x_0)$. Then according to the definition (exercise 3.10),

$$\delta A^2{}_{|\bar{A}_1(x_0)} \langle p\bar{K}_3(x_0)\rangle = d_3 A^2 \langle \bar{K}_3(x_0)\rangle. \tag{4.3}$$

Identifying K-sectors such as $\bar{K}_3(x_0)$ with ordered pairs of *tangent vectors* to $T_1(M)$ at $\bar{A}_1(x_0)$, we see that δA^2 is a *bilinear* and *symmetric* covariant 2-tensor with matrix:

$$\begin{vmatrix} \dfrac{\partial a_{ij}}{\partial x^k}_{|z_0} Z^k & a_{it}(z_0) \\ a_{sj}(z_0) & 0 \end{vmatrix}$$

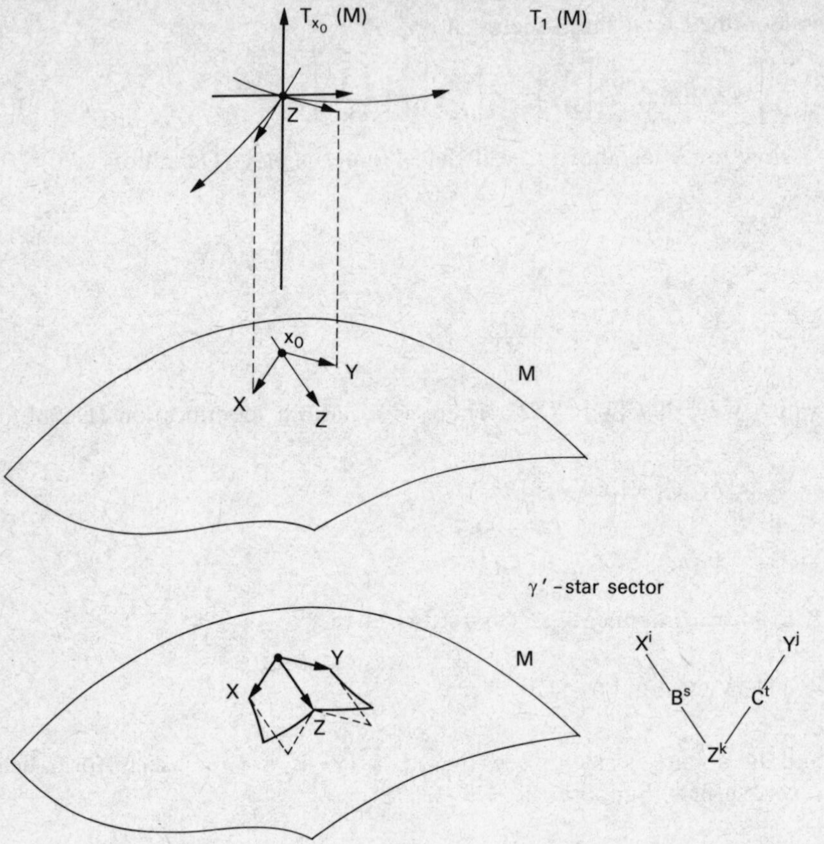

Fig. 4.1

with each block an $m \times m$ submatrix. Thus non-degeneracy follows from the non-degeneracy of $A^2(x)$. It is clear, however, that this metric will not be definite, in general.

Now naturality follows directly from the corresponding naturality of the differential operators (in particular, d_3); appeal to the fact that $d_3 f * A^2 = f * d_3 A^2$. ■

Actually, the main use of the pointing operation at this stage is heuristic. In reality, the *fundamental* object is the *sectorform field* d_3A^2. The reason for this is that it is difficult to assimilate the richer structure of the bundles $T_2(M)$ and $T_3(M)$ to the familiar structure of the tangent bundle. It is more natural to study them on their own terms, and this will mean coming to grips with their own characteristic structures. It will be useful from time to time to observe that the differential d_3A^2 is a metric tensor on the tangent bundle in order to *motivate* certain constructions, and to interpret certain classical results. For example, we shall see that the Euler–Lagrange flow on the tangent bundle for a conservative mechanical system is a 'gradient' (the gradient of the Lagrangian) with respect to the metric δA^2 on $T_1(M)$. This explains some of the naturality properties of Lagrangian (and Hamiltonian) systems. And it will be possible to *iterate* certain constructions on the metric A^2 for the metric δA^2 leading to such notions as curvature on a Riemannian manifold. But restricting language to tensors will avoid certain symmetries in the higher-order cases which can lead to new refinements. Thus we shall refer *primarily* to the 3 sectorform field d_3A^2 (and its images under even permutation: d_1A^2 and d_2A^2), leaving it to the reader to apply the interpretation as metric on the tangent bundle: δA^2. We return eventually to the latter interpretation *explicitly* only to make such applications as were mentioned above.

The following construction, which plays a central role in the elaboration of the metric structure on a Riemannian manifold, illustrates the difficulty which attends a restriction to tensor constructions. We have defined in the previous chapter the *coboundary* of an antisymmetric 2-tensor. An important fact for those considerations was that this coboundary was itself a tensor; as we observed, this leads to some of the major results of vector analysis via Stokes's theorem. Now *symmetric* tensors such as metrics can be treated in a similar way but the result will be a 3 sectorform (with some rather interesting properties) instead of a tensor. We make the definition below.

Definition 4.2. Suppose that $A^2(x)$ is a metric tensor on M with local components $\tilde{a}_{ij}(z)$ in frame ϕ. Define the *cycle* of $A^2(x)$ denoted DA^2 to be the 3 sectorform field $d_1A^2+d_2A^2-d_3A^2$.

In local components, we may write (using Convention 3.9)

$$DA^2(x)=2\tilde{a}_{rk}(z)+\frac{\partial\tilde{a}_{jk}}{\partial x^i}\bigg|_z+\frac{\partial\tilde{a}_{ki}}{\partial x^j}\bigg|_z-\frac{\partial\tilde{a}_{ij}}{\partial x^k}\bigg|_z. \tag{4.4}$$

Again, it is clear that $DA^2(x)$ is a 3 sectorform but not a tensor. The cycle can be defined for *any* symmetric 2-tensor field and we have the naturality property: for $f:N\to M$ smooth and A^2 as above, $f*A^2$ is (at worst) a covariant 2-tensor field, and

$$f*[DA^2]=D[f*A^2]. \tag{4.5}$$

The 'Riemann–Christoffel symbols' make their natural appearance in this generalization of the coboundary operator. ■

The 3 sectorform field DA^2 will be of fundamental importance for the study of the local structure of a Riemannian manifold. We shall use it to define a 'pairing', analogous to the pairing of tangent vectors given by the metric, which associates a real number to each pair consisting of a 2-sector and 1-sector at a point.

Theorem 4.3. *Let $D \subset \Gamma(2)$ be the subcomplex*

————*

*

where $\Gamma(2)$ has the standard ordering. Suppose that $B^2(x)$ is a symmetric covariant tensor *field of type* $\binom{0}{2}$ *on smooth manifold M. Then DB^2 (the cycle of B^2) is a 3 D-sectorform field. Thinking of it as a map*

$$\begin{array}{l} T_2(M) \times_M T_1(M) \to \mathbb{R} \\ \quad \downarrow q_1 \times M q_2 \\ \quad M \end{array}$$

it gives a pairing *called the* cycle product *of 2-sectors with 1-sectors.*

For x_0 in M let an element of the fiber over x_0 be denoted $(\bar{A}_2(x_0), \bar{A}_1(x_0))$. Abusing language slightly, denote the image of this point $DB^2\langle \bar{A}_2(x_0), \bar{A}_1(x_0)\rangle \subset \mathbb{R}$. Call this image the cycle product *of $\bar{A}_2(x_0)$ with $\bar{A}_1(x_0)$.*

1) *Holding $\bar{A}_2(x_0)$ fixed, the mapping $\bar{A}_1(x_0) \to DB^2\langle \bar{A}_2(x_0), \bar{A}_1(x_0)\rangle$ gives a 1 sectorform (a 1-form) on the tangent fiber.*

2) *Holding $\bar{A}_1(x_0)$ fixed, the mapping $\bar{A}_2(x_0) \to DB^2\langle \bar{A}_2(x_0), \bar{A}_1(x_0)\rangle$ gives a 2 sectorform on the fiber $T_2(x_0)$.*

3) *There is the following naturality property. For $f: N \to M$ smooth and B^2 as above, let $f * B^2$ be the pull-back 2-tensor field on N, and let $f(w_0) = x_0$ with $T_2(f)\bar{C}_2(w_0) = \bar{A}_2(x_0)$ and $T_1(f)\bar{C}_1(w_0) = \bar{A}_1(x_0)$ for sectors $\bar{C}_2(w_0)$ and $\bar{C}_1(w_0)$ at $w_0 \in N$. Then*

$$D(f * B^2)\langle \bar{C}_2(w_0), \bar{C}_1(w_0)\rangle = DB^2\langle \bar{A}_2(x_0), \bar{A}_1(x_0)\rangle. \tag{4.6}$$

Proof. First observe that there is a natural bundle mapping $T_3(M) \xrightarrow[j]{} T_2(M) \times_M T_1(M) = T_3[M; D]$ (on total spaces, over M)

$$\begin{array}{c} [\phi_3, X^i \text{——} A^r \text{——} Y^j] \xrightarrow[j]{} ([\phi_2, X^i \text{——} A^r \text{——} Y^j], [\phi_1, Z^k]). \\ \diagdown \qquad\qquad \diagup \\ B^s \;\; E^u \;\; C^t \\ \diagdown \;\; \diagup \\ Z^k \end{array}$$

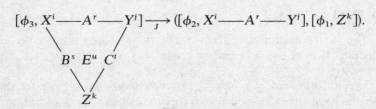

Now define the *cycle product mapping* with respect to B^2 in the following way. For $(\bar{A}_2(x_0), \bar{A}_1(x_0)) \in T_2(x_0) \times_M T_1(x_0)$ let $\bar{A}_3(x_0)$ be *any* element of $T_3(x_0)$ with the property that:

$$J[\bar{A}_3(x_0)] = (\bar{A}_2(x_0), \bar{A}_1(x_0)).$$

Then (abusing language) define

$$DB^2\langle \bar{A}_2(x_0), \bar{A}_1(x_0)\rangle = DB^2(\bar{A}_3(x_0)).$$

It is easy to see that this prescription is independent of the choice of $\bar{A}_3(x_0)$ in the J pre-image.

In fact, in these components, with respect to frame ϕ taking the origin to x_0,

$$DB^2\langle [\phi_2, X^i \text{---} A^r \text{---} Y^j], [\phi_1, Z^k]\rangle$$

$$= 2b_{rk}(0)A^rZ^k + \left[\frac{\partial b_{jk}}{\partial x^i} + \frac{\partial b_{ki}}{\partial x^j} - \frac{\partial b_{ij}}{\partial x^k}\right] X^iY^jZ^k. \tag{4.7}$$

Here the components for $B^2(x)$ are $\tilde{b}_{ij}(z)$ and we follow Convention 3.9 for notation. Assertions 1) and 2) are clear from the component formula and assertion 3) follows immediately from naturality formula (4.5). ■

The cycle product is thus seen to be a generalization of the metric pairing. It plays an important role in the local formulation of the Euler–Lagrange equations, and, as we shall see, its use is implicit in many of the other constructions of Riemannian geometry. By means of this pairing, each 2-sector at a point gives rise to a *covector* at that point (assertion 1). This allows us to formulate a classical notion in local terms as follows.

Definition 4.4. Suppose M is a smooth manifold and $A^2(x)$ is a *metric* tensor field. Let $\bar{A}_2(x_0)$ be a 2-sector at x_0. Then there is a *unique* 1-sector (tangent vector) $\bar{A}_1(x_0)$ at x_0 with the property that

$$\tfrac{1}{2}DA^2\langle \bar{A}_2(x_0), \bar{B}_1(x_0)\rangle = A^2\langle \bar{A}_1(x_0), \bar{B}_1(x_0)\rangle \tag{4.8}$$

for *all* tangent vectors $\bar{B}_1(x_0)$ at x_0. Call this tangent vector the *force* of $\bar{A}_2(x_0)$, for picturesque reasons upon which we elaborate later. When $\bar{A}_2(x_0)$ is thought of as the suspension of a vector field by a tangent vector, the force of $\bar{A}_2(x_0)$ equals the *covariant derivative* of the vector field in the direction of the tangent vector which suspends it. Refer to the *covector* dual to the force of $\bar{A}_2(x_0)$ as the *dual force* of $\bar{A}_2(x_0)$. ■

Observation 4.5. Let σ be the transposition of $S(2)$. Then the force of $\bar{A}_2(x_0)$ is equal to the force of $[\bar{A}_2(x_0)]\sigma$. Recall that the projection

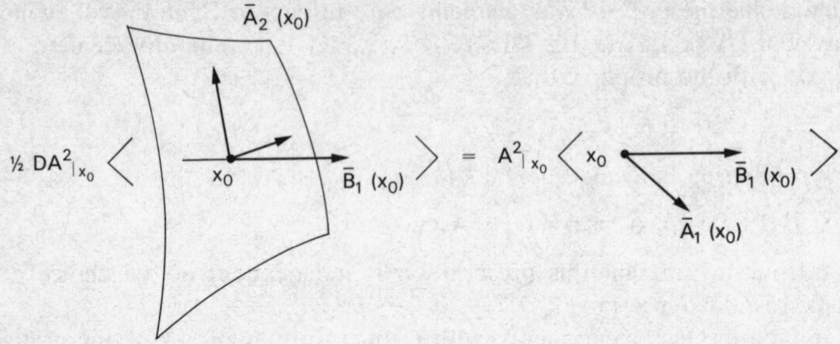

Fig. 4.2. Determination of the 'force' ($\bar{A}_1$) of $\bar{A}_2$ via the cycle product pairing

$D(2): T_2(M) \to T_1^2(M) = B^2(M)$. This is the projection of an 'affine bundle' over a certain vector bundle. Two sectors with the same boundary (in the same $D(2)$ fiber) have a *bracket* which can be thought of as a tangent vector on M.

Suppose that $\bar{A}_2(x_0)$ and $\bar{B}_2(x_0)$ are in the same $D(2)$ fiber. Then

$$\text{force of } \bar{B}_2(x_0) - \text{force of } \bar{A}_2(x_0) = [\bar{A}_2(x_0); \bar{B}_2(x_0)]. \tag{4.9}$$

Proof. All that needs to be proved here is the formula (4.9). But the corresponding formula for the dual forces follows immediately from formula (4.7) since in the difference on the right-hand side of the equation the terms involving derivatives cancel. ■

The next notion to be studied in this section gives an interpretation of the classical idea of 'affine connection'. A smooth manifold with a metric tensor field inherits, as we have seen, additional structure. The 'fundamental lemma of Riemannian geometry' (as it has been called) asserts the existence of a unique *symmetric affine connection compatible with a given metric.* We need not worry here about the signification of the italicized phrase; that will be given its interpretation in the following discussion. From the point of view which is being developed here, the existence and the uniqueness of this structure should appear natural. At least, this is the aim of the approach. There are many equivalent formulations of the notion of connection, some axiomatic, some constructive. These may lead to more or less general classes of structures (symmetric or non-symmetric, derived from a metric or given independently, etc.). The structure to be defined here is necessarily symmetric, although it may be formulated in such a way as not to be necessarily bound to a metric; however, the formulation we give does derive it from one.

We begin by recalling the *tower of frame bundles* discussed in Chapter 1. We observed that for M a smooth manifold, the projection $\pi_{2,1}: F_2(M) \to F_1(M)$ is the projection of an *affine bundle*. The fiber is an *affine space* of dimension $m\binom{m+1}{2}$ for M of dimension m. In fact, following Proposition 1.17, we may think of the fiber as the affine space of homogeneous degree-2 polynomials taking $(E, 0) \to (E, 0)$. This fiber is actually the kernel of the projection $G_2(m) \to G_1(m)$ given by truncation, and will be denoted, as it was earlier, $\gamma_2(m) \subseteq G_2(m)$.

Now a smooth cross section for $\pi_{2,1}$ gives a *choice of origin* in each fiber and thus provides a (non-natural) equivalence of fiber bundles $F_2(M) \cong \gamma_2(m) \times_M F_1(M)$ where, abusing notation, $\gamma_2(m)$ denotes the trivial *vector* bundle over M with fiber the vector space $\gamma_2(m)$. There should be no confusion with this notation which harmonizes with the construction, given later, of an *affine bundle over a vector bundle* (Definition 2.5). Now the data of a metric tensor on M give such a cross section for $\pi_{2,1}$ and we shall call it the *fundamental section for the metric tensor.* This section will be the data of the unique symmetric connection cited above which is 'compatible' with the metric.

To begin, we represent an element of $\gamma_2(m)$ by its 'polynomial representative' following formula (1.14). Recall that $g_2 \in \gamma_2(m)$ is an invertible 2-jet of a frame-germ $(E, 0) \to (E, 0)$ with identity 1-jet. Thus the polynomial representation for such a g_2 is:

$$\delta_i^r x^i + \sum_{|\alpha|=2} A_\alpha^r \frac{x^\alpha}{\alpha!}$$

with δ_i^r the Kronecker delta, and the A_α^r arbitrary scalars. Now a simpler notation, and one we shall adopt, is the following:

$$g_2 = \delta_i^r x^i + \tfrac{1}{2} A_{ij}^r x^i x^j \tag{4.10}$$

with A_{ij}^r an $m \times m \times m$ matrix symmetric in i and j.

The following definition gives the criterion which provides the section.

Definition 4.6. Suppose that smooth manifold M is equipped with metric tensor $A^2(x)$. Let $x_0 \in M$, and let ϕ_2 be a 2-frame at x_0. Then ϕ_2 is a *null frame* if each 2-sector of the form $[\phi_2, X^i$——0——$Y^i]$ has *zero force.* ■

Of course, it must be shown that null frames exist.

Proposition 4.7. For M a smooth manifold with metric tensor $A^2(x)$ there is, at $x_0 \in M$, *one and only one* null frame in *each* $\pi_{2,1}$ fiber.

Proof. First, we argue existence. This argument, which might be called the *Riemann–Christoffel shuffle*, gives the classical Riemann–Christoffel symbols. Suppose then that ϕ_1 is a 1-frame at x_0. We want to show that there is a 2-frame ϕ_2 such that (of course) $\pi_{2,1}(\phi_2)=\phi_1$ *and* such that ϕ_2 is a null frame.

Suppose then that θ_2 is an arbitrary 2-frame with $\pi_{2,1}(\theta_2)=\phi_1$ and let Λ^r_{ij} be the $m\times m\times m$ matrix satisfying: Λ^r_{ij} *symmetric* in i and j, and

$$DA^2\langle[\theta_2, X^i\text{——}0\text{——}Y^j], [\phi_1, Z^k]\rangle + 2a_{rk}\Lambda^r_{ij}X^iY^jZ^k = 0$$

for the ϕ_1 components of $A^2(x_0)$ equal to a_{rk}. Such a matrix Λ^r_{ij} can be found: inspection of formula (4.7) shows that the expression on the left $DA^2\langle[\theta_2, X^i\text{——}0\text{——}Y^j], [\phi_1, Z^k]\rangle$ is *trilinear* in the components X^i, Y^j, and Z^k and is *symmetric* in i and j. One appeals also to the non-degeneracy of a_{rk}.

Now if Λ^r_{ij} is zero, then θ_2 is, by definition, a null frame. In any case, let $g_2\in\gamma_2(m)$ be the quadratic polynomial

$$g_2=\delta^r_i x^i+\tfrac{1}{2}\Lambda^r_{ij}x^ix^j$$

for the $m\times m\times m$ matrix Λ^r_{ij} determined (for frame θ_2) above.

Let $\phi_2=\theta_2 g_2$. Then $\pi_{2,1}(\phi_2)=\pi_{2,1}(\theta_2)=\phi_1$. Also

$$\begin{aligned}&DA^2\langle[\phi_2, X^i\text{——}0\text{——}Y^j], [\phi_1, Z^k]\rangle\\ &\quad= DA^2\langle[\theta_2, X^i\text{——}\Lambda^r_{ij}X^iY^j\text{——}Y^j], [\theta_1, Z^k]\rangle = 0.\end{aligned}$$

Therefore, $\phi_2=\theta_2 g_2$ is a null frame.

Computationally, $\Lambda^r_{ij}=-\Gamma^r_{ij}$ (−'connection coefficients') depends on the frame θ_2 at x_0. In particular, for $\theta: O\to U\subset M$ a local frame, one computes smooth mapping $\Lambda^r_{ij}(z): O\to L(E, E; E)$ by the prescription:

$$\Lambda^r_{ij}(z)=\tfrac{1}{2}\tilde{a}^{rk}(z)\left[\frac{\partial \tilde{a}_{ij}}{\partial x^k}\Big|_z-\frac{\partial \tilde{a}_{ik}}{\partial x^j}\Big|_z-\frac{\partial \tilde{a}_{kj}}{\partial x^i}\Big|_z\right] \tag{4.11}$$

with $\tilde{a}_{ij}(z)$ the θ components of $A^2(x)$, that is, the components with respect to frame $(\theta\circ t_z)_2$ and where $a^{ik}(z)a_{kj}(z)=\delta^i_j$ (Kronecker delta).

Next, we argue uniqueness. Suppose that ϕ_2 and θ_2 are null frames in the $\pi_{2,1}$ fiber over ϕ_1. Then for some $g_2\in\gamma_2(m)$, $\phi_2=\theta_2 g_2$ since $F_2(M)\xrightarrow[\pi_{2,1}]{} F_1(M)$ is a *principal* bundle with fiber $\gamma_2(m)$ as we saw in Proposition 1.17. Now suppose $g_2=\delta^r_i x^i+\frac{1}{2}A^r_{ij}x^ix^j$. Then reversing the reasoning above, it is easy to see that $A^r_{ij}=0$. Thus $\phi_2=\theta_2$. ∎

The last proposition gives the prescription for defining the *fundamental section* of $\pi_{2,1}$ for $A^2(x)$. Associate to each 1-frame ϕ_1 the unique 2-frame ϕ_2 which is the null frame in the fiber over ϕ_1. Now the data of a cross section for the principal bundle

$$\begin{array}{c} F_2(M) \\ \downarrow \pi_{2,1} \\ F_1(M) \end{array}$$

are equivalent to the data of a smooth map $f: F_2(M) \to \gamma_2(m)$ with the property that $f(\theta_2 g_2) = f(\theta_2) \cdot g_2$ for $g_2 \in \gamma_2(m)$ (see, for example, Steenrod [23] or Sternberg [24]). This amounts to associating to each frame θ_2 the element $g_2 \in \gamma_2(m)$ such that $\theta_2 = \phi_2 g_2$ with ϕ_2 the 2-frame in the section in the fiber of θ_2. Then $f(\theta_2) = g_2$.

The proposition above essentially defined the map f. In the notation of that proposition, $\phi_2 g_2^{-1} = \theta_2$ and $g_2^{-1} = \delta_i^r x^i - \frac{1}{2}\Lambda_{ij}^r x^i x^j$ with Λ_{ij}^r defined from formula (4.11). Given the data of f, the pre-image of $1 \in \gamma_2(m)$ gives the cross section.

Definition 4.8. For M a smooth manifold with metric tensor $A^2(x)$, the *fundamental section* of the principal $\gamma_2(m)$ bundle

$$\begin{array}{c} F_2(M) \\ \downarrow \pi_{2,1} \\ F_1(M) \end{array}$$

associates to each 1-frame the unique null frame in its fiber. This unambiguously defines (and is defined by) $\gamma_2(m)$-equivariant map $f: F_2(M) \to \gamma_2(m)$ given by the prescription

$$f(\theta_2) = \delta_i^r x^i - \tfrac{1}{2}\Lambda_{ij}^r x^i x^j$$

with

$$DA^2\langle[\theta_2, X^i \text{---}0\text{---} Y^j], [\phi_1, Z^k]\rangle + 2a_{rk}\Lambda_{ij}^r X^i Y^j Z^k = 0$$

for all X^i, Y^j, Z^k.

The fact that f is equivariant follows a simple calculation. Smoothness of f is given, essentially, in formula (4.11). ■

A 2-sector $\bar{A}_2(x_0)$ determines in the presence of a metric tensor the tangent vector at x_0 which we have called its 'force'.

Definition 4.9. For M a smooth manifold with metric tensor $A^2(x)$, say that 2-sector $\bar{A}_2(x_0)$ is *flat* if it has *zero force.* This means that $DA^2\langle\bar{A}_2(x_0), \bar{A}_1(x_0)\rangle = 0$ for *all* tangent vectors $\bar{A}_1(x_0)$ at x_0. Thus, in a null frame, a 2-sector with zero middle (11) component is flat. ■

Before undertaking the closer study of flat 2-sectors and their relation to *covariant derivative,* we make the following observation which provides an alternative characterization of null frames.

Observation 4.10. For M a smooth manifold with metric tensor $A^2(x)$ and for ϕ_2 a frame-jet at x_0, the following are equivalent:

1) ϕ_2 is a null frame,

2) $\left.\dfrac{\partial \tilde{a}_{ij}}{\partial x^k}\right|_0 = 0$

$(A^2(x) = \tilde{a}_{ij}(z)$ in components with respect to frame-germ $\phi:(E, 0) \rightarrow (M, x_0))$,

3) δA^2, the *promotion* of A^2, has matrix (following Theorem 4.1) at any tangent vector $\bar{A}_1(x_0) = [\phi_1, Z^k]$:

$$\begin{vmatrix} 0 & a_{it}(z_0) \\ a_{sj}(z_0) & 0 \end{vmatrix}.$$

Proof. This observation will be quite useful from time to time since it provides an extremely simple form for the promoted metric δA^2 on $T_1(M)$. It is obvious that 2)$\leftrightarrow$3). From formula (4.7), it follows that 2)$\rightarrow$1). Now if ϕ_2 is a null frame, then

$$\left[\left.\frac{\partial \tilde{a}_{ij}}{\partial x^k}\right|_0 - \left.\frac{\partial \tilde{a}_{ik}}{\partial x^j}\right|_0 - \left.\frac{\partial \tilde{a}_{kj}}{\partial x^i}\right|_0\right] X^i Y^j Z^k = 0$$

since

$$DA^2\langle [\phi_2, X^i \text{---} 0 \text{---} Y^j], [\phi_1, Z^k] \rangle = 0.$$

Also, since

$$DA^2\langle [\phi_2, X^i \text{---} 0 \text{---} Z^j], [\phi_1, Y^k] \rangle = 0$$

we have

$$\left[\left.\frac{\partial \tilde{a}_{ik}}{\partial x^j}\right|_0 - \left.\frac{\partial \tilde{a}_{jk}}{\partial x^i}\right|_0 - \left.\frac{\partial \tilde{a}_{ij}}{\partial x^k}\right|_0\right] X^i Y^j Z^k = 0$$

for all X^i, Y^j, Z^k. Adding, we have

$$-2 \left.\frac{\partial \tilde{a}_{jk}}{\partial x^i}\right|_0 X^i Y^j Z^k = 0$$

for all X^i, Y^j, Z^k. From this, the implication 1)$\rightarrow$2) follows easily. ■

Now in order to develop the next element of structure we recall the projection

$$\begin{array}{c} T_2(M) \\ \downarrow D(2) \\ T_1^2(M) \end{array}$$

This is the projection of an *affine bundle over the vector bundle*

$$\begin{array}{c} T_1^3(M) \\ \downarrow q_1^* \\ T_1^2(M) \end{array}$$

—see Theorem 2.7 for notation. (Here we denote $B_2(M)$ by $T_1^2(M)$, the Whitney sum of the tangent bundle with itself.)

Earlier, we observed that any smooth cross section of $D(2)$ would establish a 'choice of origin' in each fiber, transforming the affine bundle into a vector bundle, one (non-naturally) isomorphic with

$$\begin{array}{c} T_1^3(M) \\ \downarrow q_1^* \\ T_1^2(M) \end{array}$$

The data of a metric tensor provide such a section; we shall modify classical notation and call this section the *affine connection* associated to $A^2(x)$. In a sense, the *fundamental section* for $\pi_{2,1}$ is the primary construction from which the affine connection can be immediately derived. It will be easier, however, to distinguish the fundamental section from the affine connection (as here interpreted) from the point of view of understanding its *transformation laws.*

Theorem 4.11. *Let M be a smooth manifold with metric tensor $A^2(x)$. And let*

$$\begin{array}{c} T_2(M) \\ \downarrow D(2) \\ T_1^2(M) \end{array}$$

be the 2-boundary transformation. So in components,

$$D(2)[\phi_2, X^i \text{——} Z^r \text{——} Y^j] = [\phi_1, (X^i, Y^j)]$$

in the notation of Theorem 2.7.

There is, in each $D(2)$ fiber a unique flat 2-*sector. The association that takes each element of $T_1^2(M)$ to the flat 2-sector in its fiber is a smooth cross section for $D(2)$, denoted Λ, and will be called the* affine connection *associated to the metric $A^2(x)$.*

[This takes a certain liberty with the conventional usage of the term but gives an intuitive interpretation to the 'transformation laws' for classical connections and provides the explanation for the emergence of the *tensor*, the *covariant derivative* from the standard algebraic manipulations.]

Proof. First we show that there is, in each $D(2)$ fiber, a *unique* flat 2-sector. Recall that $D(2)$ is the projection of an *affine bundle* over

$$\begin{array}{c} T_1^3(M) \\ \downarrow q_1^* \\ T_1^2(M) \end{array}$$

with $q_1^*[\phi_1, (W^r, X^i, Y^j)] = [\phi_1, (X^i, Y^j)]$. The action of the q_1^* fiber on the $D(2)$ fiber is given in Theorem 2.7. We need only recall that each $D(2)$ fiber at say $x_0 \in M$ is an affine space over the tangent space at x_0. In Observation 4.5 we mentioned that the *bracket* of two 2-sectors in a given $D(2)$ fiber is the difference of the *forces* of those sectors (formula (4.9)). But the bracket of two sectors is the unique translation that carries one to the other. Thus, given a 2-sector with non-zero force, then formula (4.9) can be solved *uniquely* to give a 2-sector with zero force. Specifically, suppose that $\bar{B}_2(x_0)$ is given arbitrarily in some $D(2)$ fiber. There is then a *unique* $\bar{A}_2(x_0)$ in that $D(2)$ fiber such that

$$\text{force of } \bar{B}_2(x_0) = [\bar{A}_2(x_0); \bar{B}_2(x_0)]$$

from the affine structure of the $D(2)$ fiber. Then from formula (4.9), the force of $\bar{A}_2(x_0)$ will be 0.

An easier way to arrive at the same conclusion is to work in a null frame. For a given element (say) $[\phi_1, (X^i, Y^j)]$ of $T_1^2(M)$, for the null frame ϕ_2 over ϕ_1 the only flat 2-sector in the fiber is $[\phi_2, X^i \text{——} 0 \text{——} Y^j]$ since any other middle (11) component clearly yields a 2-sector which is not flat. Now it is fairly easy to see in general that for *arbitrary* 2-frame θ_2 over ϕ_1 the flat 2-sector in this $D(2)$ fiber will be $[\theta_2, X^i \text{——} \Lambda_{ij}^r X^i Y^j \text{——} Y^j]$ with

$$\Lambda_{ij}^r = \tfrac{1}{2}\tilde{a}^{rk}\left[\frac{\partial \tilde{a}_{ij}}{\partial x^k} - \frac{\partial \tilde{a}_{ik}}{\partial x^j} - \frac{\partial \tilde{a}_{kj}}{\partial x^i}\right]$$

(formula (4.11), the partial derivatives calculated in frame-germ θ). Smoothness of the fundamental section then easily guarantees that the affine connection is smooth. ■

The next topic that we take up is the *transformation law*. Classically, the affine connection is the association to *each* frame $\theta: O \to U \subset M$ of the function

$$-\Lambda_{ij}^r(z) = \tfrac{1}{2}\tilde{a}^{rk}(z)\left[\frac{\partial \tilde{a}_{ik}}{\partial x^j}\bigg|_z + \frac{\partial \tilde{a}_{kj}}{\partial x^i}\bigg|_z - \frac{\partial \tilde{a}_{ij}}{\partial x^k}\bigg|_z\right] = \Gamma_{ij}^r(z).$$

The 'connection coefficients', despite their appearance, are *not* tensors. It was discovered, however, that the law by which they 'transform' under changes of frame could be used as the basis for the definition of 'abstract' connections (variable $m \times m \times m$ matrices which 'transform' according to the same law).

These transformation laws are less mysterious from the point of view taken here than they have been classically. From our viewpoint, they amount to giving the expression in various 2-frames at x_0 of the components of the flat 2-sector in a fixed $D(2)$ fiber. This reduces then to applying the ordinary transformation laws for 2-sectors. We make this precise in the following observation. (See also Dodson and Poston [8].)

Observation 4.12. Let M be a smooth manifold with metric tensor $A^2(x)$. Let Λ be the affine connection associated with the metric. So $\Lambda: T_1^2(M) \to T_2(M)$ is the cross section which takes each *pair of tangent vectors at a point* to the flat 2-sector in their $D(2)$ fiber. Let $x_0 \in M$ be fixed, and let $[\phi_1, (X^i, Y^i)] \in T_1^2(x_0)$.

Suppose that ψ_2 and θ_2 are *arbitrary* 2-frames at x_0 with $\psi_2 g_2 = \theta_2$ for $g_2 \in G_2(m)$. Let $\bar{A}_2(x_0) = \Lambda[\phi_1, (X^i, Y^i)]$. Then if

$$\bar{A}_2(x_0) = [\theta_2, X^i \text{——} \Lambda^r_{ij} X^i Y^j \text{——} Y^j]$$

(with Λ^r_{ij} computed in frame-germ θ using formula (4.11)) *then*

$$\bar{A}_2(x_0) = \left[\psi_2, \frac{\partial g^i}{\partial x^\alpha} X^\alpha \text{——} \frac{\partial g^r}{\partial x^\gamma} \Lambda^\gamma_{\alpha\beta} X^\alpha Y^\beta + \frac{\partial^2 g^r}{\partial x^\alpha \partial x^\beta} X^\alpha Y^\beta \text{——} \frac{\partial g^j}{\partial x^\beta} Y^\beta\right].$$

Proof. This is simply the transformation law for 2-sectors. Flatness is obviously not a property of frame or components, but a property of the sector itself. ■

Notice the simple form that this transformation law takes when for θ_2 in the above observation we take a null frame. Also, it is worth remarking that the transformation of middle (11) components under frame change g_2 is (as was observed in Chapter 2) an *affine transformation* $E \to E$, a *translation* if $g_2 \in \gamma_2(m)$.

Now the affine connection associated with the metric tensor $A^2(x)$ which we call Λ is essentially an intermediate construction: its *fundamental consequence* is the trivialization of the 2-boundary bundle

$$\begin{array}{c} T_2(M) \\ \downarrow D(2) \\ T_1^2(M) \end{array}$$

Recall that this bundle is an affine bundle over $T_1^3(M) \xrightarrow{q_1^*} T_1^2(M)$, a *vector* bundle with fiber E.

The cross section Λ then gives a *choice of origin* in each fiber for $D(2)$ by means of which a bundle isomorphism can be established between q_1^*

and $D(2)$. Actually, the construction has already been made via the notion of the 'force' of a sector; the digression to affine connections was made for the sake of completeness, to explain the relation between the very powerful and successful classical construction and the present view. Of course, another advantage in repeating to the extent that we have the historical development of the subject is that many of the calculations which we shall come to make (for example, the determination of $\Lambda^r_{ij}(z)$ in a frame) simply reproduce the classical results.

Recall the action of the q_1^* fiber on the $D(2)$ fiber at a fixed pair $[\phi_1, (X^i, Y^j)]$ at $x_0 \in M$:

$$[\phi_1, (W^r; X^i, Y^j)] \cdot [\phi_2, X^i \text{---} Z^r \text{---} Y^j] = [\phi_2, X^i \text{---} W^r + Z^r \text{---} Y^j]$$

(formula (2.3)). We take as origin in the $D(2)$ fiber the flat 2-sector $\Lambda[\phi_1, (Y^j, X^i)]$, which we may denote $[\phi_2, X^i \text{---} \Lambda^r_{ij} X^i Y^j \text{---} Y^j]$. Then the bundle isomorphism associates to $[\phi_1, (W^r; X^i, Y^j)]$ in $(q_1^*)^{-1}[\phi_1, (Y^j, X^i)]$ the 2-sector

$$[\phi_2, X^i \text{---} W^r + \Lambda^r_{ij} X^i Y^j \text{---} Y^j] \tag{4.12}$$

(we assume here $\pi_{2,1}\phi_2 = \phi_1$). This gives $D(2)^{-1}[\phi_1, (X^i, Y^j)]$ its vector space structure. In particular, if ϕ_2 is a *null frame*, we have the simple association:

$$[\phi_1, (W^r; X^i, Y^j)] \to [\phi_2, X^i \text{---} W^r \text{---} Y^j]. \tag{4.13}$$

We now make simpler interpretation of the bundle isomorphism which derives directly from the notion of 'force'.

Theorem 4.13. *Suppose that M is a smooth manifold with metric tensor $A^2(x)$. Then the pair (F, id) is a fiber bundle isomorphism:*

$$\begin{array}{ccc} T_2(M) & \xrightarrow{F} & T_1^3(M) \\ {\scriptstyle D(2)}\downarrow & & \downarrow{\scriptstyle q_1^*} \\ T_1^2(M) & \xrightarrow{\mathrm{id}} & T_1^2(M) \end{array}$$

It provides the bundle with projection $D(2)$ with vector bundle structure with fiber E, and it corresponds the cross section of $D(2)$ which we call the affine connection *associated to $A^2(x)$: Λ to the zero section of the bundle with projection q_1^*. Finally, the restriction of F to each fiber is a translation with respect to the affine structures on the fibers of $D(2)$ and q_1^*.*

The map F is defined in the following way:

$$F: [\phi_2, X^i \text{---} Z^r \text{---} Y^j] \to [\phi_1, (W^r; X^i, Y^j)]$$

with $[\phi_1, W^r]$: (force *of* $[\phi_2, X^i \text{---} Z^r \text{---} Y^j]$).

Proof. This is clearly enough to give (F, id) a bundle mapping over the identity. Also, from the definition of the affine connection, it carries the section Λ to the 0-section, since the force of a flat sector is zero.

We show that the restriction of F to a fiber is a translation to finish the proof. From formula (2.3) we have

$$[\phi_1, (W^r; X^i, Y^j)] \cdot [\phi_2, X^i \text{---} Z^r \text{---} Y^j] = [\phi_2, X^i \text{---} W^r + Z^r \text{---} Y^j].$$

Now it follows from formula (4.7) that

$$DA^2\langle[\phi_2, X^i \text{---} W^r + Z^r \text{---} Y^j], [\phi_1, U^k]\rangle = DA^2\langle[\phi_2, X^i \text{---} Z^r \text{---} Y^j], [\phi_1, U^k]\rangle + 2A^2\langle[\phi_1, W^r], [\phi_1, U^k]\rangle.$$

From this, the conclusion follows. ■

Corollary 4.14. *In the notation of Theorem* 4.13, *if*

$$F[\phi_2, X^i \text{---} Z^r \text{---} Y^j] = F(\bar{B}_2(x_0)) : [\phi_1, (W^r; X^i, Y^j)],$$

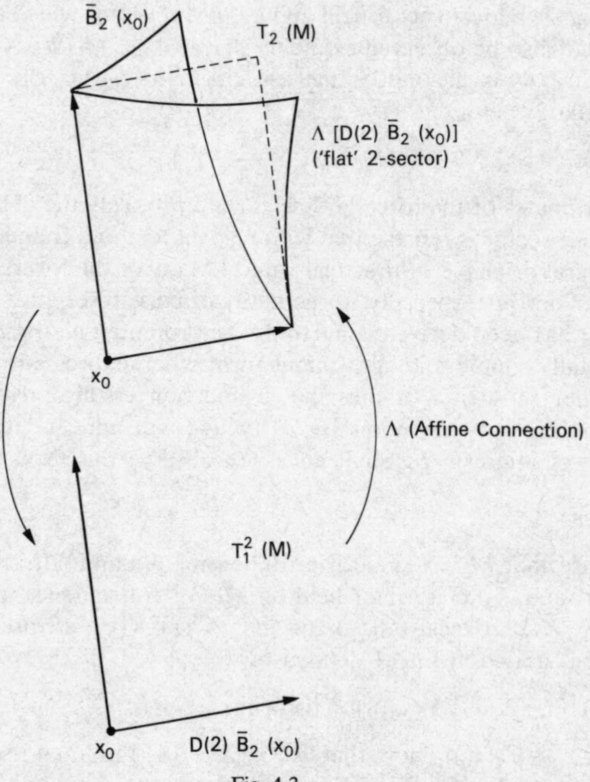

Fig. 4.3

and if

$$[\phi_2, X^i \text{——} \Lambda^r_{ij} X^i Y^j \text{——} Y^j] = \bar{A}_2(x_0)$$

is the flat 2-sector in the $D(2)$ *fiber, then*

$$[\phi_1, W^r] = [\bar{A}_2(x_0); \bar{B}_2(x_0)]$$

(bracket of 2-sectors).

Proof. Formula (4.9). ■

In a later section, we shall introduce a notation for the force of a 2-sector which recalls the classical notion of 'covariant differentiation'. We shall in fact be interested in the special case of a local vector field on M, call it $\tilde{Y}$, and a tangent vector at a point, say X. Then $X \cup \tilde{Y}$ has the interpretation that it is the 2-sector generated by 'pushing' X along the flow for $\tilde{Y}$. We shall denote the force of $X \cup \tilde{Y}$ by the symbol $\nabla(X \cup \tilde{Y})$ and this will have the same meaning as the standard symbol $\nabla_X \tilde{Y}$. Of course, it makes sense to define, for an arbitrary 2-sector $\bar{A}_2(x_0)$, its force and to call that force $\nabla \bar{A}_2(x_0)$; for arbitrary $\bar{A}_2(x_0)$ it requires little work to materialize a local vector field and a tangent vector which produce it.

It should also be observed explicitly that if $\bar{A}_2(x_0) = [\phi_2, X^i \text{——} A^r \text{——} Y^j]$ then, if θ_2 is the null frame associated to ϕ_2 via the *fundamental section*, the 2-sector

$$\bar{A}_2(x_0) = [\theta_2, X^i \text{——} A^r - \Lambda^r_{ij} X^i Y^j \text{——} Y^j].$$

The coordinates of the force of 2-sector are precisely the (11) component when the 2-sector is represented with respect to a null frame. [This points up a general principle of practical value.] Many of the formulas which we shall later derive (especially in the study of curvature) and most of what will be or has been derived concerning the promoted metric δA^2 will take an especially simple and convenient form when expressed in terms of a null frame. In such a frame, the 'connection coefficients' Λ^k_{ij} and the Riemann–Christoffel symbols $[ij; k]$ (which will appear later) vanish at the point of interest. We shall not often observe this explicitly.

Exercises

1. Suppose that M is a smooth m-dimensional manifold, $A^2(x)$ a metric tensor, and $\bar{A}_1(x)$ a vector field on M. For convenience, denote $\bar{A}_1(x)$ simply $\tilde{X}(x)$. Recall the *reduction* $A^2(x)[\tilde{X}(x)]$ (Proposition 3.22) which is a 1-form on M defined by

$$A^2(x_0)[\tilde{X}(x)]\langle Y(x_0)\rangle = A^2(x_0)\langle Y(x_0) \cup \tilde{X}(x)\rangle$$

for $Y(x_0) \in T_1(x_0)$. Show that $A^2(x)[\tilde{X}(x)]$ is the 'dual form' for $\tilde{X}(x)$ and that its definition is independent of frame.

2. With notation as above, use the contraction formula and Proposition 3.22 to show that the following relations hold between differential and reduction operators:

   a) $d_i(A^2(x)[\tilde{X}(x)]) = (d_i A^2(x))[\tilde{X}(x)]$ for $1 \leqslant i \leqslant 2$,

   b) $d(A^2(x)[\tilde{X}(x)]) = (d_1 A^2(x) - d_2 A^2(x))[\tilde{X}(x)]$,

   (that is, the first two differential operators commute with the reduction operator).

3. Let $A^r(x)$ be a sectorform field on M. Show that in general, if we define $s_k \in S(k)$ to be the permutation

$$\begin{array}{cccccc} 1 & 2 & 3 & \dots & (k-1) & k \\ \downarrow & \downarrow & \downarrow & & \downarrow & \downarrow \\ 2 & 3 & 4 & & 1 & k \end{array}$$

   (s_k *fixes* k but 'shifts' the first $k-1$ entries), then we have the following relation connecting differential and reduction operators:

$$d_j(A^r(x)[\tilde{X}(x)]) \underset{\text{def}}{\equiv} \sigma^j \circ d_r(A^r(x)[\tilde{X}(x)]) = (s_{r+1}^{j-1} \circ \sigma \circ d_{r+1} A^r(x))[\tilde{X}(x)]. \qquad (4.14)$$

4. Returning to the case $A^2(x)$ a metric tensor on M, and $\tilde{Z}(x)$ a vector field on M, define map $T_1^2(M) \to \mathbb{R}$ by $(X(x_0), Y(x_0)) \to \frac{1}{2} DA^2(x_0)\langle X(x_0) \cup \tilde{Z}(x), Y(x_0)\rangle$ and call the image simply $W_{\tilde{Z}}(X(x_0), Y(x_0))$.

   Show that $W_{\tilde{Z}}$ defines a covariant tensor field on M of type $\binom{0}{2}$, in general neither symmetric nor antisymmetric.

5. Let B be the subcomplex of $\Gamma(2)$

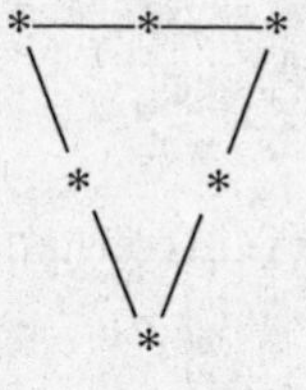

   (standard ordering) and for $A^2(x)$ a metric on M define the following B-sectorform fields (order 3):

   a) $d_3 A^2(x)$,

   b) $\frac{1}{2}\sigma^1 DA^2(x)$,

   c) $\frac{1}{2}\sigma^2 DA^2(x)$,

   d) $d_1 A^2(x) - d_2 A^2(x)$.

   Show that a) and d) reduce to sectorform fields on *star-complex* at the last vertex. Pointing these fields at the last vertex yields for a) the promoted metric δA^2. Show that pointing d) at the last vertex yields a

2-*form* on $T_1(M)$. Finally, show that

$$d_1A^2 - d_2A^2 = \tfrac{1}{2}(\sigma^2 DA^2 - \sigma^1 DA^2).$$

6. Under the conditions of exercise 5 show that if $\tilde{Z}(x)$ is a vector field on M (smooth section $M \to T_1(M)$), then the *reduction* $(d_1A^2 - d_2A^2)[\tilde{Z}]$ is the *pull-back* under $\tilde{Z}(x)$ of the pointed B-sectorform $d_1A^2 - d_2A^2$, and this is a 2-form on M. Show also that this 2-form is $d(A^2(x)[\tilde{Z}(x)])$.

7. Given an ordered pair of vector fields on M $(\tilde{X}(x), \tilde{Y}(x))$ and tangent vector $Z(x_0)$, define the following B-sector at x_0 (B as in exercise 5): 3-side, $\tilde{X}(x) \cup Y(x_0)$; 1-side, $\tilde{Y}(x) \cup Z(x_0)$; 2-side, $Z(x_0) \cup \tilde{X}(x)$. Indicate it pictorially as

$$\begin{array}{ccc} \tilde{X} & \longrightarrow & \tilde{Y} \\ & \searrow \quad \swarrow & \\ & Z(x_0) & \end{array}$$

letting arrows indicate *suspension* by the tangent vector at the *head of the arrow.* Suppose now that $(\tilde{X}(x), \tilde{Y}(x))$ is a fixed pair of vector fields. Show that the correspondence $Z(x_0) \to W_{\tilde{X}}(Z(x_0), \tilde{Y})$ gives a 1-form on M, the contraction of $\frac{1}{2}\sigma^2 DA^2$ with the B-sector above. The 1-form $W_{\tilde{X}}(\ \ , \tilde{Y})$ we shall call $W_{(\tilde{X}, \tilde{Y})}(\ \ )$ to indicate its dependence on the pair of vector fields $\tilde{X}(x)$, $\tilde{Y}(x)$.

Show that

$$W_{(\tilde{X}, \tilde{Y})}\langle Z(x_0)\rangle + W_{(\tilde{Y}, \tilde{X})}\langle Z(x_0)\rangle = d_3A^2(x_0)\left\langle \begin{array}{ccc} \tilde{X} & \longrightarrow & \tilde{Y} \\ & \searrow \quad \swarrow & \\ & Z(x_0) & \end{array} \right\rangle$$

and in particular is equal to *zero* if $A^2\langle X(x), Y(x)\rangle$ is constant.

8. For fixed $\tilde{X}$, show that

$$W_{\tilde{X}}\langle Y(x_0), Z(x_0)\rangle - W_{\tilde{X}}\langle Z(x_0), Y(x_0)\rangle = d(A^2(x)[\tilde{X}(x)])\langle Y(x_0), Z(x_0)\rangle. \quad (4.15)$$

The form $W_{(\tilde{X}, \tilde{Y})}$ and the tensor $W_{\tilde{X}}$ are important constructions for Cartan's *method of moving frames.* We return to them later. Next we investigate some consequences of the fact that the 'force' and 'affine connection' are not natural for smooth immersions. Let $(M, A^2(x))$ and $(N, B^2(y))$ be Riemannian manifolds. If $A^2 = j * B^2$ for $j: M \to N$, say j is an *isometric immersion.* Denote respective connections Λ_M and Λ_N.

Now j gives rise to a bundle mapping $(T_2(j), T_1^2(j))$

$$\begin{array}{ccc} T_2(M) & \xrightarrow{T_2(j)} & T_2(N) \\ {\scriptstyle D(2)}\downarrow \uparrow{\scriptstyle \Lambda_M} & & {\scriptstyle \Lambda_N}\uparrow \downarrow{\scriptstyle D(2)} \\ T_1^2(M) & \xrightarrow{T_1^2(j)} & T_1^2(N) \end{array} \tag{4.16}$$

We are interested in the behavior of the cross sections Λ_M and Λ_N in this diagram.

Perhaps it is surprising that $\Lambda_N \circ T_1^2(j)$ is *not* equal to $T_2(j) \circ \Lambda_M$ in general. For intuition, it is useful to think of j as an immersion of, say, a 2-dimensional manifold in $\mathbb{R}^3$ (with the usual metric). A 2-sector which is flat in M need not be flat with respect to the trivial metric on $\mathbb{R}^3$.

Theorem 4.15. *Given* $(M, A^2(x))$ *and* $(N, B^2(y))$ *as above, and* $j: M \to N$ *is an isometric immersion. Consider the pull-back diagram:*

$$\begin{array}{ccc} T_1^2(M) \times_{T_1^2(N)} T_1^3(N) & \longrightarrow & T_1^3(N) \\ \downarrow {\scriptstyle q_1^{**}} & & \downarrow {\scriptstyle q_1^*} \\ T_1^2(M) & \xrightarrow[T_1^2(j)]{} & T_1^2(N) \end{array} \tag{4.17}$$

Here, q_1^{**} is the projection of a vector bundle. Its fiber at a pair in $T_1^2(x_0)$ is the vector space $T_1[j(x_0)]$.

Now the association to each $[\phi_1, (X^i, Y^j)] \in T_1^2(M)$ of the bracket

$$[\Lambda_N \circ T_1^2(j)[\phi_1, (X^i, Y^j)]; T_2(j) \circ \Lambda_M[\phi_1, (X^i, Y^j)]] \in T_1(N)$$

gives a smooth cross section of the vector bundle with projection q_1^{**}. For simplicity we denote that bracket $\Omega(j)[\phi_1, (X^i, Y^j)]$ or $\Omega(j)(X, Y)$. This cross section is symmetric *and* bilinear *in (X, Y).*

Proof. This cross section associates to each pair of tangent vectors at $x_0 \in M$ the *force* of $T_2(j)\Lambda_M[\phi_1, (X^i, Y^j)]$. This force is computed with respect to $B^2(y)$ in N. All that really needs be shown here is that $\Omega(j)$ is symmetric and bilinear in (X, Y) at each $x_0 \in M$, since that force is simply the translation in the fiber of $T_2(N)$ carrying the flat 2-sector to the one in question. By smoothness of the action of the fibers of q_1^* on the fibers of $D(2)$ and by smoothness of the sections Λ_M and Λ_N and the bundle maps induced by j, the smoothness of $\Omega(j)$ follows.

For x_0 fixed in M, let ϕ_2 be a *null frame* at x_0 (with respect to $A^2(x)$) and let θ_2 be a *null frame* at $j(x_0) \in N$ (with respect to $B^2(y)$). Suppose that (X, Y) is equal in components to $[\phi_1, (X^i, Y^j)]$. Then $\Lambda_M[\phi_1, (X^i, Y^j)]$ is the 2-sector $[\phi_2, X^i \text{---} 0 \text{---} Y^j]$ and

$$T_2(j)\Lambda_M[\phi_1, (X^i, Y^j)] = \left[\theta_2, \frac{\partial g^i}{\partial x^\alpha} X^\alpha \text{---} \frac{\partial^2 g^r}{\partial x^\alpha \partial x^\beta} X^\alpha Y^\beta \text{---} \frac{\partial g^j}{\partial x^\beta} Y^\beta\right],$$

where g is the germ $\theta^{-1} \circ j \circ \phi : (E, 0) \to (E', 0)$, $E = \mathbb{R}^m$ and $E' = \mathbb{R}^n$.

Then the *force* of $T_2(j)\Lambda_M[\phi_1, (X^i, Y^i)]$ is

$$\left[\theta_1, \frac{\partial^2 g^r}{\partial x^\alpha \partial x^\beta} X^\alpha Y^\beta\right].$$

This is easily symmetric and bilinear in (X, Y). ■

Observation 4.16. Let $j:(M, A^2(x)) \to (N, B^2(y))$ be an isometric immersion and suppose that $j(x_0) = y_0$. Then for (X, Y) a pair of tangent vectors to M at x_0, $\Omega(j)(X, Y)$ is, for *any* frame $\phi:(E, 0) \to (M, x_0)$ and any *null frame* $\theta_2:(E', 0) \to (N, y_0)$, the '*normal component*' of the *second derivative* of j computed with respect to those frames in the following sense. If g is the germ $\theta^{-1} \circ j \circ \phi:(E, 0) \to (E', 0)$ and if $(X, Y) = [\phi_1, (X^i, Y^i)]$, then $\Omega(j)(X, Y)$ is the tangent vector

$$\left[\theta_1, \frac{\partial g^r}{\partial x^\gamma} \Lambda^\gamma_{\alpha\beta} X^\alpha Y^\beta + \frac{\partial^2 g^r}{\partial x^\alpha \partial x^\beta} X^\alpha Y^\beta\right]$$

(here it is important that θ_2 be a *null frame*) where $\Lambda^\gamma_{\alpha\beta}$ are computed for $A^2(x)$ with respect to ϕ_2, and the second derivative is computed with respect to ϕ_2 and θ_2. Then if the θ-components of $B^2(y_0)$ are b_{rk}, for any tangent vector at x_0 $[\phi_1, Z^\gamma]$ we have:

$$b_{rk} \frac{\delta g^k}{\delta x^\gamma} Z^\gamma \left[\frac{\partial g^r}{\partial x^\gamma} \Lambda^\gamma_{\alpha\beta} X^\alpha Y^\beta + \frac{\partial^2 g^r}{\partial x^\alpha \partial x^\beta} X^\alpha Y^\beta\right] = 0. \tag{4.18}$$

Proof. Formula (4.18) follows directly from naturality of the cycle product DA^2. We have immediately that

$$DA^2\langle [\phi_2, X^i \text{——} \Lambda^r_{ij} X^i Y^j \text{——} Y^j], [\phi_1, Z^k] \rangle = 0$$

which then implies that:

$$0 = DB^2 \Big\langle \Big[\theta_2, \frac{\partial g^i}{\partial x^\alpha} X^\alpha \text{——} \frac{\partial g^r}{\partial x^\gamma} \Lambda^\gamma_{\alpha\beta} X^\alpha Y^\beta + \frac{\partial^2 g^r}{\partial x^\alpha \partial x^\beta} X^\alpha Y^\beta \text{——} \frac{\partial g^j}{\partial x^\beta} Y^\beta\Big], \Big[\theta_1, \frac{\partial g^k}{\partial x^\gamma} Z^\gamma\Big]\Big\rangle.$$

The fact that θ_2 is a null frame gives then formula (4.18).

Now the 'second derivative' of j computed with respect to frame ϕ_2 and null frame θ_2 is the symmetric bilinear map

$$(X^\alpha, Y^\beta) \to \frac{\partial^2 g^r}{\partial x^\alpha \partial x^\beta} X^\alpha Y^\beta$$

and from formula (4.18) it is easy to see that its component 'along' $T_1(j)[T_1(x_0)]$ is

$$-\frac{\partial g^r}{\partial x^\gamma}\Lambda^\gamma_{\alpha\beta}X^\alpha Y^\beta.$$

In particular, if ϕ_2 is a null frame, that second derivative is equal to $\Omega(j)(X, Y)$ *and* is 'normal' to $T_1(j)[T_1(x_0)] \subset T_1[j(x_0)]$. We make this precise in the following statement. ■

Observation 4.17. Under the conditions of the previous observation, for $\phi_2:(E, 0) \to (M, x_0)$ arbitrary and $\theta_2:(E', 0) \to (N, j(x_0))$ a null frame with $g = \theta^{-1} \circ j \circ \phi$, we have orthogonal decomposition in $T_1[j(x_0)]$

$$\left[\theta_1, \frac{\partial^2 g^r}{\partial x^\alpha \partial x^\beta} X^\alpha Y^\beta\right] = \Omega(j)[\phi_1, (X^i, Y^j)] - \left[\theta_1, \frac{\partial g^r}{\partial x^\gamma}\Lambda^\gamma_{\alpha\beta}X^\alpha Y^\beta\right] \quad (4.19)$$

expressing the 'second derivative' as the sum of a tangent vector *normal* to $T_1(j)[T_1(x_0)]$ and one *in* $T_1(j)[T_1(x_0)]$. ■

In particular, if j is a local diffeomorphism as well as an isometric immersion, then $\Omega(j)$ vanishes identically. The section $\Omega(j)$ bears a close and simple relationship with the so-called *second fundamental form* in the case that $N = \mathbb{R}^n$, equipped with the usual Euclidean metric (see Abraham and Marsden [1], p. 156). We shall not pursue that relationship here, but leave it to the interested reader to derive it. The second fundamental form is not really a form at all; rather it is a symmetric covariant tensor field of type $\binom{0}{2}$ associated with j.

We shall be interested in the case that $N = \mathbb{R}^n = E'$ and is equipped with the Euclidean metric which will be denoted B^2. In this case, if $\theta: E' \to E'$ is the identity frame, then the frame-jets $(\theta \circ t_w)_2$ are *all* null frames. In order to give the geometric interpretation of $\Omega(j)$ in this setting, we recall the *generalized differentiation* which was introduced in Definition 3.18. Suppose that $f: M \to E'$ is a smooth map and let $\bar{A}_2(x_0)$ be a 2-sector at $x_0 \in M$. Then we formulate the following definition.

Definition 4.18. For smooth $f: M \to E'$, and $\bar{A}_2(x_0)$ a 2-sector at $x_0 \in M$, we define the *derivative* of f *with respect to* $\bar{A}_2(x_0)$ to be the following *tangent vector* at $y_0 = f(x_0) \in N$. Let $\theta: E' \to E'$ be the global identity frame on E' and $(y^1, y^2, \ldots, y^n)$ the linear coordinates associated with E'. Let $T_2(f)\bar{A}_2(x_0) = [(\theta \circ t_{y_0})_2, Y^i_\gamma]$ for $\gamma \in \Gamma(1)$. Let e be the 1-face (11) of $\Gamma(1)$.

Then the derivative of f with respect to $\bar{A}_2(x_0)$ is the tangent vector $[(\theta \circ t_{y_0})_2, Y^i_e]$. The components that we denote Y^i_e are just the 'derivatives' given in Definition 3.18, and the correspondence, for fixed f, $\bar{A}_2(x_0) \to Y^i_e$ for $1 \leqslant i \leqslant n$ gives a 2 sectorform at x_0. In particular, this differentiation is *linear* on $T_{2,1}(\bar{A}_2(x_0))$ *and* on $T_{2,2}(\bar{A}_2(x_0))$, for all $\bar{A}_2(x_0)$. In analogy with the notation of Observation 3.19, denote this $d_{2,1}f(x_0)\langle\bar{A}_2(x_0)\rangle=$ derivative of f with respect to $\bar{A}_2(x_0)$. ■

Thus for a given *isometric immersion* $j:(M, A^2(x))\to(E', B^2)$, there is defined an n-tuple of 2 sectorform fields on M, or simply a *vector-valued* 2 sectorform field denoted $d_{2,1}(j)$ whose value on $\bar{A}_2(x_0)$ is the contraction $d_{2,1}j(x_0)\langle\bar{A}_2(x_0)\rangle$ which is the (11) component of $T_2(j)\bar{A}_2(x_0)$. We consider it to take its value in the tangent space $T_1(j(x_0))$. Thus, abusing the notation slightly, we denote that tangent vector at $f(x_0)$ simply $d_{2,1}j(x_0)\langle\bar{A}_2(x_0)\rangle$. We may now give the geometric interpretation for $\Omega(j)$.

Theorem 4.19. *Let $j:(M, A^2(x))\to(E', B^2)$ be an isometric immersion and let $d_{2,1}(j)$ be the associated* vector-valued 2 *sectorform field on M. Then for $x_0\in M$, let $[\phi_1, (X^i, Y^i)]=(X, Y)$ be a pair of tangent vectors at x_0. Let $\bar{A}_2(x_0)$ belong to the $D(2)$ fiber over (X, Y): $\bar{A}_2(x_0)$ has 'boundary' (X, Y). Then $\Omega(j)(X, Y)$ is the component of $d_{2,1}j(x_0)\langle\bar{A}_2(x_0)\rangle$* normal *to $T_1(j)[T_1(x_0)]\subset T_1(j(x_0))$. This normal component is the same for* all $\bar{A}_2(x_0)$ *in the $D(2)$ fiber.*

The unique flat 2-*sector in that $D(2)$ fiber is* characterized *by the fact that (if it is denoted $\bar{C}_2(x_0)$), then $d_{2,1}j(x_0)\langle\bar{C}_2(x_0)\rangle$ is normal to $T_1(j)$ $[T_1(x_0)]$; that is, it is* equal *to* $\Omega(j)(X, Y)$.

Proof. Keeping formula (4.19) in mind, all that needs to be shown is that if $d_{2,1}j(x_0)\langle\bar{A}_2(x_0)\rangle$ is normal to $T_1(j)[T_1(x_0)]$, then $\bar{A}_2(x_0)$ is flat. In components, let $\bar{A}_2(x_0)=[\phi_2, X^i$——$Z^r$——$Y^i]$. Then the derivative of j with respect to $\bar{A}_2(x_0)$ is

$$\frac{\partial^2 g^k}{\partial x^\alpha \partial x^\beta}X^\alpha Y^\beta+\frac{\partial g^k}{\partial x^\gamma}Z^\gamma,$$

in standard components, $g=(\theta\circ t_{y_0})^{-1}\circ j\circ\phi$. Substituting from formula (4.19), we have this equal to the vector with $(\theta\circ t_{y_0})$ components:

$$\Omega(j)(X, Y)+\left[(\theta\circ t_{y_0})_1, \frac{\partial g^k}{\partial x^\gamma}[Z^\gamma-\Lambda^\gamma_{\alpha\beta}X^\alpha Y^\beta]\right].$$

Injectivity of the derivative of g then gives the result. ■

It should be observed that the last calculation gives the following geometric interpretation of the 'covariant derivative' or the *force* of a 2-sector at $x_0\in M$. It is the tangent vector at x_0 which transforms under $T_1(j)$ to the component of the derivative of j with respect to the 2-sector *along the tangent space of the image manifold.* What is remarkable is that, while the force of the sector is determined by $A^2(x)$, this interpretation

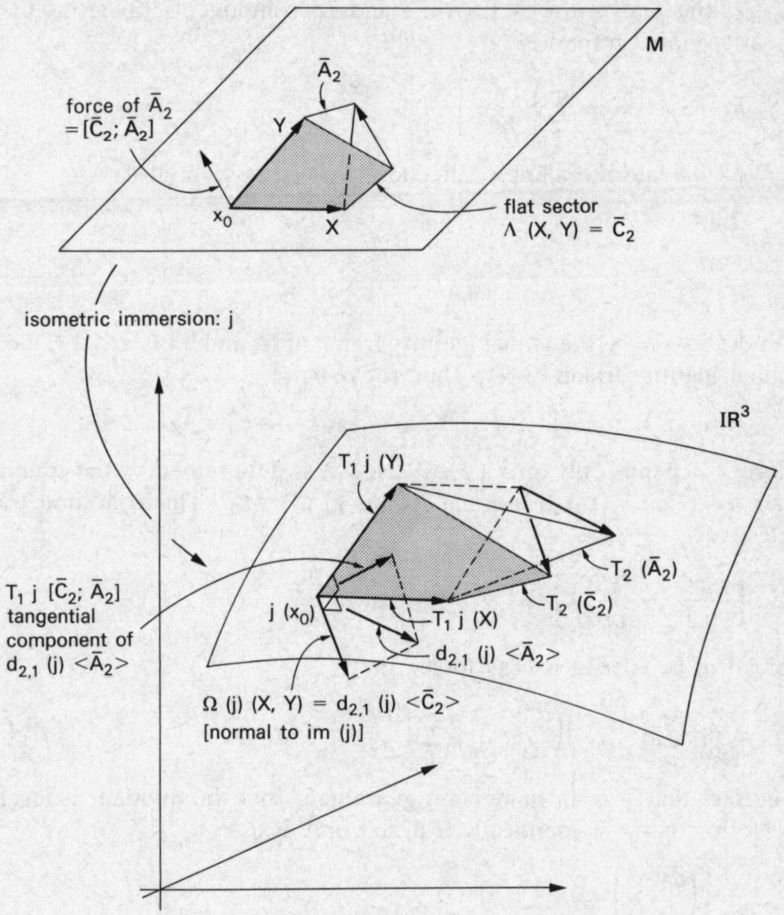

Fig. 4.4

applies for *all* isometric immersions j. In particular, it is possible to determine the affine connection *directly* (without the intermediary of DA^2 or the Riemann–Christoffel symbols) by finding for each pair (X, Y) the 2-sector in its fiber with derivative of j normal to the tangent space of the image manifold.

The last topic in this section will be the calculation of the *Gaussian curvature* of an immersion of a two-dimensional manifold in $\mathbb{R}^3$ *and* of the *curvature* of an immersion of a one-dimensional manifold in $\mathbb{R}^2$. These will, of course, be interpreted in terms of $\Omega(j)$.

We begin by considering a smooth immersion $\gamma : \mathbb{R} \to E' = \mathbb{R}^n$. Suppose that the matrix for B^2 in standard components is at each point b_{ij}, $b_{ij} = 0$ if $i \neq j$ and $b_{ii} = 1$. Then γ is an isometric immersion if and only if for each

$x_0 \in \mathbb{R}$, the matrix for $A^2(x_0)$ in standard components (in terms of the identity global frame) is

$$b_{ij} \frac{d\gamma^i}{dx}_{|x_0} \frac{d\gamma^j}{dx}_{|x_0} = \sum_i \left[\frac{d\gamma^i}{dx}_{|x_0}\right]^2.$$

We calculate the affine connection Λ as cross section of

$$\begin{array}{l} T_2(\mathbb{R}) \\ \quad\downarrow D(2) \\ T_1^2(\mathbb{R}) \end{array}$$

Let $\phi : E \to E$ be the global identity frame of $\mathbb{R}$, and let $\theta : E' \to E'$ be the global identity frame of $\mathbb{R}^n$. Then for $a, b \in \mathbb{R}$,

$$\Lambda[(\phi \circ t_{x_0})_1, (a, b)] = [(\phi \circ t_{x_0})_2, a \text{——} \lambda ab \text{——} b] = \bar{A}_2(x_0).$$

Here λ depends only on x_0, $\lambda \in \mathbb{R}$. Now λ is determined by the condition that $d_{2,1}\gamma(x_0)\bar{A}_2(x_0)$ is 'normal' to $\text{im}(\gamma)$ at $\gamma(x_0)$. This condition translates to:

$$\left[\frac{d^2\gamma^i}{dx^2}_{|x_0} + \frac{d\gamma^i}{dx}_{|x_0} \lambda\right] ab \qquad \text{normal to } \frac{d\gamma^i}{dx}_{|x_0}$$

for all a, b. Then it is easy to see that

$$\lambda = -\sum_i \left(\frac{d\gamma^i}{dx}_{|x_0}\right)\left(\frac{d^2\gamma^i}{dx^2}_{|x_0}\right) \Big/ \sum_i \left(\frac{d\gamma^i}{dx}_{|x_0}\right)^2. \tag{4.20}$$

The fact that γ is an immersion guarantees that the quotient is legal.

Notice that λ is identically 0 if and only if

$$\frac{d}{dx}\left[\left(\frac{d\gamma^i}{dx}\right)^2\right] = 0$$

if and only if ϕ is a 'multiple' of the arc length parametrization. This shows incidentally that even if λ does not vanish identically on a local frame, the curve can be reparametrized on that frame (by arc length) in such a way that for the new frame (parametrization) λ *does* vanish identically there. If we call this new frame ϕ', then *each* $(\phi' \circ t_{z_0})_2$ is a *null frame* for the metric. This property of immersed *curves* does not generalize, as we shall see, to *surfaces*: in general, there is no two-dimensional 'arc length' parametrization in the sense that $(\phi \circ t_{z_0})_2$ is a null frame for all $z_0 \in O$, $\phi : O \to U \subset M$. The analysis of this problem plays a central role in the development of differential geometry (Spivak [22]).

Now it is easy to determine $\Omega(\gamma)[\phi \circ t_{x_0})_1, (a, b)]$. This is, from (4.20), the tangent vector:

$$\left[(\theta \circ t_{\gamma(x_0)})_1, \frac{d^2\gamma^j}{dx^2}_{|x_0} ab - \left\{\sum_i \left(\frac{d\gamma^i}{dx}_{|x_0}\right)\left(\frac{d^2\gamma^i}{dx^2}_{|x_0}\right) \Big/ \sum_i \left(\frac{d\gamma^i}{dx}_{|x_0}\right)^2\right\} \frac{d\gamma^j}{dx}_{|x_0} ab\right]. \tag{4.21}$$

This is the *normal component* of

$$\left[(\theta \circ t_{\gamma(x_0)})_1, \frac{d^2\gamma^i}{dx^2}_{|x_0} ab\right]$$

and will play an important role in the calculation of the classical 'curvature' of γ at x_0 when $E'=\mathbb{R}^2$.

Now in order to motivate the definition for Gaussian curvature for immersed surfaces in $\mathbb{R}^3$, we study first the less interesting case of the 'curvature' of an immersed curve in $\mathbb{R}^2$.

Let $\mathbb{R}^2 = E'$ be equipped with an orientation. This can be taken as the global 2-form with constant matrix

$$\begin{vmatrix} 0 & 1 \\ -1 & 0 \end{vmatrix}.$$

Its contraction with the ordered pair of tangent vectors $[(\theta \circ t_{y_0})_1, (X^i, Y^i)]$ (for θ the global *identity* frame as usual) yields $X^1Y^2 - X^2Y^1$. A pair of tangent vectors (X, Y) at y_0 is *positively oriented* if and only if the above contraction is positive, is *negatively oriented* if negative, and is *degenerate* if the contraction is 0. The above contraction will be recognized as the usual determinant of the matrix

$$\begin{vmatrix} X^1 & Y^1 \\ X^2 & Y^2 \end{vmatrix}$$

the components being given in θ coordinates. Abusing language, we shall often refer to this contraction of tangent vectors at y_0 as $\det|X, Y|$. This choice of orientation for E' amounts to the selection of this standard determinant function for each tangent space $T_1(y_0)$, the selection being made via the global trivialization of $T_1(E')$ induced by θ.

Definition 4.20. Suppose that $\gamma:\mathbb{R} \to E' = \mathbb{R}^2$ is a smooth immersion. A *unit normal* 1-*variation* of γ is a 1-variation $G:(\mathbb{R}\times J, \mathbb{R}\times t_0) \to (E', \mathrm{im}(\gamma))$ with $G_{|\mathbb{R}\times t_0} = \gamma$, G an *equivalence class of ribbon-germs* with respect to the relation $F \sim G$ iff $\bar{F}_2^{\#}(x, t_0) = \bar{G}_2^{\#}(x, t_0)$ for all $x \in \mathbb{R}$ (see Definition 2.13) with the properties:

1) $B^2\langle D^1(2)\bar{G}_2^{\#}(x, t_0), D^1(2)\bar{G}_2^{\#}(x, t_0)\rangle = 1$ for all $x \in \mathbb{R}$.
2) $B^2\langle D^1(2)\bar{G}_2^{\#}(x, t_0), D^2(2)\bar{G}_2^{\#}(x, t_0)\rangle = 0$ for all $x \in \mathbb{R}$.
3) The pair $(D^1(2)\bar{G}_2^{\#}(x, t_0), D^2(2)\bar{G}_2^{\#}(x, t_0))$ is positively oriented for all $x \in \mathbb{R}$. ■

For simplicity, we denote the tangent vectors $D^2(2)\bar{G}_2^{\#}(x, t_0)$ as $X(x)$ for $X(x) = [(\theta \circ t_{\gamma(x)})_1, X^i(x)]$, *and* $D^1(2)\bar{G}_2^{\#}(x, t_0)$ as $Y(x)$ for $Y(x) = [(\theta \circ t_{\gamma(x)})_1, Y^i(x)]$.

Now for $\gamma:\mathbb{R} \to E'$ a smooth immersion, the *Gauss map* associated to

γ is the map $\hat{\gamma}:\mathbb{R}\to S^1$ given by

$$\hat{\gamma}(x)=Y^j(x)=\begin{pmatrix}Y^1(x)\\ Y^2(x)\end{pmatrix}.$$

This map is uniquely determined for any unit normal variation G of γ.

Definition 4.21. For $\gamma:\mathbb{R}\to E'=\mathbb{R}^2$ a smooth immersion, the *curvature* of γ at x_0, written as $K_\gamma(x_0)$, is the ratio

$$\frac{\det\left|Y^j(x_0),\dfrac{dY^j}{dx}_{|x_0}\right|}{\det|Y(x_0),X(x_0)|},\tag{4.22}$$

where the determinant in the denominator is calculated on $T_1(\gamma(x_0))$ and the determinant in the numerator is the determinant of a pair of vectors in $T_1(\hat{\gamma}(x_0))\subseteq T_1(E')$. (So $Y^j(x_0)$ are interpreted as the components of a vector normal to the unit circle in E' at $\hat{\gamma}(x_0)$, and $(dY^j/dx)_{|x_0}$ the components of a tangent vector to the circle at $\hat{\gamma}(x_0)$.)

It is easy to see that $K_\gamma(x_0)$ is the *Jacobian determinant* of the *germ* $\hat{\gamma}\circ\gamma^{-1}:(\gamma(\mathbb{R}),\gamma(x_0))\to(\hat{\gamma}(\mathbb{R}),\hat{\gamma}(x_0))$ at the point $\gamma(x_0)$ with respect to the one-dimensional determinant forms inherited from E'. ■

We shall see that in spite of its awkward form, this definition gives the classical curvature of an immersion at a point. It is easy to see that $|K_\gamma(x_0)|$ is *independent of parametrization,* although the sign depends upon the direction in which the curve is traversed. We relate this classical curvature with $\Omega(\gamma)(x_0)$. It is fairly easy to see that under the conditions of the above definition:

$$K_\gamma(x_0)=\frac{\det\left|Y^j(x_0),\dfrac{dY^j}{dx}_{|x_0}\right|\cdot\det|Y(x_0),X(x_0)|}{[\det|Y(x_0),X(x_0)|]^2}.\tag{4.23}$$

For further simplicity, denote $\det|Y(x_0),X(x_0)|$ by $w(x_0)$. Then $|w(x_0)|=|X(x_0)|$. Now using the multiplicative property of determinants together with their invariance under the transpose operation, (4.23) reduces to

$$\begin{aligned}K_\gamma(x_0)&=\frac{1}{w(x_0)^2}\left[\sum_j\frac{dY^j}{dx}_{|x_0}\cdot X^j(x_0)\right]\\&=-\frac{1}{w(x_0)^2}\left[\sum_j Y^j(x_0)\frac{dX^j}{dx}_{|x_0}\right]\\K_\gamma(x_0)&=-\frac{1}{w(x_0)^2}\left[\sum_j Y^j(x_0)\frac{d^2\gamma^j}{dx^2}_{|x_0}\right].\end{aligned}\tag{4.24}$$

Formula (4.24) expresses the curvature in terms of the 'normal component' (with respect to $Y(x_0)$) of the second derivative of γ at x_0; this derivative being computed, as usual, with respect to the global identity frame for $\mathbb{R}$. As a final simplification, let $A^2(x)$ be the pull-back metric on E induced by γ. So $\gamma:(E, A^2(x)) \to (E', B^2)$ is an isometric immersion. Let u be the *positively oriented unit vector* in $T_1(x_0)$,

$$u = \left[(\phi \circ t_{x_0})_1, \frac{1}{w(x_0)}\right].$$

Then

$$A^2(x_0)\langle u, u\rangle = 1 = \det|Y(x_0), T_1(\gamma)(u)|.$$

Then we have

$$K_\gamma(x_0) = -B^2\langle Y(x_0), \Omega(\gamma)(u, u)\rangle, \tag{4.25}$$

and also

$$K_\gamma(x_0) = \det|T_1(\gamma)(u), \Omega(\gamma)(u, u)|. \tag{4.26}$$

These formulas have especially simple geometric interpretations in the case that the curve is parametrized by arc length. Then

$$\Omega(\gamma)(u, u) = \pm\left[(\theta \circ t_{y_0})_1, \left.\frac{d^2\gamma^i}{dx^2}\right|_{x_0}\right]$$

with the sign depending on the direction of traversal; $y_0 = \gamma(x_0)$. In fact, for γ a parametrization by arc length, the data of $K_\gamma: E \to \mathbb{R}$ entirely determine the *shape* of $\mathrm{im}(\gamma)$ in the sense that if γ and γ' are two parametrizations of curves by arc length such that $K_\gamma = K_{\gamma'}$ as smooth maps from $E \to \mathbb{R}$, then there is a rigid orientation-preserving motion of the plane (a translation followed by a rotation) which 'carries' γ to γ' (see Spivak [22] for a discussion). This idea is made more precise with the following definitions.

Definition 4.22. Suppose that $E = \mathbb{R}^m$ with the standard basis $(e_1, \ldots, e_m)$. B^2 denotes the (constant) Euclidean metric on E. Let $\det_E$ denote the *unique* antisymmetric m-linear map in $L(E, \ldots, E; \mathbb{R})$ with the property that its contraction with $(e_1, e_2, \ldots, e_m)$ is 1. Abusing language, we denote the global m-form on E whose contraction with $[(\phi \circ t_{x_0})_1, (X_1^i, \ldots, X_m^i)]$ is $\det_E(X_1^i e_i, \ldots, X_m^i e_i)$ for ϕ the global identity frame on E, simply $\det_E$.

For $E' = \mathbb{R}^{m+1}$ with its standard basis, let B^2 again denote the Euclidean metric (there should be no confusion) and let $\det_{E'}$ denote the global $(m+1)$ form constructed as above.

The forms $\det_E$ and $\det_{E'}$ serve as 'orientations' for E and E'. ■

Definition 4.23. Let $E' = \mathbb{R}^{m+1}$ as above, and let $E = \mathbb{R}^m$. Suppose that $j:(E, A^2(x)) \to (E', B^2)$ and $\hat{j}:(E, \hat{A}^2(x)) \to (E', B^2)$ are *isometric immersions* (so $A^2(x)$ and $\hat{A}^2(x)$ are pull-back metrics).

a) Say that j is *weakly equivalent* to $\hat{j}$ if there is an *isometry* (diffeomorphism preserving the metrics) $f:(E, A^2(x)) \to (E, \hat{A}^2(x))$.
b) Say that j is *strongly equivalent* to $\hat{j}$ if there is an isometry $f:(E, A^2(x)) \to (E, \hat{A}^2(x))$ satisfying a further condition to be formulated below. Suppose, to fix ideas, that $x_0 \in E$ and denote an element of $T_1^m(x_0)$ as $(X_1, X_2, \ldots, X_m)$ in $T_1^m(E)$. For *each* choice $1 \le r, s \le m$, let $W_{rs}(j)$ $(X_1, X_2, \ldots, X_m)$ be defined by

$$W_{rs}(j)(X_1, X_2, \ldots, X_m) = \det_{E'}[T_1(j)(X_1), \ldots, T_1(j)(X_m), \Omega(j)(X_r, X_s)].$$

It is easy to see that each $W_{rs}(j)$ is a well-defined map from $T_1^m(x_0) \to \mathbb{R}$. We may therefore define an $m \times m$ matrix with rs entry $W_{rs}(X_1, \ldots, X_m)$ for each element of $T_1^m(x_0)$. Identifying the columns of that matrix with vectors in E (with respect to the standard basis for E), define the *new* mapping

$$U(j): T_1^m(x_0) \to \mathbb{R}$$

by the rule:

$$U(j)(X_1, \ldots, X_m) = \det_E[W_{rs}(j)(X_1, \ldots, X_m)]. \tag{4.27}$$

Now the further condition that f should satisfy is this:

$$U(\hat{j}) \circ T_1^m(f)(X_1, \ldots, X_m) = U(j)(X_1, \ldots, X_m)$$

at each $x_0 \in E$, for each $(X_1, \ldots, X_m) \in T_1^m(x_0)$. ■

Obviously, strongly equivalent immersions are weakly equivalent. Further, while weak equivalence requires only the data of the induced metrics on E, strong equivalence apparently makes use of further information which is implicit in j. Some justification for the clumsy formulation of strong equivalence may be found in the following.

Observation 4.24. Suppose that $j:(E, A^2(x)) \to (E', B^2)$ is an isometric immersion as above ($E = \mathbb{R}^m$, $E' = \mathbb{R}^{m+1}$). Let $h: E \to E$ be a diffeomorphism and let $h*(A^2(x)) = \hat{A}^2(x)$ on E. Then h is a 'reparametrization' of j; the map $h:(M, \hat{A}^2(x)) \to (M, A^2(x))$ is an isometry and gives a weak equivalence between j and $\hat{j} = j \circ h$.

In fact, h is a *strong equivalence.*

Proof. We show in fact that if $(X_1, \ldots, X_m) \in T_1^m(x_0)$, then $W_{rs}(\hat{j})(X_1, \ldots, X_m)$ is *equal to* $W_{rs}(j)T_1^m(h)(X_1, \ldots, X_m)$ for each pair rs.

Now

$$\det_{E'}[T_1(\hat{j})(X_1), \ldots, T_1(\hat{j})(X_m), \Omega(\hat{j})(X_r, X_s)]$$
$$= \det_{E'}[T_1(j)T_1(h)(X_1), \ldots, T_1(j)T_1(h)(X_m), \Omega(\hat{j})(X_r, X_s)]$$

as is evident from naturality of T_1. Next we show that $\Omega(j) \circ T_1^2(h) = \Omega(\hat{j})$ in the case that h is a diffeomorphism.

For Λ and $\hat{\Lambda}$ the affine connections associated to the metrics $A^2(x)$ and $\hat{A}^2(x)$ respectively, we know that $\Omega(h)$ vanishes; that is $\Lambda \circ T_1^2(h) = T_2(h) \circ \hat{\Lambda}$. (This follows from Observation 4.16.) Now composing with $d_{2,1}(j)$ gives the result. ∎

Notice that the functions $\det_E$ and $\det_{E'}$ played only an incidental role in this proof. But if we had attempted to formulate the definition of strong equivalence in terms of the functions W_{rs} alone, we would have lost the ability to identify immersions which differ by a linear isometry on E' as 'equivalent'. In any case, strong equivalence does not distinguish reparametrizations. Now let us return to the one-dimensional case.

Theorem 4.25. *Suppose that* $j:(E, A^2(x)) \to (E', B^2)$ *and* $\hat{j}:(E, \hat{A}^2(x)) \to (E', B^2)$ *are isometric immersions (in the notation of Definition* 4.23) *and that* $E = \mathbb{R}$, *and* $E' = \mathbb{R}^2$. *Let* $f:(E, A^2(x)) \to (E, \hat{A}^2(x))$ *be a* strong equivalence. *Then say that f is* orientation-preserving *if at one (hence every) point it has positive derivative in terms of the fixed basis for E; and say that f is* orientation-reversing *if that derivative is negative. Then*

1) *if f is orientation preserving,* $K_j(x) = K_{\hat{j}}[f(x)]$ *for all* $x \in E$,
2) *if f is orientation reversing,* $K_j(x) = -K_{\hat{j}}[f(x)]$ *for all* $x \in E$.

Proof. To say that f is a strong equivalence is to say that it is an isometry and that

$$\det_{E'}[T_1(j)(X), \Omega(j)(X, X)]$$
$$= \det_{E'}[T_1(\hat{j}) \circ T_1(f)(X), \Omega(\hat{j})(T_1(f)(X), T_1(f)(X))]$$

for all $X \in T_1(x)$, $x \in E$.

Now choosing X to be the *positively oriented unit vector* u at x and letting v be the positively oriented unit vector at $f(x) \in E$ (with respect to $\hat{j}$), then in case 1) we have $T_1(f)(u) = v$, and in case 2) we have $T_1(f)(u) = -v$.

Returning to the expression for curvature in formula (4.26) then gives the conclusion of the theorem. ∎

This theorem has a converse. Suppose that under the conditions of the theorem $f:(E, A^2(x)) \to (E, \hat{A}^2(x))$ is an orientation-preserving *weak* equivalence (isometry). Then if for each $x \in E$, $K_j(x) = K_{\hat{j}}[f(x)]$, we may conclude that f is a *strong equivalence*. The proof simply reverses the reasoning in the proof of the theorem.

Now a definition as elaborate and as far removed from intuition as the one given for strong equivalence of isometric immersions could hardly be justified on the grounds that it 'captures' the notion of 'having the same curvature' for curves. After all, the definition of curvature itself is easy to work with, fairly intuitive, and already parametrization-independent. It will transpire that the geometric definition which Gauss gave for the curvature of an immersed *surface* in $\mathbb{R}^3$ generalizes directly the one with which we are working. The notion of strong equivalence applies to that case also, and again it coincides with the idea of having the same curvature. This formulation of strong equivalence then shows in a formal way how the *conditions* for equal curvature change as one passes from dimension 1 to dimension 2. In particular, the *determinant* functions play a much more prominent role in the case of surfaces as we shall see. We shall also have the statement of Gauss' 'remarkable' theorem: *two isometric immersions of* $\mathbb{R}^2$ *in* $\mathbb{R}^3$ *are strongly equivalent if and only if they are weakly equivalent.* While we shall prove that result in this section, it will be the goal of later sections to develop with sector calculus a notion of curvature which is defined *entirely* by the metric with no reference to an immersion. This is the notion of Riemannian curvature, adumbrated by Gauss and given its most familiar expression in the language of tensor calculus. It should be observed here that the Riemannian notion of curvature does not coincide with the one which we have been discussing for curves. Since all curves are isometric, any notion of curvature which derives entirely from the metric must fail to distinguish them. The Riemannian curvature of a curve is identically 0. It is perhaps fair to argue that there is a semantic imprecision in the identification with one and the same term: *curvature* of both immersion-dependent and immersion-independent constructions. Conceptually, the Riemannian notion is the satisfying one; our presentation of it will be entirely in terms of the metric, although we shall discuss, via the section $\Omega(j)$, what its relation with codimension-1 immersions is. Our justification for having begun this study with the case of *curves* is twofold: it completes, in a very modest way, the historic progression of the ideas, from curves, to surfaces, to manifolds. Secondly, it provides an intuitive introduction to the notion of Gaussian curvature, which will be the next topic to be taken up here.

Suppose now that $\gamma: E \to E'$ is a smooth immersion with $E = \mathbb{R}^2$ and $E' = \mathbb{R}^3$. As usual, let B^2 denote the Euclidean metric for E'. We want to define a notion of a 'unit normal 1-variation' (Definition 4.20) for γ. Recall from the discussion immediately preceding Definition 2.13 that a map such as γ has been called a *ribbon* in E'. If we denote the coordinate functions for $E=\mathbb{R}^2$ as (x, y), then for each point $(x_0, y_0)\in E$, let $\bar{\gamma}_2^{\#}(x_0, y_0)$ denote the 2-sector in E' at $\gamma(x_0, y_0)$, θ the identity frame

$$\bar{\gamma}_2^{\#}(x_0, y_0)=\left[(\theta\circ t_{\gamma(x_0, y_0)})_2, \frac{\partial\gamma^i}{\partial x}_{|(x_0, y_0)} \;\text{——}\; \frac{\partial^2\gamma^r}{\partial x\delta y}_{|(x_0, y_0)} \;\text{——}\; \frac{\partial\gamma^j}{\partial y}_{|(x_0,y_0)}\right].$$

We generalize this in our next definition.

Definition 4.26. Let $G:\mathbb{R}^3\to O\subset E=\mathbb{R}^m$ and $\phi:O\to U\subset M$ a frame, and let the linear coordinates for $\mathbb{R}^3$ be denoted (x, y, z). Call such a G a *block* in U. G gives rise to a block in $T_3(U)$ by assigning to each $(x_0, y_0, z_0)\in\mathbb{R}^3$ the 3-sector at $\phi(G(x_0, y_0, z_0))\in U$ represented with respect to frame ϕ:

$$\begin{array}{ccccccc}
\Big[(\phi\circ t_{G(x_0,y_0,z_0)})_3, \dfrac{\partial G^i}{\partial x}_{|(x_0,y_0,z_0)} & \text{——} & & \dfrac{\partial^2 G^r}{\partial x\partial y}_{|(x_0,y_0,z_0)} & & \text{——} & \dfrac{\partial G^j}{\partial y}_{|(x_0,y_0,z_0)}\Big] \\
\diagdown & & & \dfrac{\partial^3 G^u}{\partial x\partial y\partial z}_{|(x_0,y_0,z_0)} & & & \diagup \\
 & \dfrac{\partial^2 G^s}{\partial x\partial z}_{|(x_0,y_0,z_0)} & & & & \dfrac{\partial^2 G^t}{\partial y\partial z}_{|(x_0,y_0,z_0)} & \\
 & \diagdown & & & & \diagup & \\
 & & & \dfrac{\partial G^k}{\partial z}_{|(x_0,y_0,z_0)} & & &
\end{array}$$

This is just the F-related class of $\phi\circ G\circ t_{(x_0, y_0, z_0)}:(\mathbb{R}^3, 0)\to(U, \phi\circ G(x_0, y_0, z_0))$. We denote this 3-sector $\bar{G}_3^{\#}(x_0, y_0, z_0)$.

Now suppose that $\gamma:\mathbb{R}^2\to O$ is a smooth ribbon in $U\subset M$ with respect to frame $\phi:O\to U$. A 1-*variation* of γ is an equivalence class of germs of blocks $G:(\mathbb{R}^2\times\mathbb{R}, \mathbb{R}^2\times z_0)\to(O, \operatorname{im}\gamma)$ with $G_{|\mathbb{R}^2\times z_0}=\gamma$ ($\mathbb{R}^2\times z_0$ is naturally identified with $\mathbb{R}^2$) with respect to the following equivalence relation: $F\simeq G$ if $\bar{F}_3^{\#}(x, y; z_0)=\bar{G}_3^{\#}(x, y; z_0)$ for *all* $(x, y)\in\mathbb{R}^2$. For such blocks it follows that $D^3(3)\bar{G}_3^{\#}(x, y; z_0)=\bar{\gamma}_2^{\#}(x, y)$ for all $(x, y)\in\mathbb{R}^2$. ■

Now returning to the case $\gamma:E\to E'$ a smooth immersion with $E=\mathbb{R}^2$ and $E'=\mathbb{R}^3$, we denote the respective determinant functions (defined in Definition 4.22) $\det_E$ and $\det_{E'}$. A *unit normal* 1-*variation of immersion* γ is a 1-variation $G:(\mathbb{R}^2\times\mathbb{R}, \mathbb{R}^2\times z_0)\to(E', \operatorname{im}(\gamma))$ with $G_{|(\mathbb{R}^2\times z_0)}=\gamma$ satisfying:

1) $B^2\langle Z(x, y), Z(x, y)\rangle=1$,
2) $B^2\langle Z(x, y), X(x, y)\rangle=B^2\langle Z(x, y), Y(x, y)\rangle=0$,
3) $\det_{E'}[Z(x, y), X(x, y), Y(x, y)]$ *positive* for all $(x, y)\in\mathbb{R}^2$.

(Here we write $X(x, y)=[(\theta\circ t_{\gamma(x, y)})_1, X^i(x, y)]$ the 100 vertex of $\bar{G}_3^{\#}(x, y; z_0)$, $Y(x, y)$ the 010 vertex, and $Z(x, y)$ the 001 vertex with components defined in terms of global identity frame θ for E'.)

Definition 4.27. Let $\gamma: E \to E'$ be a smooth immersion $\mathbb{R}^2 \to \mathbb{R}^3$ and let G be a *unit normal* 1-*variation* of γ. The *Gauss map* associated with γ is the smooth map $\mathbb{R}^2 \to S^2$ defined in the following way:

$$\hat{\gamma}(x, y) = Z^k(x, y) = \begin{bmatrix} Z^1(x, y) \\ Z^2(x, y) \\ Z^3(x, y) \end{bmatrix}.$$

This map is uniquely determined by γ. ■

Definition 4.28. Under the conditions of the above definition for $\gamma: E \to E'$ a smooth immersion, let $w: E \to \mathbb{R}$ be defined by $w(x, y) = \det_{E'}[Z(x, y), X(x, y), Y(x, y)]$ for unit normal 1-variation G. For $(x_0, y_0) \in E$, define the *Gaussian curvature* of γ at (x_0, y_0), denoted $K_\gamma(x_0, y_0)$, to be the ratio of determinants:

$$\frac{\det_{E'}\left[\hat{\gamma}(x_0, y_0), \frac{\partial \hat{\gamma}}{\partial x}_{|(x_0, y_0)}, \frac{\partial \hat{\gamma}}{\partial y}_{|(x_0, y_0)}\right]}{\det_{E'}[Z(x_0, y_0), X(x_0, y_0), Y(x_0, y_0)]} \tag{4.28}$$

where the determinant in the numerator is calculated in $T_1[\hat{\gamma}(x_0, y_0)]$ (so $\hat{\gamma}(x_0, y_0)$ is interpreted as a vector tangent to E' *at* $\hat{\gamma}(x_0, y_0)$) and the determinant in the denominator is calculated in $T_1[\gamma(x_0, y_0)]$.

As in the one-dimensional case, it is easy to see that $K_\gamma(x_0, y_0)$ is defined *locally* as the *Jacobian determinant* of the germ $\hat{\gamma} \circ \gamma^{-1}: (\gamma(E), \gamma(x_0, y_0)) \to (\hat{\gamma}(E), \hat{\gamma}(x_0, y_0))$. This makes sense as γ is locally a diffeomorphism onto its image and the volume 2-forms are inherited in the manner indicated in (4.28) from $\det_{E'}$. ■

Now

$$K_\gamma(x_0, y_0) = \frac{\det_{E'}\left[\hat{\gamma}(x_0, y_0), \frac{\partial \hat{\gamma}}{\partial x}_{|(x_0, y_0)}, \frac{\partial \hat{\gamma}}{\partial y}_{|(x_0, y_0)}\right] \det_{E'}[Z(x_0, y_0), X(x_0, y_0), Y(x_0, y_0)]}{w(x_0, y_0)^2}$$

and if we translate tangent vectors to $T_1[\gamma(x_0, y_0)]$ in $T_1(E')$ this is equal to

$$\frac{1}{w(x_0, y_0)^2}\left(B^2\left\langle\frac{\partial \hat{\gamma}}{\partial x}_{|(x_0, y_0)}, X(x_0, y_0)\right\rangle B^2\left\langle\frac{\partial \hat{\gamma}}{\partial y}_{|(x_0, y_0)}, Y(x_0, y_0)\right\rangle - B^2\left\langle\frac{\partial \hat{\gamma}}{\partial y}_{|(x_0, y_0)}, X(x_0, y_0)\right\rangle B^2\left\langle\frac{\partial \hat{\gamma}}{\partial x}_{|(x_0, y_0)}, Y(x_0, y_0)\right\rangle\right).$$

It is not difficult to see that in light of condition 2) for the unit normal variation G, we have:

$$B^2\left\langle \frac{\partial \hat{\gamma}}{\partial x}_{|(x_0, y_0)}, X(x_0, y_0) \right\rangle + B^2\left\langle \hat{\gamma}(x_0, y_0), \frac{\partial^2 \gamma}{\partial x^2}_{|(x_0, y_0)} \right\rangle = 0,$$

$$B^2\left\langle \frac{\partial \hat{\gamma}}{\partial y}_{|(x_0, y_0)}, X(x_0, y_0) \right\rangle + B^2\left\langle \hat{\gamma}(x_0, y_0), \frac{\partial^2 \gamma}{\partial x \partial y}_{|(x_0, y_0)} \right\rangle = 0,$$

and two other equations, essentially the *Weingarten equations*.

Let $u = \partial/\partial x$, $v = \partial/\partial y$ give in each tangent fiber of $T_1(E)$ the basis (u, v) dual to the linear coordinate functions (dx, dy) induced from linear coordinates (x, y) on E. Thus,

$$u = \left[(\phi \circ t_{(x_0, y_0)})_1, \begin{pmatrix} 1 \\ 0 \end{pmatrix} \right]$$

and

$$v = \left[(\phi \circ t_{(x_0, y_0)})_1, \begin{pmatrix} 0 \\ 1 \end{pmatrix} \right]$$

in the fiber at (x_0, y_0) with respect to global identity frame ϕ on E. We shall not make reference to the particular fiber when referring to the basis given in this way. There should be no confusion with this notational liberty. In particular, $T_1(\gamma)(u) = X(x_0, y_0)$ for u in the (x_0, y_0) fiber and $T_1(\gamma)(v) = Y(x_0, y_0)$ in that fiber. With this notation, and substituting from the Weingarten equations above, we have the following expression for $K_\gamma(x_0, y_0)$:

$$K_\gamma(x_0, y_0) = \frac{1}{w(x_0, y_0)^2} \times [B^2\langle \Omega(\gamma)(u, u), \Omega(\gamma)(v, v)\rangle - B^2\langle \Omega(\gamma)(u, v), \Omega(\gamma)(u, v)\rangle] \qquad (4.29)$$

where B^2 is computed in $T_1[\gamma(x_0, y_0)]$ and u and v are in the tangent fiber at (x_0, y_0). This formula follows from Observation 4.16 which interprets the values of $\Omega(\gamma)$ as 'normal components' of the second derivative in a frame-independent way. In this form, it is clear that $K_\gamma(x_0, y_0)$ depends on the particular immersion γ *and* on the system of coordinates chosen for E. Eventually, we shall show that it is independent of both choices, but depends only on the induced metric $\gamma * (B^2)$ on E. We now prove the analog for this case of Theorem 4.25.

Theorem 4.29. *Let* $E = \mathbb{R}^2$ *and* $E' = \mathbb{R}^3$ *and suppose that* $j : (E, A^2(x, y)) \to (E', B^2)$ *is an* isometric immersion, *and let* $\hat{j} : (E, \hat{A}^2(x, y)) \to (E', B^2)$ *be another isometric immersion. Let* $f : (E, A^2(x, y)) \to (E, \hat{A}^2(x, y))$ *be a* weak equivalence (*isometry*), *then* f *is a* strong equivalence if and only if *for each* $(x, y) \in E$, $K_j(x, y) = K_{\hat{j}}[f(x, y)]$.

Proof. First suppose that f is a strong equivalence. Then for (u, v) in the (x, y) fiber

$$\det_E\begin{bmatrix}\det_{E'}[T_1(j)(u), T_1(j)(v), \Omega(j)(u, u)] & \det_{E'}[T_1(j)(u), T_1(j)(v), \Omega(j)(u, v)] \\ \det_{E'}[T_1(j)(u), T_1(j)(v), \Omega(j)(v, u)] & \det_{E'}[T_1(j)(u), T_1(j)(v), \Omega(j)(v, v)]\end{bmatrix}$$
$$= w(x, y)^4 K_j(x, y)$$

and this is equal to the following for $\hat{u} = T_1(f)(u)$ and $\hat{v} = T_1(f)(v)$:

$$\det_E\begin{bmatrix}\det_{E'}[T_1(\hat{j})(\hat{u}), T_1(\hat{j})(\hat{v}), \Omega(\hat{j})(\hat{u}, \hat{u})] & \det_{E'}[T_1(\hat{j})(\hat{u}), T_1(\hat{j})(\hat{v}), \Omega(\hat{j})(\hat{u}, \hat{v})] \\ \det_{E'}[T_1(\hat{j})(\hat{u}), T_1(\hat{j})(\hat{v}), \Omega(\hat{j})(\hat{v}, \hat{u})] & \det_{E'}[T_1(\hat{j})(\hat{u}), T_1(\hat{j})(\hat{v}), \Omega(\hat{j})(\hat{v}, \hat{v})]\end{bmatrix}$$
$$= w(x, y)^2[B^2\langle\Omega(\hat{j})(\hat{u}, \hat{u}), \Omega(\hat{j})(\hat{v}, \hat{v})\rangle - B^2\langle\Omega(\hat{j})(\hat{u}, \hat{v}), \Omega(\hat{j})(\hat{v}, \hat{u})\rangle]$$

since f is an isometry.

Now suppose that $\hat{u} = au + bv$ and $\hat{v} = cu + dv$. Computing

$$\begin{aligned}&\hat{A}^2[f(x, y)]\langle\hat{u}, \hat{u}\rangle\hat{A}^2[f(x, y)]\langle\hat{v}, \hat{v}\rangle - \hat{A}^2[f(x, y)]\langle\hat{u}, \hat{v}\rangle\hat{A}^2[f(x, y)]\langle\hat{v}, \hat{u}\rangle \\ &\quad = A^2(x, y)\langle u, u\rangle A^2(x, y)\langle v, v\rangle - A^2(x, y)\langle u, v\rangle A^2(x, y)\langle v, u\rangle \\ &\quad = (ad - bc)^2\{\hat{A}^2[f(x, y)]\langle u, u\rangle\hat{A}^2[f(x, y)]\langle v, v\rangle \\ &\qquad - \hat{A}^2[f(x, y)]\langle u, v\rangle\hat{A}^2[f(x, y)]\langle v, u\rangle\}\end{aligned}$$

we conclude that

$$w(x, y)^2 = (ad - bc)^2\hat{w}[f(x, y)]^2$$

for $\hat{w}$ defined for the immersion $\hat{j}$ in the manner analogous to that in which w was defined for immersion j.

A similar calculation shows that

$$\begin{aligned}&[B^2\langle\Omega(\hat{j})(\hat{u}, \hat{u}), \Omega(\hat{j})(\hat{v}, \hat{v})\rangle - B^2\langle\Omega(\hat{j})(\hat{u}, \hat{v}), \Omega(\hat{j})(\hat{v}, \hat{u})\rangle] \\ &\quad = (ad - bc)^2\{B^2\langle\Omega(\hat{j})(u, u), \Omega(\hat{j})(v, v)\rangle - B^2\langle\Omega(\hat{j})(u, v), \Omega(\hat{j})(v, u)\rangle\},\end{aligned} \tag{4.30}$$

so

$$\begin{aligned}w(x, y)^4K_j(x, y) &= (ad - bc)^4\hat{w}[f(x, y)]^2[B^2\langle\Omega(\hat{j})(u, u), \Omega(\hat{j})(v, v)\rangle \\ &\quad - B^2\langle\Omega(\hat{j})(u, v), \Omega(\hat{j})(v, u)\rangle] \\ &= w(x, y)^4K_{\hat{j}}[f(x, y)].\end{aligned}$$

It follows then that $K_j(x, y) = K_{\hat{j}}[f(x, y)]$. Notice that no 'orientability condition' was here necessary.

The reverse implication is proved by reversing the algebraic steps. ■

Observation 4.30. Suppose that $j:(E, A^2(x, y)) \to (E', B^2)$ is an *isometric immersion* with $E' = \mathbb{R}^3$ and $E = \mathbb{R}^2$. Suppose also that $h: E \to E$ is a diffeomorphism with $h * A^2 = \hat{A}^2$. Let $\hat{j} = j \circ h$. Then for all $(x, y) \in E$, $K_{\hat{j}}(x, y) = K_j[h(x, y)]$. That is, the Gaussian curvature is *independent of parametrization*. Further, if at $(x, y) \in E$, $(\hat{u}, \hat{v})$ is an *arbitrary* basis for $T_1(x, y)$, then

$$K_j(x, y) = \frac{B^2\langle \Omega(j)(\hat{u}, \hat{u}), \Omega(j)(\hat{v}, \hat{v})\rangle - B^2\langle \Omega(j)(\hat{u}, \hat{v}), \Omega(j)(\hat{v}, \hat{u})\rangle}{A^2(x, y)\langle \hat{u}, \hat{u}\rangle A^2(x, y)\langle \hat{v}, \hat{v}\rangle - A^2(x, y)\langle \hat{u}, \hat{v}\rangle A^2(x, y)\langle \hat{u}, \hat{v}\rangle}. \tag{4.31}$$

Proof. The first assertion follows from the fact that a reparametrization gives a strong equivalence. The second follows fairly easily from formula (4.30). ■

For the calculation of Gaussian curvature, formula (4.31) is very convenient since it is independent of the choice of basis for the tangent space and need make no direct reference to the 'unit normal variation' which, while it is uniquely determined by the immersion j, can obscure the invariance properties of curvature. In fact, the only real dependence of $K_j(x, y)$ on the immersion (over and above the metric which it induces on E) is given by $\Omega(j) = d_{2,1}j\langle\Lambda\rangle$. And since Λ itself depends only on the metric, that dependence resides in the application of $d_{2,1}j$ to the flat 2-sector. Our discussion in §4.3 of Gauss' *Theorema Egregium*, that weak equivalence implies strong equivalence for surfaces immersed in $\mathbb{R}^3$, will exhibit the dependence of the numerator in formula (4.26) *on the metric alone* via the invariance of a certain sectorform field naturally derived from the metric. From this point of view, the result, while no less remarkable, will perhaps appear less surprising than it will after the *direct* calculation which follows. The procedure we follow here will use the Gram–Schmidt technique to calculate $\Omega(j)(X, Y)$ via its characterization as the 'normal component' of the 'second derivative' of j with respect to X and Y computed in any 2-frame. The alternative to this makes the direct calculation of Λ via the Riemann–Christoffel symbols and grinds the result out.

We begin with the following observation. Suppose that X and Y are *tangent vectors* at some fixed point in $E' = \mathbb{R}^3$. Let $V_X(Y)$ denote the vector:

$$Y - \frac{B^2\langle X, Y\rangle}{B^2\langle X, X\rangle} X.$$

[Here and throughout this discussion we assume that X and Y are linearly independent.] The vector $V_X(Y)$ has this interpretation:

$$B^2\langle X, V_X(Y)\rangle = 0 \qquad \text{and} \qquad Y = V_X(Y) + \frac{B^2\langle X, Y\rangle}{B^2\langle X, X\rangle} X.$$

Thus $V_X(Y)$ is the 'component' of Y in the plane spanned by X and Y 'normal' to X.

Now let Δ denote the scalar $B^2\langle X, X\rangle B^2\langle Y, Y\rangle - B^2\langle X, Y\rangle^2$. This is non-zero and is in fact the square of the area of the parallelogram spanned by X and Y. Then for an arbitrary tangent vector based at the source of X and Y, the component of Z *normal* to the plane spanned by X and Y is given by the following simple formula:

normal component of Z

$$= Z - \frac{1}{\Delta}[B^2\langle X, X\rangle B^2\langle Z, Y\rangle V_X(Y) - B^2\langle Y, Y\rangle B^2\langle Z, X\rangle V_Y(X)]. \tag{4.32}$$

This is immediately verified by showing that the inner product with X and Y gives 0.

Observation 4.31. Let $\gamma: E \to E'$ be an immersion for $E = \mathbb{R}^2$ and $E' = \mathbb{R}^3$ and let $(x_0, y_0) \in E$. Then in the notation of formula (4.29) for (u, v) the standard basis for $T_1(x_0, y_0)$ and for $T_1(\gamma)(u) = X$ and $T_1(\gamma)(v) = Y$, then

1)

$$\Omega(\gamma)(u, u) = \frac{\partial^2\gamma}{\partial x^2}_{|(x_0,y_0)} - (1/\Delta)\left[B^2\langle X, X\rangle B^2\left\langle \frac{\partial^2\gamma}{\partial x^2}_{|(x_0,y_0)}, Y\right\rangle V_X(Y) + B^2\langle Y, Y\rangle B^2\left\langle \frac{\partial^2\gamma}{\partial x^2}_{|(x_0,y_0)}, X\right\rangle V_Y(X)\right]$$

2)

$$\Omega(\gamma)(u, v) = \Omega(\gamma)(v, u) = \frac{\partial^2\gamma}{\partial x\partial y}_{|(x_0,y_0)} - (1/\Delta)\left[B^2\langle X, X\rangle B^2\left\langle \frac{\partial^2\gamma}{\partial x\partial y}_{|(x_0,y_0)}, Y\right\rangle V_X(Y) + B^2\langle Y, Y\rangle B^2\left\langle \frac{\partial^2\gamma}{\partial x\partial y}_{|(x_0,y_0)}, X\right\rangle V_Y(X)\right]$$

3)

$$\Omega(\gamma)(v, v) = \frac{\partial^2\gamma}{\partial y^2}_{|(x_0,y_0)} - (1/\Delta)\left[B^2\langle X, X\rangle B^2\left\langle \frac{\partial^2\gamma}{\partial y^2}_{|(x_0,y_0)}, Y\right\rangle V_X(Y) + B^2\langle Y, Y\rangle B^2\left\langle \frac{\partial^2\gamma}{\partial y^2}_{|(x_0,y_0)}, X\right\rangle V_Y(X)\right].$$

Proof. Use formula (4.32) together with Observation 4.16. ■

Now suppose that we set

$$E = B^2\langle X, X\rangle = B^2\left\langle \frac{\partial\gamma}{\partial x}_{|(x_0,y_0)}, \frac{\partial\gamma}{\partial x}_{|(x_0,y_0)}\right\rangle,$$

$$F = B^2\langle X, Y\rangle = B^2\left\langle \frac{\partial\gamma}{\partial x}_{|(x_0,y_0)}, \frac{\partial\gamma}{\partial y}_{|(x_0,y_0)}\right\rangle,$$

and

$$G = B^2\langle Y, Y\rangle = B^2\left\langle \frac{\partial\gamma}{\partial y}_{|(x_0,y_0)}, \frac{\partial\gamma}{\partial y}_{|(x_0,y_0)}\right\rangle$$

and let $\gamma * B^2 = A^2(x, y)$. Then the matrix for $A^2(x_0, y_0)$ with respect to the global identity frame on E is

$$\begin{bmatrix} E & F \\ F & G \end{bmatrix}.$$

Clearly Δ is equal to $\det_E$ applied to this matrix.

Theorem 4.32 (**Gauss' Theorema Egregium**). (First form.) *Suppose that under the conditions of Observation* 4.31, $\gamma : E \to E'$ *is an immersion of* $\mathbb{R}^2 \to \mathbb{R}^3$ *with* $A^2(x, y) = \gamma * B^2$. *Letting E, F, and G be taken as smooth functions of* (x, y) *in* $\mathbb{R}^2$ *for* (x_0, y_0) *fixed,* $K_\gamma(x_0, y_0)$ *can be expressed as a function of E, F, G and their first- and second-order partial derivatives at* (x_0, y_0). *This specifies the curvature in terms of the expression of the induced metric in* standard coordinates.

Now suppose that $\hat{\gamma} : E \to E'$ *is another immersion with* $\hat{\gamma} * B^2 = \hat{A}^2(x, y)$ *and suppose that* $f : (E, A^2(x, y)) \to (E, \hat{A}^2(x, y))$ *is an* isometry. *Then for all* (x_0, y_0), $K_\gamma(x_0, y_0) = K_{\hat{\gamma}}[f(x_0, y_0)]$. *Therefore, in this setting,* every weak equivalence is a strong equivalence. *This means that the curvature depends* only *on the induced metric; it is, to this extent, independent of the particular immersion.*

Proof. First we express $K_\gamma(x_0, y_0)$ in terms of E, F, and G. Next, let $\hat{\gamma} \circ f : E \to E'$ and conclude from Observation 4.24 that $\hat{\gamma}$ is *strongly equivalent* via reparametrization f to $\hat{\gamma} \circ f$. Thus $K_{\hat{\gamma}}[f(x_0, y_0)] = K_{\hat{\gamma}\circ f}(x_0, y_0)$. Now we have assumed that $\gamma * B^2 = (\hat{\gamma} \circ f) * B^2$ for all $(x, y) \in E$. Since the curvatures $K_\gamma(x_0, y_0)$ and $K_{\hat{\gamma}\circ f}(x_0, y_0)$ have the same expression in terms of E, F, and G, they are equal. This will give the conclusion of the theorem.

Now to obtain the expression for the curvature $K_\gamma(x_0, y_0)$ in terms of E, F, and G, return to 1), 2), and 3) of Observation 4.31 and to formula

(4.29). It is then fairly easy to see that in $T_1[\gamma(x_0, y_0)]$,

$$\begin{aligned}
\Delta \cdot K_\gamma(x_0, y_0) = {} & B^2\left\langle \frac{\partial^2\gamma}{\partial x^2}_{|(x_0,y_0)}, \frac{\partial^2\gamma}{\partial y^2}_{|(x_0,y_0)} \right\rangle \\
& - B^2\left\langle \frac{\partial^2\gamma}{\partial x\partial y}_{|(x_0,y_0)}, \frac{\partial^2\gamma}{\partial x\partial y}_{|(x_0,y_0)} \right\rangle \\
& - \frac{1}{\Delta} B^2\left\langle \frac{\partial^2\gamma}{\partial x^2}_{|(x_0,y_0)}, B^2\langle X, X\rangle B^2\left\langle \frac{\partial^2\gamma}{\partial y^2}_{|(x_0,y_0)}, Y\right\rangle V_X(Y) \right. \\
& \left. + B^2\langle Y, Y\rangle B^2\left\langle \frac{\partial^2\gamma}{\partial y^2}_{|(x_0,y_0)}, X\right\rangle V_Y(X) \right\rangle \\
& + \frac{1}{\Delta} B^2\left\langle \frac{\partial^2\gamma}{\partial x\partial y}_{|(x_0,y_0)}, B^2\langle X, X\rangle B^2\left\langle \frac{\partial^2\gamma}{\partial x\partial y}_{|(x_0,y_0)}, Y\right\rangle V_X(Y) \right. \\
& \left. + B^2\langle Y, Y\rangle B^2\left\langle \frac{\partial^2\gamma}{\partial x\partial y}_{|(x_0,y_0)}, X\right\rangle V_Y(X) \right\rangle. \qquad (4.33)
\end{aligned}$$

In this form, it is not quite obvious that E, F, and G determine K. We make a few calculations:

$$\begin{aligned}
& B^2\left\langle \frac{\partial^2\gamma}{\partial x^2}_{|(x_0,y_0)}, \frac{\partial^2\gamma}{\partial y^2}_{|(x_0,y_0)} \right\rangle - B^2\left\langle \frac{\partial^2\gamma}{\partial x\partial y}_{|(x_0,y_0)}, \frac{\partial^2\gamma}{\partial x\partial y}_{|(x_0,y_0)} \right\rangle \\
& = -\frac{1}{2}\frac{\partial^2 E}{\partial y^2}_{|(x_0,y_0)} + \frac{\partial^2 F}{\partial x\partial y}_{|(x_0,y_0)} - \frac{1}{2}\frac{\partial^2 G}{\partial x^2}_{|(x_0,y_0)},
\end{aligned}$$

$$\begin{aligned}
& B^2\left\langle \frac{\partial^2\gamma}{\partial x^2}_{|(x_0,y_0)}, B^2\langle X, X\rangle B^2\left\langle \frac{\partial^2\gamma}{\partial y^2}_{|(x_0,y_0)}, Y\right\rangle V_X(Y) \right\rangle \\
& = \tfrac{1}{2} E \frac{\partial G}{\partial y}_{|(x_0,y_0)} \left\{ \left[\frac{\partial F}{\partial x}_{|(x_0,y_0)} - \frac{1}{2}\frac{\partial E}{\partial y}_{|(x_0,y_0)} \right] \right. \\
& \qquad \left. - \frac{1}{2}\frac{\partial E}{\partial x}_{|(x_0,y_0)} (F/E) \right\} \\
& = \tfrac{1}{2} E \frac{\partial G}{\partial y}\frac{\partial F}{\partial x} - \tfrac{1}{4} E \frac{\partial G}{\partial y}\frac{\partial E}{\partial y} - \tfrac{1}{4} F \frac{\partial G}{\partial y}\frac{\partial E}{\partial x}
\end{aligned}$$

(omitting evaluation points),

$$B^2\left\langle \frac{\partial^2\gamma}{\partial x\partial y}, B^2\langle X, X\rangle B^2\left\langle \frac{\partial^2\gamma}{\partial x\partial y}, Y\right\rangle V_X(Y) \right\rangle = \tfrac{1}{4} E\left(\frac{\partial G}{\partial x}\right)^2 - \tfrac{1}{4} F \frac{\partial G}{\partial x}\frac{\partial E}{\partial y},$$

$$B^2\left\langle \frac{\partial^2\gamma}{\partial x\partial y}, B^2\langle Y, Y\rangle B^2\left\langle \frac{\partial^2\gamma}{\partial x\partial y}, X\right\rangle V_Y(X) \right\rangle = \tfrac{1}{4} G\left(\frac{\partial E}{\partial y}\right)^2 - \tfrac{1}{4} F \frac{\partial G}{\partial x}\frac{\partial E}{\partial y},$$

and finally,

$$B^2\left\langle\frac{\partial^2\gamma}{\partial x^2}, B^2\langle Y, Y\rangle B^2\left\langle\frac{\partial^2\gamma}{\partial y^2}, X\right\rangle V_Y(X)\right\rangle$$
$$=\tfrac{1}{2}G\frac{\partial E}{\partial x}\left[\frac{\partial F}{\partial y}-\frac{1}{2}\frac{\partial G}{\partial x}\right]-F\left[\frac{\partial F}{\partial y}-\frac{1}{2}\frac{\partial G}{\partial x}\right]\left[\frac{\partial F}{\partial x}-\frac{1}{2}\frac{\partial E}{\partial y}\right].$$

Therefore we have the following (everything evaluated at (x_0, y_0), $\Delta = EG - F^2$):

$$\Delta^2 K_\gamma(x_0, y_0) = (-\Delta)\left[\frac{1}{2}\frac{\partial^2 E}{\partial y^2}-\frac{\partial^2 F}{\partial x \partial y}+\frac{1}{2}\frac{\partial^2 G}{\partial x^2}\right]$$
$$+\frac{1}{4}\left[\frac{\partial E}{\partial y} \quad \frac{\partial G}{\partial x}\right]\begin{bmatrix} G & -F \\ -F & E\end{bmatrix}\begin{bmatrix}\dfrac{\partial E}{\partial y} \\ \dfrac{\partial G}{\partial x}\end{bmatrix}$$
$$-\left[\left(\frac{\partial F}{\partial y}-\frac{1}{2}\frac{\partial G}{\partial x}\right), \frac{1}{2}\frac{\partial G}{\partial y}\right]\begin{bmatrix} G & -F \\ -F & E\end{bmatrix}\begin{bmatrix}\dfrac{1}{2}\dfrac{\partial E}{\partial x} \\ \dfrac{\partial F}{\partial x}-\dfrac{1}{2}\dfrac{\partial E}{\partial y}\end{bmatrix}. \tag{4.34}$$

■

It is obvious at this point that there are more questions raised than answers given in this amazing calculation of Gauss. The formula (4.34) for curvature hardly gives any insight into its invariance (although one knows *a priori* that if E, F, G are replaced by E', F', G' obtained from representing the metric with respect to a different global frame, then K_γ remains invariant). A much more satisfactory description of K_γ (following Riemann's interpretation of curvature) will follow in §4.3. From this point of view, the curvature will be shown *ab initio* to depend only on the metric. Its invariance will follow from the 'naturality' of the *second differential operator* on sectorform fields.

§4.2. The geodesic flow and critical energy curves

It will be necessary to digress briefly in this section to take up a circle of ideas that have played a critical role in the development of the relations between geometry and mechanics. The notion that the configurations of a mechanical system should evolve along 'straightest paths' connects the work of Lagrange, Hertz, and Einstein. These straightest paths, or *geodesics*, should be determined by the 'metrical' structure of the continuum (whether space or space-time) as Riemann suggested. From the viewpoint of mechanics, that metrical structure is intimately related to the idea of 'energy'. In particular, the *kinetic energy* is maintained constant when a particle moves in a Cartesian coordinate system according to the Galilean dictum: in a straight line with constant velocity. Some deviation from this 'natural motion' must in the eyes of Newton and Galileo have a 'cause', classically formulated in terms of 'force'. (See Bruter [4].)

The properties of geodesics and their cognates, 'parallel translations', give a clear and intuitive picture of the structure of a Riemannian manifold. In fact, a Riemannian manifold with a positive definite metric can be assimilated to a classical mechanical system by defining from the metric $A^2(x)$ a *kinetic energy function* $T: T_1(M) \to \mathbb{R}$ by $T[X(x)] = \frac{1}{2}A^2(x)\langle X(x), X(x)\rangle$. Then the Euler–Lagrange equations determine the geodesics.

There are many ways to define *geodesics* for a manifold equipped with a metric tensor. They all make use, in one form or another, of the following picture. Let $(M, A^2(x))$ be given. Let $\gamma: \mathbb{R} \to M$ be a smooth curve. Then $T_1(\gamma): T_1(\mathbb{R}) \to T_1(M)$ in the usual way. Now let $\mathbb{R} = E$ be equipped with the standard basis (e) and dual coordinate function (t), and let $\phi: E \to \mathbb{R}$ be the global identity frame (ϕ can be thought of as the inverse of the coordinate function t). Now if we define the 'constant' vector field on $\mathbb{R}$:

$$[(\phi \circ t_\tau), 1] = u(\tau) = \frac{d}{dt_{|\tau}}\ .$$

We shall denote it simply $u(\tau)$ or u. Then γ defines a *curve* in $T_1(M)$ by the rule $\gamma_*: \tau \to T_1(\gamma)(u(\tau))$. This curve, denoted γ_*, is essentially the '1-jet extension' of γ (though not precisely). It is easily seen to be smooth. We have the following relation between γ and γ_*. The following diagram commutes:

$$\begin{array}{ccc} & & T_1(M) \\ & \overset{\gamma_*}{\nearrow} & \downarrow q_1 \\ \mathbb{R} & \xrightarrow[\gamma]{} & M \end{array}$$

Now for a curve γ to be a geodesic, certain conditions must be satisfied by the curve γ_*. Locally, we shall insist that γ_* be a solution curve for a certain *vector field* defined on $T_1(M)$ called the *geodesic flow* on $T_1(M)$. We define this vector field now.

Definition 4.33. Let $(M, A^2(x))$ be a smooth manifold with a metric tensor. The *geodesic vector field* on $T_1(M)$ associated with $A^2(x)$ is defined in the following way: for each $x_0 \in M$ and frame $\theta: (E', 0) \to (M, x_0)$ (for $E' = \mathbb{R}^m$), associate to $[\theta_1, X^i] \in T_1(x_0)$ the 2-sector

$\Lambda_M[\theta_1, (X^i, X^i)]$ for Λ_M the affine connection associated to the metric and $[\theta_1, (X^i, X^i)] \in T_1^2(x_0)$ the 'diagonal' pair (X, X) at x_0.

With respect to a 2-frame at x_0, this 2-sector has representation $[\theta_2, X^i$——$\Lambda^r_{ij}X^iX^j$——$X^i]$ with the 'coefficients' Λ^r_{ij} determined by the *fundamental section* (Definition 4.8) for $\pi_{2,1}$, and so depending on the choice of θ_2 over θ_1.

Now this 2-sector has, via Theorem 1.27, the identification as an element of $T_1[T_1(M)]$ using natural isomorphism ν. Thus the geodesic vector field associates to $[\theta_1, X^i]$ the tangent vector *at* $[\theta_1, X^i]$:

$$\nu \circ \Lambda_M[\theta_1, (X^i, X^i)]. \tag{4.35}$$

In simple language, the geodesic vector field associates to each tangent vector the unique *flat* 2-sector both of whose vertices are equal to the tangent vector. Denote this vector field $G(X)$. ■

The geodesic vector field $G(X)$ on $T_1(M)$ then determines the geodesics in M by the stipulation:

Definition 4.34. For $(M, A^2(x))$ a Riemannian manifold (smooth manifold with metric tensor), a curve $\gamma: \mathbb{R} \to M$ is a *geodesic* if $\gamma_*: \mathbb{R} \to T_1(M)$ is a *solution curve* for the vector field $G(X)$.

Analytically, this means the following: If $\gamma: \mathbb{R} \to M$ is in general a smooth curve, for $\tau_0 \in \mathbb{R}$ consider the 2-sector $[(\phi \circ t_{\tau_0})_2, 1$——$0$——$1] \in T_2(\tau_0)$ for ϕ as usual the global identity frame on $\mathbb{R}$. Call this sector $U(\tau_0)$ in analogy with the 1-sector $u(\tau_0)$. So $T_2(\gamma)(U(\tau_0))$ is a tangent vector *at* $\gamma_*(\tau_0)$ in $T_1(M)$ (via ν). We shall denote it $\gamma_{**}(\tau_0)$ and call it the *natural ribbon* associated to γ at τ_0. In components, if θ is a frame-germ at $(M, \gamma(\tau_0))$ and if $g = \theta^{-1} \circ \gamma \circ \phi$ is the germ $(E, 0) \to (E', 0)$, then

$$\gamma_{**}(\tau_0) = \left[\theta_2, \frac{dg^i}{dt} \text{——} \frac{d^2g^r}{dt^2} \text{——} \frac{dg^j}{dt}\right].$$

Just as γ_* does, γ_{**} has a coordinate-independent meaning. Returning to the notation of Theorem 1.27 and with respect to frame $\tilde{\theta}^1$ on $T_1(M)$ *at* $\gamma_*(\tau_0)$, $\gamma_{**}(\tau_0)$ is easily seen to be associated via ν to the tangent vector $T_1(\gamma_*)(u(\tau_0)) = (\gamma_*)_*(\tau_0)$. This justifies the notation $\gamma_{**} = (\gamma_*)_*$.

We thus have the interesting characterization of geodesics in M, namely $\gamma: \mathbb{R} \to M$ is a geodesic if and only if $\gamma_{**}(\tau)$ is *flat* for all $\tau \in \mathbb{R}$ and this is true if and only if the *force* of $\gamma_{**}(\tau)$ is 0 for all $\tau \in \mathbb{R}$. ■

Observe that in the above notation we have the following: with respect to frame $\theta: O \to U \subset M$ with $\gamma: I \to U$ for I some open interval in $\mathbb{R}$, $(\gamma_*)_{|I}$ is a solution curve for $G(X)$ on I if and only if for

$g = \theta^{-1} \circ \gamma \circ \phi : I \to O$ we have the classical

$$\frac{d^2 g^r}{dt^2}_{|\tau} = \Lambda^r_{ij}(g(\tau)) \frac{dg^i}{dt}_{|\tau} \frac{dg^j}{dt}_{|\tau}$$

$$= \tfrac{1}{2}\tilde{a}^{rk}(g(\tau)) \left[\frac{\partial \tilde{a}_{ij}}{\partial x^k}_{|g(\tau)} - 2 \frac{\partial \hat{a}_{ik}}{\partial x^j}_{|g(\tau)} \right] \frac{dg^i}{dt}_{|\tau} \frac{dg^j}{dt}_{|\tau} \tag{4.36}$$

for $\tilde{a}_{ij}(z)$ the local representation for $A^2(x)$ in frame θ. This follows directly from formula (4.11) and using symmetry of the coefficients in i and j. This gives a second-order differential equation for geodesics.

We develop now an interpretation of the vector field $G(X)$ as the 'gradient' of the kinetic energy with respect to the (indefinite) metric δA^2 on $T_1(M)$. Here δA^2 is the *promotion* of the metric tensor (Theorem 4.1). In light of the naturality property given in formula (4.1), namely $T_1(f)*[\delta A^2] = \delta[f*A^2]$ for $f: N \to M$ a smooth immersion with $f*A^2$ a metric, the interpretation of $G(X)$ as a gradient flow establishes certain naturality properties for geodesics and gives a clear rational procedure for passing the geodesic flow on a manifold M to the geodesic flow on an immersed submanifold N (of course, we are abusing language slightly, these flows are on the tangent bundles). Among the naturality properties for geodesics there is one which, in the context of mechanics with conservative forces introduced, will be recognized as *D'Alembert's principle.*

Recall from the proof of Theorem 4.1 that for Riemannian manifold $(M, A^2(x))$ and local frame $\theta: O \to U \subset M$ the matrix for $\delta A^2_{|\bar{A}_1(x_0)}$ for $\bar{A}_1(x_0) = [(\theta \circ t_{z_0})_1, Z^k]$ has the form

$$\begin{bmatrix} \dfrac{\partial a_{ij}}{\partial x^k}_{|z_0} Z^k & a_{it}(z_0) \\ a_{sj}(z_0) & 0 \end{bmatrix}$$

when it is represented in terms of the local frame $\tilde{\theta}^1: O \times E' \to T_1(U)$.

Definition 4.35. Let $(M, A^2(x))$ as a Riemannian manifold, allowing that $A^2(x)$ is a possibly indefinite metric. Let $Z = \bar{A}_1(x_0) \in T_1(x_0)$ and let $\theta: O \to U \subset M$ be a local frame, $\theta(z_0) = x_0$.

Now let $T: T_1(M) \to \mathbb{R}$ be the kinetic energy function associated to the metric; thus $T(X) = \tfrac{1}{2}A^2\langle X, X\rangle$. Also, let $\tilde{\theta}^1: O \times E' \to T_1(U)$ be the local frame associated to θ defined in Theorem 1.27. Suppose that $Z = [(\theta \circ t_{z_0})_1, Z^k]$ and suppose that the general 2-sector $[(\theta \circ t_{z_0})_2, Z^k$——$B^s$——$X^i]$ (violating our notational convention) is interpreted as the element of $T_1[T_1(M)]$ at Z with $\tilde{\theta}^1$ representation:

$$\left[(\tilde{\theta}^1 \circ t_{(z_0, Z^k)})_1, \begin{pmatrix} X^i \\ B^s \end{pmatrix} \right].$$

(This representation is explained in formula (4.2) at the beginning of §4.1.)

Then for

$$\bar{A}_2(x_0) = [(\theta \circ t_{z_0})_2, Z^k \text{---} B^s \text{---} X^i]$$

and

$$\bar{B}_2(x_0) = [(\theta \circ t_{z_0})_2, Z^k \text{---} C^t \text{---} Y^j]$$

let the corresponding tangent vectors at $Z \in T_1(x_0)$ be denoted $X_2(Z)$ and $Y_2(Z)$ with $\tilde{\theta}^1$ components respectively

$$\begin{pmatrix} X^i \\ B^s \end{pmatrix} \quad \text{and} \quad \begin{pmatrix} Y^j \\ C^t \end{pmatrix}.$$

Then the contraction $\delta A^2{}_{|Z}\langle X_2(Z), Y_2(Z)\rangle$ is given in Theorem 4.1 as the matrix product:

$$[X^i \quad B^s] \begin{bmatrix} \dfrac{\partial a_{ij}}{\partial x^k}{}_{|z_0} Z^k & a_{it}(z_0) \\ a_{sj}(z_0) & 0 \end{bmatrix} \begin{bmatrix} Y^j \\ C^t \end{bmatrix}. \tag{4.37}$$

Next, define the *differential of the kinetic energy*, $dT_{|Z}$, to be the covector at Z defined in terms of $\tilde{\theta}^1$ components by the contraction formula:

$$dT_{|Z}\langle Y_2(Z)\rangle = \frac{1}{2} \frac{\partial a_{ij}}{\partial x^k}{}_{|z_0} Y^k Z^i Z^j + a_{it}(z_0) Z^i C^t. \tag{4.38}$$

This is the ordinary differential of real-valued function T on $T_1(U)$.

Then the *gradient of T at Z with respect to metric* δA^2 is defined in the usual way: it is the *unique* tangent vector $X_2(Z)$ at Z with the property that, for all $Y_2(Z)$ tangent vectors at Z, the following identity holds:

$$\delta A^2{}_{|Z}\langle X_2(Z), Y_2(Z)\rangle = dT_{|Z}\langle Y_2(Z)\rangle. \tag{4.39}$$

Denote the gradient $\nabla_{\delta A^2}[T](Z)$. Letting Z vary in $T_1(U)$, this prescription defines a smooth vector field on $T_1(U)$. It should also be observed here that this definition is *independent of frame.* Both the differential of T and the metric δA^2 have a frame-independent meaning. We have chosen the component representations above in order to facilitate the proof of the following theorem. ■

Theorem 4.36. *Under the conditions of Definition* 4.35, *let* $G(Z)$ *be the* geodesic vector field *on* $T_1(M)$ *associated to the metric* $A^2(x)$. *Then for each* $Z \in T_1(M)$ *we have* $G(Z) = \nabla_{\delta A^2}[T](Z)$. *The geodesic vector field is the gradient field for the kinetic energy with respect to the promoted metric* δA^2.

Proof. This result is perhaps not surprising in light of the fact that the 'Hamiltonian' field associated with a conservative system is a 'symplectic gradient' on the cotangent bundle (Hirsch and Smale [11] or Maclane [16]). Just as in that case, the metric is indefinite and so care must be taken in forming conclusions concerning the gradient flow. (See also Thom [25].)

Setting (4.37) equal to (4.38), we conclude immediately that $X^i = Z^i$ for all i by equating the terms involving arbitrary C^t. Next, since Y^j is arbitrary, we conclude that

$$\frac{\partial a_{ik}}{\partial x^j}\Big|_{z_0} Z^i Z^j Y^k + a_{sk}(z_0) B^s Y^k = \frac{1}{2}\frac{\partial a_{ij}}{\partial x^k}\Big|_{z_0} Y^k Z^i Z^j. \tag{4.40}$$

This easily implies that

$$B^r = \tfrac{1}{2} a^{rk}(z_0)\left[\frac{\partial a_{ij}}{\partial x^k}\Big|_{z_0} - 2\frac{\partial a_{ik}}{\partial x^j}\Big|_{z_0}\right] Z^i Z^j = \Lambda^r_{ij}(z_0) Z^i Z^j.$$

Thus $X_2(Z)$ is the flat 2-sector with both vertices equal to Z. ■

It is worthwhile to point out that, in many cases, the matrix form of δA^2 is easy to write down and the differential dT easy to describe in local coordinates on $T_1(M)$. The second-order differential equation for geodesics can then be written down directly without an explicit detour through the Riemann–Christoffel symbols (although, of course, they make their *implicit* appearance). A final remark on this interpretation is this: we have excluded 'forces' from this picture only for the sake of clarity of presentation in order first to present the 'kinematical' aspect of the strategy. Classically, in the study of the motion of conservative systems, the kinetic energy is augmented by $-V$, the negative potential, a function whose differential gives 'force' when force is thought of (as it should be) as a 1-form on M, the manifold of configurations. Thus the kinetic energy is replaced by the general 'Lagrangian', $T - V = L$. And L is interpreted as a smooth map from $T_1(M) \to \mathbb{R}$. As such, it has a differential and its gradient can be computed with respect to the metric δA^2 on $T_1(M)$. Then $\nabla_{\delta A^2}[L]$ will give a vector field on the tangent bundle with the property that the curves $\gamma : \mathbb{R} \to M$, such that $\gamma_* : \mathbb{R} \to T_1(M)$ gives a solution for that vector field, are precisely the curves that satisfy the *Euler–Lagrange conditions:* the classical 'evolution' curves for the conservative mechanical system. We shall develop this idea in a slightly different way.

We have presented two *equivalent* conditions for a curve $\gamma : \mathbb{R} \to (M, A^2(x))$ to be a geodesic: (i) that $\gamma_* : \mathbb{R} \to T_1(M)$ be a solution curve for $G(X)$, the geodesic vector field, and (ii) that $\gamma_* : \mathbb{R} \to T_1(M)$ be a solution for the gradient vector field $\nabla_{\delta A^2}[T](X)$ on $T_1(M)$. These are obviously *local conditions.* And if we examine formula (4.40), we see that they are satisfied locally *if and only if* for g as in Definition 4.34

$$\frac{\partial a_{ik}}{\partial x^j}_{|g(t_0)} \frac{dg^i}{dt}_{|t_0} \frac{dg^j}{dt}_{|t_0} + a_{sk}(g(t_0)) \frac{d^2 g^s}{dt^2}_{|t_0}$$
$$= \frac{1}{2} \frac{\partial a_{ij}}{\partial x^k}_{|g(t_0)} \frac{dg^i}{dt}_{|t_0} \frac{dg^j}{dt}_{|t_0}. \tag{4.41}$$

This formula may be written in a slightly different form:

$$\frac{d}{dt_{|t_0}}\left[a_{ik}[g(t)]\frac{dg^i}{dt}\right] = d_x T_{|[g(t_0),(dg^k/dt)_{|t_0}]} \tag{4.42}$$

where $T = T(z, Z^k)$ is a function of *position* variables and *velocity* variables in frame $\tilde{\theta}^1$ and the symbol $d_x T$ is the 'partial differential' with respect to the position variables. Formula (4.42) is the classical *Euler–Lagrange equation* and the expression $[a_{ij}[g(t)](dg^i/dt)]$ which appears on the left is interpreted as the 'momentum' of g at t_0; it is a covector on M at $g(t_0)$ (again, with a frame-independent meaning).

We shall not pursue this classical interpretation for the local conditions for a geodesic as the notation becomes cumbersome; the condition we use is the one given in Definition 4.34, namely $\gamma:\mathbb{R} \to (M, A^2(x))$ is a geodesic if $\gamma_{**}(\tau)$ is *flat* (has 0 force) for each τ. A curve in a Riemannian manifold satisfying this condition will be called an *Euler–Lagrange curve.* We conclude this section with a discussion of the equivalence of this *local* condition with a certain *global* condition which appears in the formulation of the 'principle of least action'. Let $I \subset \mathbb{R}$ be a closed interval $I = [a, b]$ and let $\gamma: I \to (M, A^2(x))$ be a smooth curve (hence with smooth extension to some interval $(a-\varepsilon, b+\varepsilon)$, $\varepsilon > 0$). We shall say that γ is a geodesic in this case if $\gamma_{**}: I \to T_2(M)$ gives flat 2-sectors for each $\tau \in I$.

Definition 4.37. For $\gamma: I \to (M, A^2(x))$ as above, define the *energy* of γ to be

$$\tfrac{1}{2}\int_a^b A^2(\gamma(t))\langle \gamma_*(t), \gamma_*(t)\rangle\, dt = \mathscr{E}(\gamma)$$

This is just the integral over the path of the kinetic energy. ■

Intuitively speaking, a curve γ is *critical for the energy* if for a smooth 1-parameter family of curves $\gamma_u: I \to (M, A^2(x))$, $u \in \mathbb{R}$ *with* $\gamma_0 = \gamma$, $\gamma_u(a) = \gamma(a)$ for all u and $\gamma_u(b) = \gamma(b)$ for all u, and if the mapping $u \to \mathscr{E}(\gamma_u)$ is differentiable at $u = 0$ *then*

$$\frac{d}{du_{|0}}[\mathscr{E}(\gamma_u)] = 0.$$

The intention of the definition is clear, although the formulation is awkward because of the difficulty of explaining quickly what the 'differential' of a function like $\mathscr{E}(\gamma)$ ought to be. It is less difficult to define, heuristically, a 'tangent vector' to the curve γ: in fact such a notion has already been met in the study of 1-variations. We use the idea of 1-variation of γ to give a precise meaning to what was formulated intuitively above. (See Arnold [3].)

Since we want a somewhat more *global* formulation than that given in Definition 2.13, we shall *not* restrict $\text{im}(\gamma)$ to belong to the image of a local frame on M.

Definition 4.38. For $\gamma: I \to (M, A^2(x))$ as above, define a *global* 1-variation of γ to be an *equivalence class* of germs $V:(I\times\mathbb{R}, I\times O)\to (M, \text{im}(\gamma))$ with $V(t,0)=\gamma(t)$ for $t\in I$ with respect to the equivalence relation $V\simeq W$ if $\bar{V}_2^{\#}(t,0)=\bar{W}_2^{\#}(t,0)$ for all $t\in I$. (Here we let (t,u) be standard coordinates in $I\times\mathbb{R}$ and $\bar{V}_2^{\#}(t,0)$ be the F-related class of the germ $V:(I\times\mathbb{R},(t,0))\to(M,\gamma(t)))$. Denote the variation V. For simplicity, denote the tangent vector $D^2(2)\bar{V}_2^{\#}(t,0)$ by $X_V(t)=\gamma_*(t)$ and denote the tangent vector $D^1(2)\bar{V}_2^{\#}(t,0)$ by $Y_V(t)$.

Finally, for $\gamma: I\to(M, A^2(x))$ there is a special 1-variation of γ called the *natural* 1-*variation* which will be distinguished. The natural 1-variation is the equivalence class of the germ $N:(I\times\mathbb{R}, I\times O)\to (M, \text{im}(\gamma))$ given (locally) by $N(t,u)=\gamma(t+u)$. Recall that γ has a smooth extension to an open interval containing I, so that this makes sense even when t is a or b. As a germ, N is well-defined, and thus it has a well-defined equivalence class with respect to the equivalence relation introduced above. The associated tangent vectors for N will be denoted $X_N(t)$ and $Y_N(t)$ for $t\in I$. It should be observed that, for a given $t\in I$, $\bar{N}_2^{\#}(t,0)=\gamma_{**}(t)$. This is most easily seen by calculating components with respect to a frame for M. ■

We are now in a position to give a clear statement of what it means to say that a curve is 'critical for the energy'.

Definition 4.39. Let $\gamma: I\to(M, A^2(x))$ be as above and let V be a 1-variation of γ. For t_0 fixed in I, consider the tangent vector $Y_V(t_0)$. We may consider $\bar{V}_2^{\#}(t_0,0)$ a tangent vector to $T_1(M)$ *at* $Y_V(t_0)$ and form the contraction $\delta A^2{}_{|Y_V(t_0)}\langle \bar{V}_2^{\#}(t_0,0), \bar{V}_2^{\#}(t_0,0)\rangle$ (we drop mention of the natural isomorphism ν, hoping that by now there is no danger of confusion). This contraction is simply the contraction of d_3A^2 with *any* 3-sector with components with respect to frame θ at $\gamma(t_0)$ in M given below (with

$\theta^{-1} \circ V = f$):

$$\begin{array}{ccccc}
\left[(\theta \circ t_{\gamma(t_0)})_3, \dfrac{\partial f^i}{\partial t}\Big|_{(t_0,0)} \right. & & \text{———— * ————} & & \left. \dfrac{\partial f^j}{\partial t}\Big|_{(t_0,0)} \right] \\
& \diagdown & * & \diagup & \\
& \dfrac{\partial^2 f^s}{\partial t \partial u}\Big|_{(t_0,0)} & & \dfrac{\partial^2 f^t}{\partial t \partial u}\Big|_{(t_0,0)} & \\
& & \diagdown \quad \diagup & & \\
& & \dfrac{\partial f^k}{\partial u}\Big|_{(t_0,0)} & &
\end{array}$$

and its value is

$$\frac{\partial a_{ij}}{\partial x^k}\Big|_{\gamma(t_0)} \frac{\partial f^i}{\partial t}\Big|_{(t_0,0)} \frac{\partial f^j}{\partial t}\Big|_{(t_0,0)} \frac{\partial f^k}{\partial u}\Big|_{(t_0,0)} + 2a_{sj}[\gamma(t_0)] \frac{\partial^2 f^s}{\partial t \partial u}\Big|_{(t_0,0)} \frac{\partial f^j}{\partial t}\Big|_{(t_0,0)} \tag{4.43}$$

(computed in frame θ) according to formula (4.3). Notice that this can be obtained also directly from formula (4.37).

Now for a given V, define

$$\Delta_V(\mathscr{E}) = \int_a^b \tfrac{1}{2}\delta A^2{}_{|Y_V(t)}\langle \bar{V}_2^{\#}(t,0), \bar{V}_2^{\#}(t,0)\rangle \, dt. \tag{4.44}$$

Thinking of V as a 1-parameter family of paths 'passing through' γ, this is the interpretation we give to $(d/du)_{|0}[\mathscr{E}(\gamma_u)]$ via the usual differentiation 'under the integral'. Checking formula (4.43) with respect to local frames, we see that it is precisely what is required and that the 1-variation V contains just enough information to give its value.

Now say that the curve γ is *critical for the energy* if for each variation V with $Y_V(a) = Y_V(b) = 0$ we have $\Delta_V(\mathscr{E}) = 0$. ■

Now if for the 3-sector which appears in formula (4.43) we take one whose 3-side is $\gamma_{**}(t_0)$, then the *cycle product* $DA^2\langle \gamma_{**}(t_0), Y_V(t_0)\rangle$ for fixed t_0 has a special interpretation.

Observation 4.40. Under the conditions of the previous definition, for $\gamma: I \to (M, A^2(x))$ and V an arbitrary 1-variation of γ, for $t_0 \in I$ fixed,

$$\begin{aligned} DA^2\langle \gamma_{**}(t_0), Y_V(t_0)\rangle &= 2\delta A^2{}_{|X_N(t_0)}\langle \bar{N}_2^{\#}(t_0,0), \bar{V}_2^{\#}(t_0,0)\rangle \\ &\quad - \delta A^2{}_{|Y_V(t_0)}\langle \bar{V}_2^{\#}(t_0,0), \bar{V}_2^{\#}(t_0,0)\rangle \end{aligned} \tag{4.45}$$

where the right-hand side is the contraction of $d_1A^2+d_2A^2-d_3A^2$ (the cycle DA^2) with *any* 3-sector at $\gamma(t_0)$ with θ components

$$\left[(\theta\circ t_{\gamma(t_0)})_3,\ \begin{array}{ccccccc} \dfrac{\partial f^i}{\partial t}\Big|_{(t_0,0)} & & \text{———} & \dfrac{\partial^2 g^r}{\partial t^2}\Big|_{(t_0,0)} & \text{———} & & \dfrac{\partial f^j}{\partial t}\Big|_{(t_0,0)} \\ & \diagdown & & & & \diagup & \\ & \dfrac{\partial^2 f^s}{\partial t\partial u}\Big|_{(t_0,0)} & & * & & \dfrac{\partial^2 f^t}{\partial t\partial u}\Big|_{(t_0,0)} & \\ & & \diagdown & & \diagup & & \\ & & & \dfrac{\partial f^k}{\partial u}\Big|_{(t_0,0)} & & & \end{array}\right]$$

This contraction is computed according to formula (4.7) and with $f=\theta^{-1}\circ V$, $g=\theta^{-1}\circ\gamma$ (notice that $\partial f^i/\partial t=\partial g^i/\partial t$).

Also it follows from the discussion leading up to formula (3.19) or from a direct calculation in components that

$$\delta A^2{}_{|X_N(t_0)}\langle \bar{N}_2^{\#}(t_0,0),\ \bar{V}_2^{\#}(t_0,0)\rangle=\frac{d}{dt}\Big|_{t_0}[A^2[\gamma(t)]\langle X_V(t),\ Y_V(t)\rangle]. \qquad (4.46)$$

■

Putting all of this together, and remembering that

$$\tfrac{1}{2}DA^2\langle\gamma_{**}(t_0),\ Y_V(t_0)\rangle=A^2\langle F(t_0),\ Y_V(t_0)\rangle$$

where $F(t_0)=\text{force of }\gamma_{**}(t_0)$, we have the *first variation formula*, as follows.

Theorem 4.41. *Let $\gamma: I\to(M, A^2(x))$ be smooth and let V be a 1-variation of γ, then (for $I=[a,b]$, and $F(t)=$* force of *$\gamma_{**}(t)$ for $t\in I$)*

$$\int_a^b A^2[\gamma(t)]\langle F(t),\ Y_V(t)\rangle\,dt=A^2[\gamma(t)]\langle X_V(t),\ Y_V(t)\rangle\Big|_a^b-\Delta_V(\mathscr{E}). \qquad (4.47)$$

Proof. There is nothing left to prove here. We think of $F(t)$ as a 'lifting' from γ to $T_1(M)$ which, of course, depends only on γ (or N). Then this formula can be interpreted as depending only on V. ■

Corollary 4.42. *γ is a geodesic in the above situation* if and only if *γ is critical for the energy: that is $\Delta_V(\mathscr{E})=0$ for all V with $Y_V(a)=Y_V(b)=0$.*

Proof. γ is a geodesic if and only if $F(t)=0$ for all $t\in I$, by definition. Thus for a geodesic, if V satisfies the condition stated above at the endpoints, then $\Delta_V(\mathscr{E})=0$. On the other hand, if $\Delta_V(\mathscr{E})=0$ for all such variations, then it is easy to see that $F(t)=0$ for all $t\in I$ by supposing that, for some $t_0\in(a, b)$, $F(t_0)\neq 0$, and then constructing an appropriate variation for which the left-hand integral would not vanish. ■

It follows immediately from Observation 4.40 that if γ is a geodesic then it is parametrized in such a way that $A_2[\gamma(t)]\langle\gamma_*(t), \gamma_*(t)\rangle$ is *constant* on I. Simply apply the result in the case that $V=N$, the natural variation.

The above theorem has another interesting application which will eventually lend itself to an interpretation of the notion of curvature.

Definition 4.43. Let $\gamma: I\to(M, A^2(x))$ be smooth. Then 1-variation V of γ is a *parallel transport* if $\bar{V}_2^{\#}(t, 0)$ is *flat* for all $t\in I$. Recalling the definition of a *linear variation equation* given in Observation 2.15 as section α of the bundle

$$\begin{array}{c} I\times_M T_2(M) \\ \alpha\uparrow\ \downarrow \mathrm{id}\times_M D^1(2) \\ I\times_M T_1(M) \end{array}$$

satisfying conditions a), b), and c) stated there, we see that if we define α by $\alpha(t; Y)=\Lambda(\gamma_*(t), Y)$ then these conditions are met. And for a *fixed* choice of $Y\in T_1[\gamma(a)]$ there is a *unique* parallel transport V with $Y=Y_V(a)$. ■

Before giving the application promised, we need a lemma.

Lemma 4.44. *Suppose that $\bar{A}_2(x_0)$ and $\bar{B}_2(x_0)$ are flat 2-sectors in $T_2(M)$ for $(M, A^2(x))$ a Riemannian manifold. And suppose that $D^2(2)\bar{A}_2=D^2(2)\bar{B}_2=X$ in $T_1(M)$. Abusing notation slightly, we form $\delta A^2{}_{|X}\langle\bar{A}_2(x_0), \bar{B}_2(x_0)\rangle$. This contraction is* 0.

Proof. It follows from Observation 4.10 that the matrix for δA^2 in a null frame is

$$\begin{vmatrix} 0 & a_{it}(z_0) \\ a_{sj}(z_0) & 0 \end{vmatrix}.$$

Now since $\bar{A}_2(x_0)$ and $\bar{B}_2(x_0)$ are flat, their components in this null frame both give middle (11) face 0. ■

This last lemma has interesting geometric interpretation. In particular, every flat 2-sector is 'orthogonal' to itself, showing again that δA^2 is indefinite.

Theorem 4.45. *If $\gamma: I \to (M, A^2(x))$ is smooth and 1-variation V is a parallel transport, then:*

1) $\Delta_V(\mathscr{E}) = 0$.
2) *For $F(t)$ the* force of $\gamma_{**}(t)$

$$\int_a^b A^2[\gamma(t)]\langle F(t), Y_V(t)\rangle\, dt = A^2[\gamma(t)]\langle X_V(t), Y_V(t)\rangle|_a^b.$$

In particular, if γ is a geodesic, then $A^2[\gamma(t)]\langle X_V(t), Y_V(t)\rangle$ is constant.

3) *If W is another parallel transport for γ, then $A^2[\gamma(t)]\langle Y_V(t), Y_W(t)\rangle$ is constant on I (whether or not γ is a geodesic).*

Proof. 1) This follows from the fact that

$$\delta A^2{}_{|Y_V(t_0)}\langle \bar{V}_2^{\#}(t_0, 0), \bar{V}_2^{\#}(t_0, 0)\rangle = 0.$$

From this 2) is immediate from formula (4.47). For 3) observe that

$$\frac{d}{dt}\Big|_{t_0} [A^2[\gamma(t)]\langle Y_V(t), Y_W(t)\rangle] = \delta A^2{}_{|X_V(t_0)}\langle \bar{V}_2^{\#}(t_0, 0), \bar{W}_2^{\#}(t_0, 0)\rangle$$

and this, by hypothesis, is 0. ■

We see that a parallel transport is 'length-preserving' and for two parallel transports the 'angles' are preserved. In particular, if $\gamma(a) = \gamma(b)$, then parallel transport around γ, as a map from $T_1[\gamma(a)] \to T_1[\gamma(a)]$, is a linear isometry (linearity follows easily from the fact that these transports are the solutions of *linear* variation equations). When γ is homotopic to the constant map, we conclude that it is a *rotation.* In the two-dimensional case, there is a simple relationship between the 'amount' of rotation, the curvature, and the area enclosed by the loop γ (see section 4.3). In order to elaborate upon these ideas, it will be necessary to develop the notion of *Riemannian curvature.*

We conclude this section with a formulation of the Euler–Lagrange equations for mechanical systems including conservative forces. This formulation will easily be seen to be equivalent to the Hamiltonian approach via the 'balance' of the metric. Begin with the data (M, A^2, h) of a smooth manifold, a Riemannian metric, and a smooth real-valued function $h: M \to \mathbb{R}$. A^2 determines a 'kinetic energy' function $T: T_1(M) \to \mathbb{R}$ as was mentioned at the beginning of the section, and h is thought of as the 'negative potential' so that dh is the force form on M.

Definition 4.46. *Given the data* (M, A^2, h) as above, define a vector field on $T_1(M)$ by the following prescription: for each $X(x)$ in $T_1(M)$

there is a *unique* 2-sector $\bar{X}_2(x)$ at x in M such that

1) $D(2)\bar{X}_2(x) = (X(x), X(x))$, and
2) $\frac{1}{2}DA^2\langle \bar{X}_2(x), Y(x)\rangle = dh\langle Y(x)\rangle$ for all tangent vectors $Y(x)$ at x.

Calculating in a null frame shows the uniqueness immediately.

The association to each $X(x)$ of the corresponding $\bar{X}_2(x)$ defines a smooth vector field on $T_1(M)$, a generalization of the 'gradient' in the obvious sense, possible because the sectorform DA^2 gives a pairing of 2-sectors with tangent vectors. With this fact in mind, call the field $\nabla_{\frac{1}{2}DA^2}(h)$ with the 2-sectors *pointed* in the usual way (Definition 3.10).

Now if $\phi: O \to U \subset M$ is a frame and $\gamma(t)$ a path in O with $\phi(\tau) = z$, let

$$\gamma_{**}(\tau) = \left[(\phi \circ t_z)_2, \frac{d\gamma^i}{dt}_{|\tau} \quad \frac{d^2\gamma^r}{dt^2}_{|\tau} \quad \frac{d\gamma^j}{dt}_{|\tau}\right]$$

then $\gamma_{**}(\tau)$ is a solution curve for the field $\nabla_{\frac{1}{2}DA^2}(h)$ if and only if for *each* τ_0 and for *each* tangent vector $Y(\gamma(\tau_0)) = [(\phi \circ t_{z_0})_1, Y^k(\gamma(\tau_0))]$,

$$\left(\frac{d}{dt}_{|\tau_0}\left[\tilde{a}_{jk}(\gamma(\tau))\frac{d\gamma^j}{dt}\right]\right)Y^k(\gamma(\tau_0))$$
$$= \left(\frac{\partial}{\partial x^k}_{|\gamma(\tau_0)}\left[\tfrac{1}{2}\tilde{a}_{ij}\frac{d\gamma^i}{dt}\frac{d\gamma^j}{dt} + h \circ \phi\right]\right)Y^k(\gamma(\tau_0)). \quad \blacksquare$$

Exercise

9. Prove D'Alembert's equation for the Euler–Lagrange flow: if $f:(M, A^2) \to (N, B^2)$ is a smooth *immersion* with $A^2 = f*B^2$, $k \circ f = f^*(k) = h$, show that the *bracket* (for $X(x) \in T_1(M)$) $[T_2(f)\nabla_{\frac{1}{2}DA^2}(h)(X); \nabla_{\frac{1}{2}DB^2}(k)(T_1(f)(X))]$ is *orthogonal* relative to B^2 to $T_1(f)[T_1(x)]$.

§4.3. Riemannian curvature

We saw in §4.1 that for a Riemannian manifold $(M, A^2(x))$ it is possible to define an order-3 sectorform field $DA^2(x)$. This field is a K-sectorform field for K the subcomplex of $\Gamma(2)$:

(*——*——*).
      *

We use this field to give a pairing of 2-sectors and 1-sectors from which the tensor $\Omega(j)$ is defined for isometric immersions j. In the case of an immersion of a surface in $\mathbb{R}^3$, this tensor is the basis for the definition of Gaussian curvature (formula (4.31)). In this section, we define a sectorform field of order 4 which we shall call the *cross* of the metric and which will give an interpretation of the invariance of the numerator in formula

(4.31). We shall interpret the 'covariant' curvature tensor in terms of the cross of the metric and develop some of its classical geometric properties by a method analogous to that of Cartan's *Repere Mobile.*

In order to motivate the definition of the cross, we introduce another 3 sectorform field which is naturally derived from the metric.

Definition 4.47. For $(M, A^2(x))$ a Riemannian manifold, define 3 sectorform field

$$BA^2(x) = d_1A^2(x) - d_2A^2(x)$$

and call it the *balance* of the metric. In components, if $A^2(x) = \tilde{a}_{ij}(z)$ then $BA^2(x)$ has the form (following Convention 3.9):

$$\frac{\partial \tilde{a}_{jk}}{\partial x^i} - \frac{\partial \tilde{a}_{ik}}{\partial x^j} + \tilde{a}_{sj} - \tilde{a}_{it}. \tag{4.48}$$

Let us agree to give the following names to subcomplexes of $\Gamma(2)$:

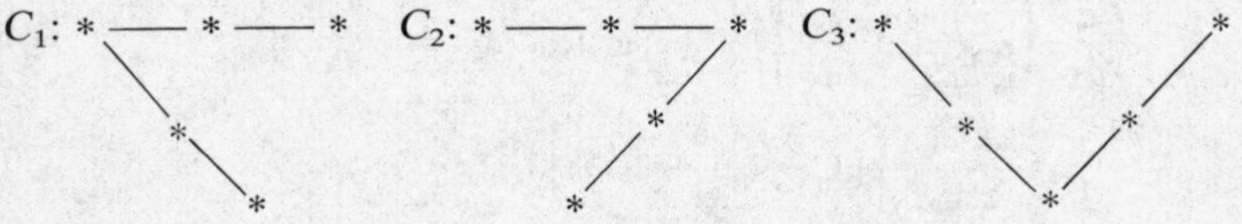

Then BA^2 is a C_3-sectorform field. ■

Clearly, on pointing, the sectorform BA^2 yields a 2-form on the tangent bundle $T_1(M)$ (exercise 5). This 2-form is the pull-back (under the Legendre diffeomorphism associated to the Kinetic Energy of the metric) of the canonical 2-form on the cotangent bundle. Recall formula (3.12) which represents $p\bar{A}_3$ for

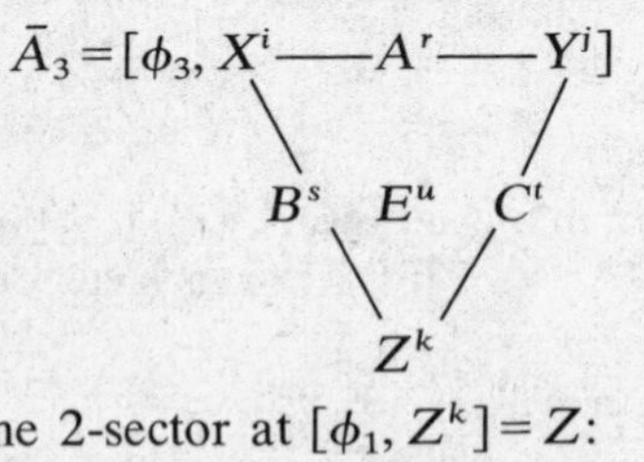

as the 2-sector at $[\phi_1, Z^k] = Z$:

$$\left[(\phi^1 \circ t_{(0,Z^k)})_2, \begin{bmatrix} X^i \\ B^s \end{bmatrix} \!-\!\!-\! \begin{bmatrix} A^r \\ E^u \end{bmatrix} \!-\!\!-\! \begin{bmatrix} Y^j \\ C^t \end{bmatrix}\right] = p\bar{A}_3.$$

We have $BA^2{}_{|Z}\langle p\bar{A}_3\rangle = BA^2\langle \bar{A}_3\rangle$ and so if $A^2(x)$ has local components $\tilde{a}_{ij}(z)$ then $BA^2{}_{|Z}$ has components:

$$\frac{\partial \tilde{a}_{jk}}{\partial x^i} Z^k - \frac{\partial \tilde{a}_{ik}}{\partial x^j} Z^k + \tilde{a}_{sj} - \tilde{a}_{it}.$$

Now the pointed *metric* $A^2{}_{|Z}$ has local components $\tilde{a}_{ik}Z^k$ as 1-form on the tangent bundle. Computing the coboundary of this 1-form in local components is slightly tricky: the contraction of $A^2{}_{|Z}$ with the tangent vector *at* Z with $\tilde{\phi}^1 \circ t_{(0,Z^k)}$ components $\begin{bmatrix} X^i \\ B^s \end{bmatrix}$ is simply $a_{ik}(z)Z^kX^i$. From this, it is fairly easy to see that $d_1(A^2{}_{|Z})$ has local components

$$\frac{\partial \tilde{a}_{jk}}{\partial x^i} Z^k + \tilde{a}_{js} + \tilde{a}_{rk}Z^k \tag{4.49}$$

and that $d_2(A^2{}_{|Z})$ has local components

$$\frac{\partial \tilde{a}_{ik}}{\partial x^j} Z^k + \tilde{a}_{it} + \tilde{a}_{rk}Z^k, \tag{4.50}$$

and thus that

$$d(A^2{}_{|Z}) = d_1(A^2{}_{|Z}) - d_2(A^2{}_{|Z}) = \frac{\partial \tilde{a}_{jk}}{\partial x^i} Z^k - \frac{\partial \tilde{a}_{ik}}{\partial x^j} Z^k + \tilde{a}_{sj} - \tilde{a}_{it} \tag{4.51}$$

in local components.

This gives the desired interpretation of the *pointed balance of the metric*, as follows.

Observation 4.48. For $(M, A^2(x))$ a Riemannian manifold, the *pointed balance* $BA^2{}_{|Z}$ is a 2-form on $T_1(M)$ with components given in (4.51). It is equal to $d(A^2{}_{|Z})$, the coboundary of the pointed metric. ■

We observed in exercise 5 that if $\tilde{Z}^k(z) = Z(x)$ is a local vector field on M, then the *reduction* $A^2[Z]$ is a 1-form on M, the *dual form with respect to the metric* defined by the contraction formula $A^2[Z]\langle X\rangle = A^2\langle X \cup Z(x)\rangle$. Thinking of $Z(x)$ as smooth section $M \to T_1(M)$ (defined locally) we see that $A^2[Z]$ is the pull-back of the 1-form $A^2{}_{|Z}$. Therefore it follows from the naturality of the coboundary operator that $d(A^2[Z])$ is the pull-back of the pointed balance $BA^2{}_{|Z}$.

This is the strategy that we intend to generalize: we start with 3-sectorform field DA^2 on Riemannian manifold $(M, A^2(x))$. Then given a 2-sector *field* $\bar{V}_2(x)$ on M, it may be thought of as a smooth section $M \to T_2(M)$, and since DA^2 is identifiable as a 1-form on $T_2(M)$ (via an extension of the pointing operation which we shall describe) that 1-form pulls back under $\bar{V}_2$ to give a 1-form on M. Then the *cross* of A^2 will be a 4-sectorform which when 'pointed' by 2-sector $\bar{V}_2$ gives the coboundary of the 1-form on $T_2(M)$, namely DA^2 'pointed' by $\bar{V}_2$. Now the cross of A^2 will give an immediate invariant definition of the covariant Riemann curvature tensor; its pull-back under $\bar{V}_2(x)$ will have some interesting geometric interpretations in the case that $D(2)\bar{V}_2(x)$ is an orthonormal pair of vector fields. In a sense, the cross of A^2 gives a number which describes the motion of a 2-sector field with respect to a fixed 2-sector: that number is the *curvature* (determined entirely by four tangent vectors, the *boundaries* of the pair of 2-sectors at the point of interest) when the motion of the sector field is 'parallel' in a sense to be made precise later.

The first thing to do is to give the appropriate generalization of the 'pointing' operator together with the corresponding notion of 'reduction'.

Definition 4.49. Let $\Gamma(k+r-1)$ be the $(k+r-1)$ simplex and let $K(r)$ be the subcomplex of faces of the form $(\underbrace{0,\ldots,0}_{k};\gamma'_1,\ldots,\gamma'_r)=\gamma'$. Agree to represent an $(r+k)$ sector $\bar{A}_{r+k}$ *with respect to* $K(r)$ in the following way:

1) γ' represents an arbitrary face of $K(r)$,
2) γ'' represents a general face of the form $(\gamma''_1,\ldots,\gamma''_k;\underbrace{0,\ldots,0}_{r})$,
3) $\gamma'\cup\gamma''$ represents a general join of faces of type 1) and 2).

Then in a fashion similar to that used in Definition 3.10, we represent $\bar{A}_{k+r}$ as $[\phi_{k+r}, X^i_{\gamma'}\text{——}Z^i_{\gamma'\cup\gamma''}\text{——}Y^i_{\gamma''}]$ and call this the representation with respect to $K(r)$.

Modify the natural isomorphism ν (introduced in Theorem 1.27) to give the following representation of $\bar{A}_{k+r}$ as an element of $T_k[T_r(M)]$ *at* $[\phi_r, X^i_{\gamma'}]$ as $[\tilde{\theta}^r\circ t_{(0,X^i_{\gamma'})})_k, W^\alpha_{\gamma''}]$ with $W^\alpha_{\gamma''}$ determined in the following way. We suppose that M is m-dimensional, and that α has the form $(i,0)$ or (i,γ') for $1\leq i\leq m$, γ' and γ'' as above. Thus α may take $m\cdot 2^r$ values and we set:

$$W^{(i,0)}_{\gamma''}=Y^i_{\gamma''}\quad\text{and}\quad W^{(i,\gamma')}_{\gamma''}=Z^i_{\gamma'\cup\gamma''}.$$

Call the modified natural isomorphism $p:T_{k+r}(M)\to T_k[T_r(M)]$ as we did in the case of Definition 3.10. In this case, $p\bar{A}_{k+r}$ is the sector $\bar{A}_{k+r}$ *pointed by the r-sector* $[\phi_r, X^i_{\gamma'}]$.

Now the map p is linear on intrinsic spaces $T_{k+r,i}(X)\to T_{k,i}(pX)$ for $1\leq i\leq k$. Therefore, if $A^{k+r}(x_0)$ is a $(k+r)$ sectorform at $x_0\in M$ and $\bar{A}_r(x_0)$ is a fixed sector at x_0 of order r, the association $T_k[\bar{A}_r(x_0)]\to\mathbb{R}$ (the fiber in $T_k[T_r(M)]$ over r-sector $\bar{A}_r(x_0)$) given by $A^{k+r}(x_0)\circ p^{-1}$ is a k sectorform at $\bar{A}_r(x_0)=[\phi_r, X^i_{\gamma'}]$. We shall call that k sectorform the *sectorform* $A^{k+r}(x_0)$ *pointed at the r-sector* $\bar{A}_r(x_0)$ and denote it $A^{k+r}(x_0)_{|\bar{A}_r(x_0)}$. ■

Exercise

10. Show the following naturality property of the pointing operation. If $f:(M,x_0)\to(N,y_0)$ is smooth and $\bar{A}_r(x_0)$ and $B^{k+r}(y_0)$ are given,

$$f^*[B^{k+r}(y_0)]_{|\bar{A}_r(x_0)}=T_r(f)^*[B^{k+r}(y_0)_{|T_r(f)(\bar{A}_r(x_0))}].$$

The next step is to introduce the idea of 'reduction' of a $(k+r)$ sectorform by an r-sector *field*. In order to do this, we need a further generalization of the suspension operation. Suppose that for $x_0 \in M$, $\bar{A}_k(x_0)$ is a fixed k-sector and $\bar{B}_r(x)$ is a local r-sector field. We want to define a $(k+r)$ sector at x_0 denoted $\bar{A}_k(x_0) \cup \bar{B}_k(x)$ which extends all previous notions of 'suspension' to the case of the suspension of a sector field by a sector.

Definition 4.50. Suppose that at $x_0 \in M$, $\bar{A}_k(x_0) = [\phi_k, Y^i_{\gamma''}]$ is a k-sector and $\bar{B}_r(x) = [(\phi \circ t_z)_r, \tilde{X}^i_{\gamma'}(z)]$ is a local r-sector *field* (z in some neighborhood of $0 \in E$ and ϕ a local frame defined on that neighborhood). Then define the *suspension* of $\bar{B}_r(x)$ *by sector* $\bar{A}_k(x_0)$ to be the following $(k+r)$ sector at x_0: identify each $\gamma' \in \Gamma(r-1)$ with a face of $\Gamma(k+r-1)$ by the association

$$(\gamma'_1, \gamma'_2, \ldots, \gamma'_r) \to (0, \underbrace{0, \ldots, 0}_{k}; \gamma'_1, \gamma'_2, \ldots, \gamma'_r)$$

and similarly identify each $\gamma'' \in \Gamma(k-1)$ with a face of $\Gamma(k+r-1)$ by

$$(\gamma''_1, \gamma''_2, \ldots, \gamma''_k) \to (\gamma''_1, \gamma''_2, \ldots, \gamma''_k; 0, \underbrace{0, \ldots, 0}_{r}).$$

Then if we represent elements of $T_{k+r}(M)$ with respect to $K(r)$ as we did in Definition 4.49, consider the $(k+r)$ sector

$$[\phi_{k+r}, X^i_{\gamma'}(0) \text{———} Z^i_{\gamma' \cup \gamma''} \text{———} Y^i_{\gamma''}]$$

with the obvious values for γ' and γ'' components and the components $Z^i_{\gamma' \cup \gamma''}$ defined in the following way:

$$Z^i_{\gamma' \cup \gamma''} = \sum_{\gamma''_{i_1} \cup \ldots \cup \gamma''_{i_s} = \gamma''} \frac{\partial^s \tilde{X}^i_{\gamma'}(z)}{\partial x^{k_1} \ldots \partial x^{k_s}} Y^{k_1}_{\gamma''_{i_1}} \ldots Y^{k_s}_{\gamma''_{i_s}}. \tag{4.52}$$

Denote this $(k+r)$ sector $\bar{A}_k(x_0) \cup \bar{B}_r(x)$. The following lemma will demonstrate that this suspension is well-defined, that is that it does not depend on the particular local frame used to compute its components. ■

Lemma 4.51. *Under the conditions of the previous definition the $(k+r)$ sector $\bar{A}_k(x_0) \cup \bar{B}_r(x)$ is well-defined.*

Proof. We suppose that θ is another local frame defined in a neighborhood of 0 in E. Say $\theta(u) = \phi(z) = x$ and let $g(z) = \theta^{-1} \circ \phi(z)$ be the germ of a frame change. Then

$$\bar{A}_k(x_0) = [\theta_k, g_k Y^i_{\gamma''}] = [\phi_k, Y^i_{\gamma''}]$$

and

$$\bar{B}_r(x) = [(\phi \circ t_z)_r, \tilde{X}^i_{\gamma'}(z)] = [(\theta \circ t_u)_r, g_r(z) \tilde{X}^i_{\gamma'}(g^{-1}(u))]$$

where $g_r(z)$ is the r-jet of $t_u^{-1} \circ g \circ t_z$.

We calculate $\bar{A}_k(x_0) \cup \bar{B}_r(x)$ with respect to local frame θ and get:

$$[\theta_{k+r}, g_r X^i_{\gamma'}(0) \text{——} \bar{Z}^i_{\gamma' \cup \gamma''} \text{——} g_k Y^i_{\gamma''}].$$

Following formula (4.52) the components $\bar{Z}^i_{\gamma' \cup \gamma''}$ are defined in the following way:

$$\bar{Z}^i_{\gamma' \cup \gamma''} = \sum_{\gamma''_{i_1} \cup \ldots \cup \gamma''_{i_s} = \gamma''} \frac{\partial^s}{\partial v^{k_1} \ldots \partial v^{k_s}} [(g_r X)^j_{\gamma'} \circ g^{-1}(u)](g_k Y^{k_1}_{\gamma''_{i_1}} \ldots (g_k Y^{k_s}_{\gamma''_{i_s}}) \tag{4.53}$$

where the symbol $(g_r X)^j_{\gamma'}$ is the smooth real-valued function of z which associates to z the number $\bar{X}^j_{\gamma'}$ for $\bar{X}^j_{\gamma'}$ the 'γ' component' of the transform of $(\tilde{X}_\gamma(z))$ by $g_r(z)$.

Now for f an arbitrary smooth real-valued function defined on the domain of ϕ, we may interpret

$$\sum_{\gamma''_{i_1} \cup \ldots \cup \gamma''_{i_s} = \gamma''} \frac{\partial^s f}{\partial x^{k_1} \ldots \partial x^{k_s}} Y^{k_1}_{\gamma''_{i_1}} \ldots Y^{k_s}_{\gamma''_{i_s}}$$

as the *derivative of f with respect to the k-sector* $\bar{A}_k(x_0)$ (Definition 3.18).

An easy argument appealing to the 'invariance' of this derivative then shows that (4.53) is equal to:

$$\sum_{\gamma''_{i_1} \cup \ldots \cup \gamma''_{i_s} = \gamma''} \frac{\partial^s (g_r X)^j_{\gamma'}}{\partial x^{k_1} \ldots \partial x^{k_s}} Y^{k_1}_{\gamma''_{i_1}} \ldots Y^{k_s}_{\gamma''_{i_s}}. \tag{4.54}$$

We must then compare this with the $\gamma' \cup \gamma''$ component of the transform of

$$[\phi_{k+r}, X^i_{\gamma'}(0) \text{——} Z^i_{\gamma' \cup \gamma''} \text{——} Y^i_{\gamma''}]$$

which, according to Theorem 1.27 has the following form:

$$\sum_{\substack{\gamma' = \gamma'_{i_1} \cup \ldots \cup \gamma'_{i_t} \cup \ldots \cup \gamma'_{i_R} \\ \gamma'' = \gamma''_{i_1} \cup \ldots \cup \gamma''_{i_t} \cup \ldots \cup \gamma''_{i_S}}} \frac{\partial^{(R+S-t)} g^j}{\partial x^{a_1} \ldots \partial x^{a_{R-t}} \partial y^{b_1} \ldots \partial y^{b_{S-t}} \partial z^{c_1} \ldots \partial z^{c_t}} (P)$$

with

$$(P) = X^{a_1}_{\gamma'_{i_{t+1}}}(0) \ldots X^{a_{R-t}}_{\gamma'_{i_R}}(0) Y^{b_1}_{\gamma''_{i_{t+1}}} \ldots Y^{b_{S-t}}_{\gamma''_{i_S}} Z^{c_1}_{(\gamma'_{i_1} \cup \gamma''_{i_1})} \ldots Z^{c_t}_{(\gamma'_{i_t} \cup \gamma''_{i_t})}.$$

The notation and conventions are explained in the proof of the theorem mentioned above.

Now writing expression (4.54) in the following form:

$$\sum_{\gamma''_{i_1} \cup \ldots \cup \gamma''_{i_s} = \gamma''} \frac{\partial^s \left[\sum_{\gamma'_{i_1} \cup \ldots \cup \gamma'_{i_t} = \gamma'} \frac{\partial^t g^j(z)}{\partial x^{k_1} \ldots \partial x^{k_t}} X^{k_1}_{\gamma'_{i_1}}(z) \ldots X^{k_t}_{\gamma'_{i_t}}(z) \right]}{\partial y^{h_1} \ldots \partial y^{h_s}} Y^{h_1}_{\gamma''_{i_1}} \ldots Y^{h_s}_{\gamma''_{i_s}}$$

gives the conclusion of the lemma after a little bookkeeping. ■

Exercise

11. Show that the suspension of a *vector field* by a 2-sector is frame-independent applying the argument above.

It is now a simple matter to define the general 'reduction' of a sectorform at a point by a local sector field and to state its relationship with the pointing operator.

Definition 4.52. For smooth manifold M, suppose that $A^{k+r}(x_0)$ is a sectorform at x_0 and suppose that $\bar{B}_r(x)$ is a local r-sector field near x_0. Then the correspondence $\bar{A}_k(x_0) \to \bar{A}_k(x_0) \cup \bar{B}_r(x)$ taking $T_k(x_0) \to T_{k+r}(x_0)$ is linear on the intrinsic spaces $T_{k,i}(x_0) \to T_{k+r,i}(x_0)$, $1 \leqslant i \leqslant k$. Therefore the composition $\bar{A}_k(x_0) \to A^{k+r}(x_0)\langle \bar{A}_k(x_0) \cup \bar{B}_r(x)\rangle$ is a k *sectorform* at x_0 which we call the *reduction* of $A^{k+r}(x_0)$ *by the r-sector field* $\bar{B}_r(x)$. We shall denote this sectorform as $A^{k+r}(x_0)[\bar{B}_r(x)]$ in a manner consistent with our notation for the reduction of a sectorform by a local *vector* field. ■

Now given a fixed r-sector field $\bar{B}_r(x)$ on a smooth manifold M, thought of as a smooth section of the bundle

$$\begin{array}{c} T_r(M) \\ \downarrow{\scriptstyle q_r} \\ M \end{array}$$

we see that if $A^{k+r}(x)$ is a $(k+r)$ sectorform field on M then pointing it yields a k sectorform field on $T_r(M)$, namely $A^{k+r}{}_{|\bar{U}_r(x)}$. We may harmlessly denote the pull-back of this k sectorform field under the section $\bar{B}_r(x)$ by $\bar{B}_r(x)*(A^{k+r}{}_{|U_r(x)})$, a k sectorform field on M. The following theorem then gives the connection between pointing and reduction.

Theorem 4.53. *Given r-sector field* $\bar{B}_r(x)$ *on M and* $(k+r)$ *sectorform field* $A^{k+r}(x)$, *then the k sectorform field on M, i.e.* $A^{k+r}(x)[\bar{B}_r(x)]$ (*the* reduction *of* A^{k+r} *by* $\bar{B}_r(x)$) *is equal to the* pull-back of the pointed sectorform *under* $\bar{B}_r(x)$, *i.e.* $\bar{B}_r(x)^*(A^{k+r}{}_{|U_r(x)})$.

Proof. It is a matter of showing that $T_k(\bar{B}_r(x))(\bar{A}_k(x_0)) = p(\bar{A}_k(x_0) \cup \bar{B}_r(x))$, where p represents the pointing operator with respect to $K(r)$ (Definition 4.49). This follows directly (but tediously) from formula (4.52) together with the coordinatization of $T_k(T_r(M)]$ given in Theorem 1.27. That verification is left as an exercise. ■

We now extend the calculation made earlier in Observation 4.48 to the next higher dimension. Let $(M, A^2(x))$ be a Riemannian manifold. There is then defined on M a 3 sectorform field, the *cycle* of the metric, DA^2. Let $K(2)$ be the subcomplex of $\Gamma(2)$ of faces of the form $(0; \gamma_1, \gamma_2)$, and letting $\bar{U}_2$ represent an arbitrary 2-sector, we wish to study the properties of $DA^2{}_{|\bar{U}_2}$, the *cycle pointed with respect to* $K(2)$.

Now $DA^2{}_{|\bar{U}_2}$ is a 1-form on $T_2(M)$. If $A^2(x) = \tilde{a}_{ij}(z)$ in local components with respect to frame-germ ϕ, and if $\phi(0) = x_0$ so that $\bar{U}_2(x_0) = [\phi_2, Y^j$——$C^t$——$Z^k]$, then we may represent $DA^2{}_{|\bar{U}_2(x_0)}$ in local components. First let

$$\begin{array}{l} \bar{A}_3(x_0) = [\phi_3, X^i \text{——} A^r \text{——} Y^j] \\ \qquad\qquad\quad \diagdown \qquad\quad \diagup \\ \qquad\qquad\quad B^s \quad E^u \quad C^t \\ \qquad\qquad\qquad \diagdown \quad \diagup \\ \qquad\qquad\qquad\quad Z^k \end{array}$$

be a 3-sector pointed at $\bar{U}_2(x_0)$ with respect to $K(2)$. Then considering a tangent vector at $\bar{U}_2(x_0)$, it has components (X^i, A^r, B^s, E^u). Thus we have

$$DA^2{}_{|\bar{U}_2(x_0)}\langle p\bar{A}_3(x_0)\rangle = \left[\frac{\partial \tilde{a}_{jk}}{\partial x^i} + \frac{\partial \tilde{a}_{ik}}{\partial x^j} - \frac{\partial \tilde{a}_{ij}}{\partial x^k}\right] Y^j Z^k + 2\tilde{a}_{rk} Z^k. \tag{4.55}$$

This is clearly linear with respect to the components (X^i, A^r, B^s, E^u).

Now we desire to do two things. First, for 2-sector field $\bar{B}_2(x)$ on M, we intend to study the pull-back

$$\bar{B}_2(x)^*(DA^2{}_{|\bar{U}_2}) = DA^2[\bar{B}_2(x)]$$

with respect to that field. This is a 1-form on M determined by $\bar{B}_2(x)$ which represents the 'infinitesimal motion' of the field with respect to the metric. It actually depends only on the *boundary* $D(2)\bar{B}_2(x)$ and corresponds to a multiple of one of the characteristic forms associated by Cartan to a pair of vectors in a 'moving frame'. For the classical presentation of the 'rotational coefficients' associated with a moving frame on a Riemannian manifold, Hermann's translation of Ricci and Levi-Civita's paper (Hermann [10]) is heartily recommended.

Secondly, we compute the coboundary of the 1-form on $T_2(M)$, that is $DA^2{}_{|\bar{U}_2(x)}$, and study the properties of its pull-back. This coboundary will be the pointed cross, a basic tool for defining curvature and demonstrating its invariance, and its pull-back, a 2-form on M associated to a 2-sector field, will give a geometric interpretation of the Gaussian curvature in terms of area and 'parallel translation'.

Thus let

$$\bar{B}_2(x) = [(\phi \circ t_z)_2, \tilde{Y}^j(z) \text{——} \tilde{C}^t(z) \text{——} \tilde{Z}^k(z)]$$

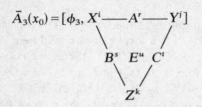

be a local 2-sector field represented in local components with respect to frame-germ ϕ. The local components for the reduction $DA^2[\bar{B}_2(x)]$ are derived from (4.55) to be

$$DA^2[\bar{B}_2(x)] = \left[\frac{\partial \tilde{a}_{jk}}{\partial x^i}\bigg|_z + \frac{\partial \tilde{a}_{ik}}{\partial x^j}\bigg|_z - \frac{\partial \tilde{a}_{ij}}{\partial x^k}\bigg|_z\right] \tilde{Y}^j(z)\tilde{Z}^k(z) + 2\tilde{a}_{rk}(z)\frac{\partial \tilde{Y}^r}{\partial X^i}\bigg|_z \tilde{Z}^k(z) \tag{4.56}$$

and this may be represented

$$DA^2[\bar{B}_2(x)]\langle \tilde{X}^i(z_0)\rangle = DA^2\langle X^i(z_0) \cup \tilde{Y}^j(z), Z^k(z_0)\rangle. \tag{4.57}$$

That is, in the notation following Corollary 4.14,

$$DA^2[\bar{B}_2(x)]\langle X^i(z_0)\rangle = 2A^2\langle \nabla(X \cup \tilde{Y}), Z\rangle.$$

For simplicity, let us agree to denote $\frac{1}{2}DA^2[\bar{B}_2(x)]$ by $W_{\bar{B}_2(x)}$. Then $W_{\bar{B}_2(x)}$ is a 1-form determined by the 2-sector field $\bar{B}_2(x)$. Again, it is easy to see that $W_{\bar{B}_2(x)}$ depends only on the boundary $D(2)\bar{B}_2(x)$.

Exercise

12. For σ the transposition of $S(2)$, show that

$$W_{\bar{B}_2(x)} + W_{\bar{B}_2(x)\sigma} = d_1A^2[\bar{B}_2(x)]$$

and show that the latter form is the coboundary (differential) of a smooth real-valued function on M. In particular, show that if the contraction $A^2\langle \bar{B}_2(x)\rangle$ is constant, then

$$W_{\bar{B}_2(x)} + W_{\bar{B}_2(x)\sigma} = 0.$$

At this level of generality, there is not much more to say about the form $W_{\bar{B}_2(x)}$. In order to develop its geometric applications, we restrict attention to the following special case. A 2-sector field on Riemannian manifold (M, A^2) is *orthonormal* if, for $D(2)\bar{B}_2(x) = (Y(x), Z(x))$, $A^2(Y, Y) = A^2(Z, Z) = 1$ for all x and $A^2(Y, Z) = 0$ for all x. We shall be interested, for reasons which will become obvious, in orthonormal 2-sector fields in the immediate sequel.

For σ the 'shift' in $S(3)$ and for $\bar{B}_2(x)$ an orthonormal 2-sector field, we see that

$$\begin{aligned} DA^2[\bar{B}_2(x)] &= d_1A^2[\bar{B}_2(x)] + d_2A^2[\bar{B}_2(x)] - d_3A^2[\bar{B}_2(x)] \\ &= d_2A^2[\bar{B}_2(x)] - d_3A^2[\bar{B}_2(x)] \\ &= \sigma BA^2[\bar{B}_2(x)]. \end{aligned} \tag{4.58}$$

Now for $\bar{B}_2(x)$ an *orthonormal* field of 2-sectors on (M, A^2), the form $W_{\bar{B}_2(x)}$ has an especially simple interpretation. In the above notation,

$W_{\bar{B}_2(x)}\langle X(x)\rangle$ is just $A^2\langle\nabla(X \cup \bar{Y}), Z(x)\rangle$. In a frame, this may be written:

$$\tilde{a}_{rk}(z)\left(\frac{\partial Y^r}{\partial x^i}_{|z} X^i - \Lambda^r_{ij}(z)X^iY^j\right)Z^k(z)$$

and has the interpretation of the *projection* on $Z(x)$ of the difference of the 'acceleration' $(\partial Y^r/\partial x^i)_{|z}\, X^i$ of Y along X and the acceleration of the *flat* 2-sector $\Lambda^r_{ij}(z)X^iY^j$ pushing Y along X. It is the infinitesimal deviation, *in the Z direction*, of the motion of Y along X, from 'parallel' motion (Definition 4.43). Now in the case that M is a two-dimensional manifold (the case that concerns us for the elucidation of Gaussian curvature), the integral of the form $W_{\bar{B}_2(x)}$ along a smooth curve has a simple geometric interpretation for $\bar{B}_2(x)$ an orthonormal 2-sector field. In that case, we assume the manifold *oriented* and the boundary $D(2)\bar{B}_2(x)$ a *positively oriented* pair of tangent vectors at each point. Recall that this means that in components the matrix

$$\begin{bmatrix} Y^1 & Z^1 \\ Y^2 & Z^2 \end{bmatrix}$$

has positive determinant $Y^1Z^2 - Y^2Z^1$, these components being computed with respect to an oriented frame.

Suppose then that $f: I \to M$ is a path with $I \subset \mathbb{R}$ some closed interval. For local frame $\phi: O \to U \subset M$ if $f(\tau) = x = \phi(z)$ say that $f_\phi(\tau) = z$. Restricting attention to a local frame, we may write

$$(f_\phi)_*(\tau) = \frac{df_\phi}{dt}_{|\tau} = X^i(z) \qquad \text{and} \qquad f_*(\tau) = X(x).$$

Now for $\bar{B}_2(x)$ an *orthonormal positively oriented 2-sector field*, we want to compute the integral:

$$\int_{\tau_0}^{\tau_1} W_{\bar{B}_2(f(t))}\langle f_*(t)\rangle\, dt. \tag{4.59}$$

For convenience set $f(\tau_0) = x_0$ and $f(\tau_1) = x_1$. Now suppose that $U(x_0)$ is an *arbitrary* unit tangent vector ($A^2\langle U, U\rangle = 1$) at $x_0 \in M$. Let V be the 1-variation of f which gives the *parallel transport* of $U(x_0)$ along f between τ_0 and τ_1 (Definition 4.43). Here let $D^2(2)\bar{V}_2^{\#}(t, 0) = X(x)$ and $D^1(2)\bar{V}_2^{\#}(t, 0) = U(x)$, deviating slightly from the notation of that definition. Then $U(x)$ is a unit vector for all $x = f(t)$ (Theorem 4.45, part 3)).

Assuming now that $I = [\tau_0, \tau_1]$, these data define a smooth map $F: I \to S^1$ (S^1 the unit circle in $\mathbb{R}^2$) by

$$F(t) = \begin{bmatrix} A^2(f(t))\langle U(f(t)), Y(f(t))\rangle \\ A^2(f(t))\langle U(f(t)), Z(f(t))\rangle \end{bmatrix} = \begin{bmatrix} a(t) \\ b(t) \end{bmatrix}$$

for $D(2)\bar{B}_2(x) = (Y(x), Z(x))$. Letting (x^1, x^2) be standard coordinates in $\mathbb{R}^2$, we consider the 1-form on I, $F^*(x^1\,dx^2 - x^2\,dx^1)$. It is clear that this form can be written

$$\left[a(t)\frac{db}{dt} - b(t)\frac{da}{dt}\right] dt$$

from the contraction formula and the integral:

$$\int_{\tau_0}^{\tau_1} \left[a(t)\frac{db}{dt} - b(t)\frac{da}{dt}\right] dt \tag{4.60}$$

has the interpretation which follows.

Letting $F(\tau_0)$ correspond to angle $\theta_0 \in \mathbb{R}$ measured in the counterclockwise sense from the positive x_1 axis, then this integral is the algebraic sum of the lengths of arcs traversed by F with positive contribution for counterclockwise traversal and negative contribution for clockwise traversal. We refer to it simply as $\Delta\theta$. Clearly it does not depend on the initial choice θ_0.

Exercise

13. Show that integral (4.60) does not depend on the initial choice $U(x_0)$. [*Hint:* Theorem 4.45.)

Now we intend to show in our next theorem that integral (4.59) + integral (4.60) = 0.

Theorem 4.54. *For (M, A^2) an oriented two-dimensional Riemannian manifold, $f: I \to M$ a smooth path as above, and $\bar{B}_2(x)$ an orthonormal positively oriented 2-sector field, if $I = [\tau_0, \tau_1]$ and $\Delta\theta$ is the integral* (4.60) *computed for this interval and some preassigned unit vector $U(x_0)$, then*

$$\int_{\tau_0}^{\tau_1} W_{\bar{B}_2(f(t))}\langle f_*(t)\rangle\, dt + \Delta\theta = 0. \tag{4.61}$$

Proof. In a local frame we may write

$$U^i(f_\phi(t)) = a(t)Y^i(f_\phi(t)) + b(t)Z^i(f_\phi(t)). \tag{4.62}$$

Then

$$\frac{d}{dt}\Big|_{\tau}[U^i(f_\phi(t))] = \frac{da}{dt}\Big|_{\tau} Y^i(f_\phi(\tau)) + a(\tau)\frac{\partial Y^r}{\partial x^i}\Big|_z X^i(z)$$
$$+ \frac{db}{dt}\Big|_{\tau} Z^i(f_\phi(\tau)) + b(\tau)\frac{\partial Z^r}{\partial x^i}\Big|_z X^i(z) \tag{4.63}$$

and this is equal to

$$\frac{\partial U^r}{\partial x^i}\Big|_z X^i(z).$$

Next for τ fixed and computing with respect to this frame we have

$$\Lambda^r_{ij}(z)X^i(z)U^j(z) = a(\tau)\Lambda^r_{ij}(z)X^i(z)Y^j(z) + b(\tau)\Lambda^r_{ij}(z)X^i(z)Z^j(z). \tag{4.64}$$

Taking (4.63) − (4.64) and remembering that V gives parallel translation, we have

$$0 = \nabla(\bar{V}_2^{\#}(\tau, 0)) = a(\tau)\nabla(X(x) \cup \tilde{Y}) + b(\tau)\nabla(X(x) \cup \tilde{Z}) + \frac{da}{dt}_{|\tau} Y(x) + \frac{db}{dt}_{|\tau} Z(x).$$

Finally computing the scalar product with respect to $A^2(x)$ with:

a) $Z(x)$ yields

$$0 = a(\tau) W_{\bar{B}_2(x)}\langle X(x)\rangle + \frac{db}{dt}_{|\tau}$$

(note: $A^2(x)\langle\nabla(X \cup \tilde{Z}), Z\rangle = 0$),

b) $Y(x)$ yields

$$0 = -b(\tau) W_{\bar{B}_2(x)}\langle X(x)\rangle + \frac{da}{dt}_{|\tau}$$

(exercise 12).

Then multiplying a) by $a(\tau)$ and b) by $-b(\tau)$ and adding yields

$$W_{\bar{B}_2(x)}\langle X(x)\rangle + a(\tau)\frac{db}{dt}_{|\tau} - b(\tau)\frac{da}{dt}_{|\tau} = 0.$$

Since this last expression is independent of frame, integration yields (4.61). ∎

We should point out that it would have been sufficient to have instead of a 2-sector field $\bar{B}_2(x)$ an 'orthonormal' 2-variation of path f defining a path of 3-sectors along path f. Contraction with $\frac{1}{2}DA^2$ and integration along f would still yield the result.

The form $W_{\bar{B}_2(x)}$ thus describes the 'rotation' of $D(2)\bar{B}_2(x)$ with respect to 'parallel' motion. Also, it can be used to describe the rotation of the vectors of $D(2)\bar{B}_2(x)$ with respect to the vectors of $D(2)\bar{C}_2(x)$ for a *pair* of orthonormal, oriented 2-sector fields $\bar{B}_2(x)$ and $\bar{C}_2(x)$.

Observation 4.55. Suppose we are given a pair of orthonormal, positively oriented 2-sector fields $\bar{B}_2(x)$ and $\bar{C}_2(x)$ on Riemannian *surface* (M, A^2) and given a path $f: I \to M$ as above with $I = [\tau_0, \tau_1]$. If $\beta(t)$ defines the angle from $D^2(2)\bar{B}_2(f(t))$ to $D^2(2)\bar{C}_2(f(t))$ measured in the same way that $\theta(t)$ was in the previous theorem, then

$$\int_{\tau_0}^{\tau_1} [W_{\bar{C}_2(f(t))} - W_{\bar{B}_2(f(t))}]\langle f_*(t)\rangle \, dt = \Delta\beta.$$

Proof. On subtracting the corresponding integrals from formula (4.61), this result follows immediately. ■

Now in order to motivate the definition of the cross of the metric, we give one of the major geometric applications of the form $W_{\bar{B}_2(x)}$ for a slightly restricted special case. This result gives the relationship between the angle traversed by parallel translation along a closed path and the integral of the Gaussian curvature enclosed by the path (assuming of course that the path is the boundary of a 2-cell). In making the calculation, we shall use a certain 2-form, $dW_{\bar{B}_2(x)}$, and we return to this form with the definition of the cross and Riemannian curvature.

Assume that (M, A^2) is a Riemannian manifold of dimension 2 and suppose that $\phi: O \to U \subset M$ is a local frame where in this case $O \subset E$ is diffeomorphic to E and ϕ defines the orientation on U with respect to which $[\binom{1}{0}, \binom{0}{1}]$ is positively oriented in each tangent fiber. Let (s, t) denote standard coordinates in $E = \mathbb{R}^2$. Finally, we suppose ϕ to be an 'orthogonal parametrization'. This means that for each $z \in O$, the pair $[(\phi \circ t_z)_1, \binom{1}{0}] = S(x)$ and $[(\phi \circ t_z)_1, \binom{0}{1}] = T(x)$ satisfy $A^2(x)\langle S(x), T(x)\rangle = 0$. This restriction guarantees that in the notation of Theorem 4.32, $F = 0$ and the matrix for A^2 in local components may be written

$$\begin{bmatrix} E(z) & 0 \\ 0 & G(z) \end{bmatrix}.$$

Finally, restrict A^2 to be *positive definite*. We observe here that, in fact, it would be possible to relax the orthogonality restriction (Do Carmo [7]). Now given these data, define the orthonormal, positively oriented 2-sector field on U as

$$\left[(\phi \circ t_z)_1 \begin{pmatrix} E^{-1/2}(z) \\ 0 \end{pmatrix} \!-\! \begin{pmatrix} 0 \\ 0 \end{pmatrix} \!-\! \begin{pmatrix} 0 \\ G^{-1/2}(z) \end{pmatrix}\right]$$

(note that $E(z)$ and $G(z)$ are *positive*) and call it $\bar{B}_2(x)$. This sector field is 'adapted' naturally to the frame.

First, we calculate the 1-form $W_{\bar{B}_2(x)}$ on U. Thus we need the coefficients $\Lambda^r_{ij}(z)$ for general $z = (s, t) \in O$. In particular, fixing z and setting

$$X(x) = \left[(\phi \circ t_z)_1, \begin{pmatrix} X^1(z) \\ X^2(z) \end{pmatrix}\right],$$

$$Y(x) = \left[(\phi \circ t_z)_1, \begin{pmatrix} Y^1(z) \\ Y^2(z) \end{pmatrix}\right],$$

$$Z(x) = \left[(\phi \circ t_z)_1, \begin{pmatrix} Z^1(z) \\ Z^2(z) \end{pmatrix}\right]$$

then

$$\Lambda(X(x), Y(x)) = \left[(\phi \circ t_z)_2, \begin{pmatrix} X^1(z) \\ X^2(z) \end{pmatrix} \!\!-\!\!\!-\!\! \begin{pmatrix} V^1(z) \\ V^2(z) \end{pmatrix} \!\!-\!\!\!-\!\! \begin{pmatrix} Y^1(z) \\ Y^2(z) \end{pmatrix}\right]$$

for $V^1(z)$, and $V^2(z)$ to be determined in terms of the metric, $X(x)$ and $Y(x)$. Writing the equation

$$\tfrac{1}{2}DA^2(x)\langle\Lambda(X(x), Y(x)), Z(x)\rangle = 0$$

for all vectors $Z(x)$, we deduce (omitting mention of the evaluation point z) that

$$\begin{aligned} &EV^1Z^1 + GV^2Z^2 \\ &\quad + \frac{1}{2}\left[\frac{\partial E}{\partial s} Y^1Z^1X^1 + \frac{\partial G}{\partial s} Y^2Z^2X^1 + \frac{\partial E}{\partial t} Y^1Z^1X^2 + \frac{\partial G}{\partial t} Y^2Z^2X^2\right. \\ &\quad + \frac{\partial E}{\partial s} Z^1X^1Y^1 + \frac{\partial G}{\partial s} Z^2X^2Y^1 + \frac{\partial E}{\partial t} Z^1X^1Y^2 + \frac{\partial G}{\partial t} Z^2X^2Y^2 \\ &\quad \left. - \frac{\partial E}{\partial s} X^1Y^1Z^1 - \frac{\partial G}{\partial s} X^2Y^2Z^1 - \frac{\partial E}{\partial t} X^1Y^1Z^2 - \frac{\partial G}{\partial t} X^2Y^2Z^2\right] = 0. \end{aligned}$$

After cancellation and elimination of Z we conclude that

$$\begin{bmatrix} V^1 \\ V^2 \end{bmatrix} = \begin{bmatrix} \tfrac{1}{2}E^{-1}\left(\dfrac{\partial G}{\partial s} X^2Y^2 - \dfrac{\partial E}{\partial s} X^1Y^1 - \dfrac{\partial E}{\partial t}(X^1Y^2 + X^2Y^1)\right) \\ \tfrac{1}{2}G^{-1}\left(\dfrac{\partial E}{\partial t} X^1Y^1 - \dfrac{\partial G}{\partial t} X^2Y^2 - \dfrac{\partial G}{\partial s}(X^1Y^2 + X^2Y^1)\right) \end{bmatrix} \tag{4.65}$$

From this, it is a simple matter to compute the contraction

$$W_{\bar{B}_2(x)}\langle X(x)\rangle = \tfrac{1}{2}(EG)^{-1/2}\left(\frac{\partial G}{\partial s} X^2 - \frac{\partial E}{\partial t} X^1\right).$$

Thus we may write

$$W_{\bar{B}_2(x)} = \tfrac{1}{2}[E(z)G(z)]^{-1/2}\left(\frac{\partial G}{\partial s}_{|z}\, dt - \frac{\partial E}{\partial t}_{|z}\, ds\right). \tag{4.66}$$

Theorem 4.56. *Suppose that (M, A^2) is a two-dimensional Riemannian manifold with positive definite metric and that $\phi : O \to U \subset M$ is an orthogonal parametrization positively orienting U in the manner described above, and suppose that O is diffeomorphic with $E = \mathbb{R}^2$. Letting $\bar{B}_2(x)$ be the positively oriented, orthonormal 2-sector field described above, then $W_{\bar{B}_2(x)}$ is given in local components by formula (4.66).*

Now for $D \subset O$ a 2-cell (a subspace diffeomorphic to the closed unit ball in $\mathbb{R}^2$) with boundary the image of a diffeomorphism $f_\phi : S^1 \to$ boundary (D) *with $(df^i_\phi/d\theta, N^i)$ positively oriented for N^i the components* *of the* inward pointing normal *at each point, then*

$$\int_{S^1} -f^* W_{\bar{B}_2(x)}$$

is the total angular variation (measured with respect to ϕ coordinates) of parallel translation of a unit vector once around S^1 along f.

Now applying Green's theorem, this integral is equal to

$$\iint_D -\left[\frac{\partial}{\partial s}\bigg|_z \left(\tfrac{1}{2}[E(z)G(z)]^{-1/2}\frac{\partial G}{\partial s}\bigg|_z\right)+\frac{\partial}{\partial t}\bigg|_z\left(\tfrac{1}{2}[E(z)G(z)]^{-1/2}\frac{\partial E}{\partial t}\bigg|_z\right)\right] ds\, dt. \tag{4.67}$$

A computation then yields

$$\iint_D -\tfrac{1}{2}[E(z)G(z)]^{-1/2}\left[\frac{\partial^2 E}{\partial t^2}\bigg|_z + \frac{\partial^2 G}{\partial s^2}\bigg|_z\right] ds\, dt$$

$$+\iint_D \tfrac{1}{4}[E(z)G(z)]^{-3/2}\left[\frac{\partial E}{\partial t}\bigg|_z\left(\frac{\partial E}{\partial t}\bigg|_z G(z)+\frac{\partial G}{\partial t}\bigg|_z E(z)\right)\right.$$

$$\left.+\frac{\partial G}{\partial s}\bigg|_z\left(\frac{\partial E}{\partial s}\bigg|_z G(z)+\frac{\partial G}{\partial s}\bigg|_z E(z)\right)\right] ds\, dt.$$

Returning to formula (4.34) (*Theorema Egregium*), *we see that this is equal to*

$$\iint_D K_\gamma(z)[E(z)G(z)]^{1/2}\, ds\, dt$$

in the case that the metric A^2 is derived from an isometric immersion $\gamma:(M, A^2)\to(\mathbb{R}^3, B^2)$. ■

Informally stated, this result says that if $\gamma:(M, A^2)\to(\mathbb{R}^3, B^2)$ is an isometric immersion giving rise to Gaussian curvature function $K_\gamma(x)$ on M and if $\phi: O\to U\subset M$ is an orthogonal parametrization with O diffeomorphic to $\mathbb{R}^2$ giving a positive orientation to U, and if $D\subset O$ is a 2-cell with the boundary of $\phi(D)$ traversed once in the positive sense by loop $f: S^1\to U$, then the total angular variation of parallel translation of a unit vector along f in U is equal to the surface integral over $\phi(D)\subset U$ of the Gaussian curvature $K_\gamma(x)$.

It is this result which Riemannian curvature explains and extends. There are two puzzling points: the invariance of the *curvature* under isometric immersions giving the same 'intrinsic' metric, and the independence of the *angular variation* of orthogonal parametrization. At a critical step, it was necessary to compute the coboundary of the 1-form $W_{\bar{B}_2(x)}$. This is the clue which leads to the definition of the *cross* of the metric (a 4 sectorform field naturally associated with the metric) and to the definition of Riemannian curvature.

Exercise

14. Use 'latitudes' and 'longitudes' to give an orthogonal parametrization of a cell in S^2 immersed in the usual way in $\mathbb{R}^3$ (exclude the 'poles'). Calculate the angular variation of parallel translation along some arcs and loops.

Now we come to the definition and interpretation of the cross of a Riemannian metric. This will be a 4 sectorform field which reduces to an S-sectorform field for S a certain subcomplex of $\Gamma(3)$ of the form

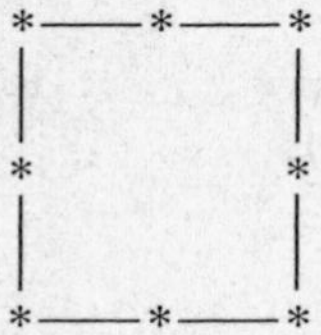

Further, it will satisfy the usual naturality properties which belong to sectorform fields.

The aim is to materialize the 4 sectorform field which when *pointed* with respect to $K(2) \subset \Gamma(3)$ ($K(2)$ the subcomplex of faces of the form $(0, 0; \gamma_1, \gamma_2)$) yields the *coboundary* of the 1-form on $T_2(M)$: DA^2 pointed with respect to $K(2) \subset \Gamma(2)$. Then its *reduction* with respect to $\bar{B}_2(x)$ will yield the coboundary of the form $W_{\bar{B}_2(x)}$. This construction begins with the *second differential operator* $d_{1,2} = d_1 \circ d_2$ applied to the metric $A^2(x)$. This notation is unambiguous and we shall give the component description of this differential as soon as the notational Convention 3.9 is extended to give a representation of 4-sectors.

It should be noticed here that $d_{1,2}A^2$ can be interpreted as the 'promotion' of a metric on $T_1(M)$ yielding a metric on $T_2(M)$. It will be possible to construct an analog of the cycle of A^2, a 'cycle' of the metric d_3A^2, to give a pairing of 1-sectors and 2-sectors *on the tangent bundle.* Many of the constructions of §4.1 will then have extensions to $T_1(M)$ as we shall see and, in particular, the 'contravariant' curvature will emerge in a natural way as bracket of 3-sectors.

Convention 4.57. Elements of E_4 will be represented in components:

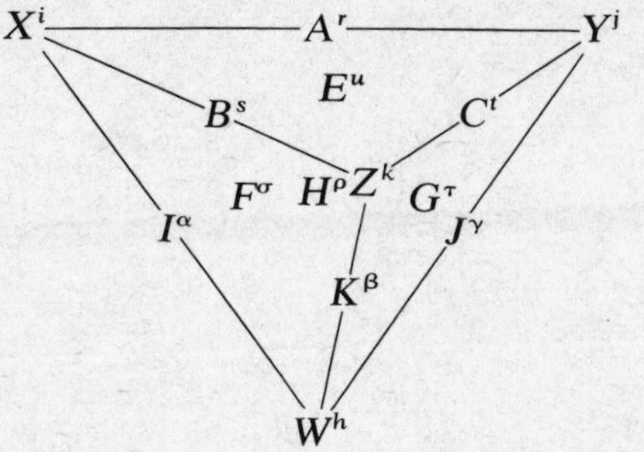

with (1111) component R^δ

The following list explains the labeling procedure:

Face	*Label*	*Face*	*Label*
1000	i	1001	α
0100	j	0011	β
0010	k	0101	γ
0001	h	1011	σ
1100	r	0111	τ
1010	s	1101	ρ
0110	t	1111	δ
1110	u		

Obviously, it would be impractical to attempt to extend this convention much beyond the 4-level. For higher sectors and sectorforms, the notation originally introduced to define them will be serviceable enough. ■

Now if $A^2(x)$ is represented in local components $\tilde{a}_{ij}(z)$, then using the above convention we may represent the 4 sectorform field $d_1[d_2A^2]$ on M in local components:

$$d_1[d_2A^2] = d_{1,2}A^2 = \frac{\partial^2 \tilde{a}_{hj}}{\partial x^k \partial x^i}\bigg|_z + \frac{\partial \tilde{a}_{\alpha j}}{\partial x^k}\bigg|_z + \frac{\partial \tilde{a}_{hr}}{\partial x^k}\bigg|_z + \frac{\partial \tilde{a}_{hj}}{\partial x^s}\bigg|_z$$

$$+ \frac{\partial \tilde{a}_{\beta j}}{\partial x^i}\bigg|_z + \frac{\partial \tilde{a}_{ht}}{\partial x^i}\bigg|_z + \tilde{a}_{\sigma j}(z) + \tilde{a}_{\beta r}(z) + \tilde{a}_{\alpha t}(z) + \tilde{a}_{hu}(z). \quad (4.68)$$

The 4 sectorform field $d_{1,2}A^2$ yields a *metric* on $T_2(M)$ which might be described by a minor modification of the pointing operator, but it is easier for our purposes to describe it in the following way. This sectorform reduces to the following subcomplex of $\Gamma(3)$

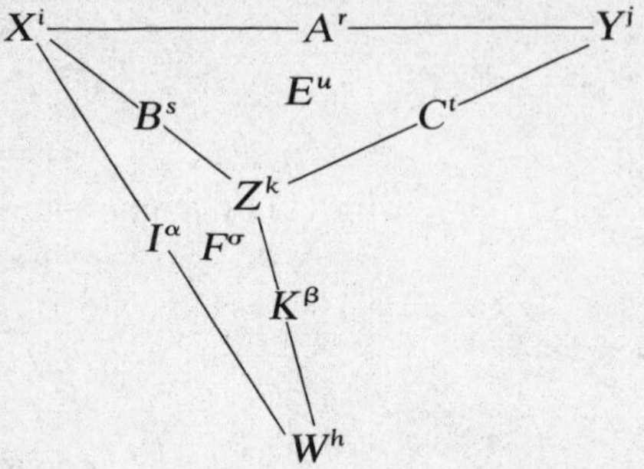

(4.69)

Thus it gives a pairing of 3-sectors whose (say) 2-sides have components Z^k——B^2——X^i. This pairing is then easily seen to be bilinear and non-degenerate and symmetric with respect to the structure of the intrinsic space $T_{3,2}$. This intrinsic space has an interpretation as a tangent fiber for the vector bundle projection

$$\begin{array}{c} T_3(M) \\ \downarrow \\ T_2(M) \end{array}$$

taking each 3-sector to its 2-side. We shall relax notation and refer to this metric as the *second promotion* of A^2, denoted $\delta^2 A^2$. [Recall that δA^2 denotes the (first) promotion of the metric, a 3 sectorform field which was obtained by pointing d_3A^2 (Theorem 4.1).]

If we let

$$[\phi_2, Z^k \text{——} B^s \text{——} X^i] = \bar{C}_2(x_0)$$

and

$$\left[\phi_3, \begin{array}{ccccc} X^i & \text{——} & A^r & \text{——} & Y^j \\ & \diagdown & & \diagup & \\ & B^s & E^u & C^t & \\ & \diagdown & & \diagup & \\ & & Z^k & & \end{array}\right] = \bar{A}_3(x_0),$$

$$\left[\phi_3, \begin{array}{ccccc} X^i & \text{——} & I^\alpha & \text{——} & W^h \\ & \diagdown & & \diagup & \\ & B^s & F^\sigma & K^\beta & \\ & \diagdown & & \diagup & \\ & & Z^k & & \end{array}\right] = \bar{B}_3(x_0)$$

then the *contraction*

$$\delta^2 A^2{}_{|\bar{C}_2}\langle \bar{A}_3(x_0), \bar{B}_3(x_0)\rangle$$

$$= \frac{\partial^2 \tilde{a}_{hj}}{\partial x^k \partial x^i} X^i Y^j Z^k W^h + \frac{\partial \tilde{a}_{\alpha j}}{\partial x^k} I^\alpha Y^j Z^k + \frac{\partial \tilde{a}_{hr}}{\partial x^k} W^h A^r Z^k$$

$$+ \frac{\partial a_{hj}}{\partial x^s} W^h Y^j B^s + \frac{\partial a_{\beta j}}{\partial x^i} K^\beta Y^j X^i + \frac{\partial a_{ht}}{\partial x^i} W^h C^t X^i$$

$$+ a_{\sigma j} F^\sigma Y^j + a_{\beta r} K^\beta A^r + a_{\alpha t} I^\alpha C^t + a_{hu} W^h E^u. \tag{4.70}$$

It is far more useful to learn the geometric pattern defining the contraction with a picture of complex (4.69) in mind than to memorize the formula. Similarly, for the contraction of δA^2 the pattern for producing the terms from complex C_3 is a simple mnemonic. Recall the naturality formula which follows (essentially) from the naturality of pointing sectorforms: suppose $f: N \to M$ is a smooth immersion and that $f^* A^2 = B^2$ is a metric, then

$$\delta^2 B^2{}_{|\bar{C}_2}\langle \bar{A}_3, \bar{B}_3\rangle = \delta^2 A^2{}_{|T_2(f)\bar{C}_2}\langle T_3(f)\bar{A}_3, T_3(f)\bar{B}_3\rangle. \tag{4.71}$$

Recall that for $C(3)$ the subgroup of $S(3)$ of even permutations generated by $\sigma = (231)$ the 'shift' (after Definition 1.25), the *cycle* of metric A^2 is the 3 sectorform field $DA^2 = \sigma d_3 A^2 + \sigma^2 d_3 A^2 - \sigma^3 d_3 A^2$. We shall define a 4 sectorform field according to this same plan which we call the *cycle* $D[\delta A^2]$ to be $d_1[DA^2]$. Now $d_1[DA^2]$ may be interpreted and defined directly in terms of a certain subgroup of $S(4)$ acting on $d_{2,1}A^2$. Since the field $d_1[DA^2]$ is visibly identifiable as the cycle cf the promoted metric δA^2, such notions as 'flatness' and 'force' can be defined for 3-sectors (when these are identified as elements of $T_2[T_1(M)]$. This will be important for our later interpretation of 'contravariant' curvature. It is clear that

$$d_1[DA^2] = d_{1,1}A^2 + d_{1,2}A^2 - d_{1,3}A^2$$

should play some role in the definition of the *cross* by analogy with our earlier treatment of the *balance*, and by the fact that the cross, pointed with respect to $K(2)$, should give the coboundary of DA^2 pointed with respect to $K(2)$. In fact, letting $S(4)$ represent the symmetric group on four letters (identified also as the group of isometries of an equilateral tetrahedron with Convention 4.57 in mind), let us identify a certain subgroup, isomorphic to the Klein 4-group, which we shall call $Kl(4) \subset S(4)$.

The group $Kl(4)$ consists of the following permutations:

$$
\begin{array}{rlrl}
e = \text{identity}: & 1 \to 1 & S: & 1 \to 2 \\
& 2 \to 2 & & 2 \to 1 \\
& 3 \to 3 & & 3 \to 3 \\
& 4 \to 4 & & 4 \to 4 \\
T: & 1 \to 1 & ST = TS: & 1 \to 2 \\
& 2 \to 2 & & 2 \to 1 \\
& 3 \to 4 & & 3 \to 4 \\
& 4 \to 3 & & 4 \to 3
\end{array}
$$

The permutations S and T are *odd* (transpositions) and the permutations e and ST are even. S, T, and ST have order 2 and the group is Abelian. Now for a first definition of the cross we have the following.

Definition 4.58. For a Riemannian manifold $(M, A^2(x))$ define the *cross* of $A^2(x)$ to be the 4-sectorform

$$d_1 DA^2 - Sd_1 DA^2 = CA^2.$$

Now this is the same as

$$d_{1,1}A^2 + d_{1,2}A^2 - d_{1,3}A^2 - d_{2,1}A^2 - d_{2,2}A^2 + d_{2,3}A^2$$
$$= d_{1,2}A^2 + d_{2,3}A^2 - d_{2,2}A^2 - d_{1,3}A^2 \qquad (4.72)$$

since $d_{1,1} = d_{2,1}$.

If the local components for $A^2(x)$ are $\tilde{a}_{ij}(z)$, we may write the local components for CA^2 (omitting mention of the evaluation point) as

$$\frac{\partial^2 \tilde{a}_{hj}}{\partial x^k \partial x^i} + \frac{\partial \tilde{a}_{\alpha j}}{\partial x^k} + \frac{\partial \tilde{a}_{hr}}{\partial x^k} + \frac{\partial \tilde{a}_{hj}}{\partial x^s} + \frac{\partial \tilde{a}_{\beta j}}{\partial x^i} + \frac{\partial \tilde{a}_{ht}}{\partial x^i} + \tilde{a}_{\sigma j} + \tilde{a}_{\beta r} + \tilde{a}_{\alpha t} + \tilde{a}_{hu}$$

$$+ \frac{\partial^2 \tilde{a}_{ik}}{\partial x^j \partial x^h} + \frac{\partial \tilde{a}_{\alpha k}}{\partial x^j} + \frac{\partial \tilde{a}_{i\beta}}{\partial x^j} + \frac{\partial \tilde{a}_{ik}}{\partial x^\gamma} + \frac{\partial \tilde{a}_{rk}}{\partial x^h} + \frac{\partial \tilde{a}_{it}}{\partial x^h} + \tilde{a}_{\rho k} + \tilde{a}_{r\beta} + \tilde{a}_{\alpha t} + \tilde{a}_{i\tau}$$

$$- \frac{\partial^2 \tilde{a}_{ih}}{\partial x^j \partial x^k} - \frac{\partial \tilde{a}_{i\beta}}{\partial x^j} - \frac{\partial \tilde{a}_{sh}}{\partial x^j} - \frac{\partial \tilde{a}_{ih}}{\partial x^t} - \frac{\partial \tilde{a}_{rh}}{\partial x^k} - \frac{\partial \tilde{a}_{i\gamma}}{\partial x^k} - \tilde{a}_{uh} - a_{\beta r} - \tilde{a}_{\gamma s} - \tilde{a}_{i\tau}$$

$$- \frac{\partial^2 \tilde{a}_{jk}}{\partial x^h \partial x^i} - \frac{\partial \tilde{a}_{rk}}{\partial x^h} - \frac{\partial \tilde{a}_{js}}{\partial x^h} - \frac{\partial \tilde{a}_{jk}}{\partial x^\alpha} - \frac{\partial \tilde{a}_{j\beta}}{\partial x^i} - \frac{\partial \tilde{a}_{\gamma k}}{\partial x^i} - \tilde{a}_{j\sigma} - \tilde{a}_{r\beta} - \tilde{a}_{s\gamma} - \tilde{a}_{\rho k}.$$

Now after cancellations this expression greatly simplifies if we admit the symbol $[ij; k]$ as shorthand for

$$\frac{1}{2}\left[\frac{\partial \tilde{a}_{ik}}{\partial x^j}+\frac{\partial \tilde{a}_{kj}}{\partial x^i}-\frac{\partial \tilde{a}_{ij}}{\partial x^k}\right]$$

these being Riemann–Christoffel symbols 'of the second kind' computed with respect to a given frame. [*Warning*: these are not tensors or even sectorforms.] Making the reduction we have that the components for $CA^2(x)$ are (dropping the tildes)

$$\frac{\partial^2 a_{hj}}{\partial x^k \partial x^i}+\frac{\partial^2 a_{ik}}{\partial x^j \partial x^h}-\frac{\partial^2 a_{ih}}{\partial x^j \partial x^k}-\frac{\partial^2 a_{jk}}{\partial x^h \partial x^i}+2a_{\alpha t}-2a_{\gamma s}$$

$$+2[ih; t]+2[jk; \alpha]-2[jh; s]-2[ik; \gamma]. \quad (4.73)$$

Finally, notice that $CA^2(x)$ is an S-sectorform field for S the subcomplex of $\Gamma(3)$ with components:

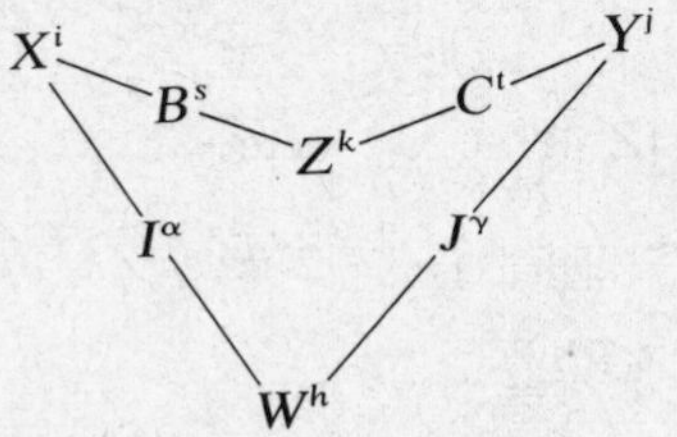

For simplicity, we shall denote this complex

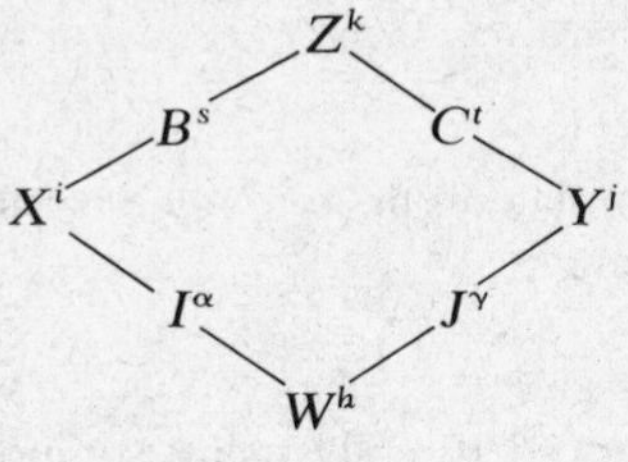

and call it the 'square'. ■

Observation 4.59. Another useful description of CA^2 is given by representing it as a signed sum of multiples of $d_{1,2}A^2$ by elements of $Kl(4)$. From this, many of the algebraic properties of the sectorform will be obvious:

$$CA^2(x)=[e-S+ST-T]d_{1,2}A^2. \quad (4.74)$$

Proof. This follows immediately from formula (4.72). ■

Given an S-sector

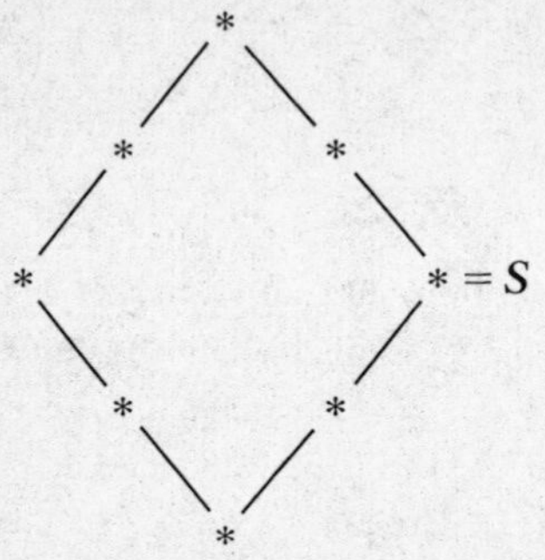

call it V_2, then we shall call the contraction $CA^2\langle V_2\rangle$ the *cross product* of V_2. Now we will want to point CA^2 with respect to $K(2)$. Thus suppose

$$\bar{U}_2=[\phi_2, Z^k \text{——} K^\beta \text{——} W^h].$$

Then the 2 sectorform on $T_2(M)$ at $\bar{U}_2$, *i.e.* $CA^2{}_{|\bar{U}_2}$, will have the local component representation:

$$\left[\frac{\partial^2 a_{hj}}{\partial x^k \partial x^i}+\frac{\partial^2 a_{ik}}{\partial x^h \partial x^j}-\frac{\partial^2 a_{ih}}{\partial x^j \partial x^k}-\frac{\partial^2 a_{jk}}{\partial x^h \partial x^i}\right]Z^k W^h+2a_{\alpha t}-2a_{\gamma s}$$
$$+2[ih;t]W^h+2[jk;\alpha]Z^k-2[jh;s]W^h-2[ik;\gamma]Z^k. \quad (4.75)$$

Exercise

15. Considering $K(2)$ simultaneously to be a subcomplex of $\Gamma(3)$ and $\Gamma(2)$ (faces of the form $(0, 0; \gamma_1, \gamma_2)$ or $(0; \gamma_1, \gamma_2)$ respectively), show that

$$d(DA^2{}_{|\bar{U}_2})=CA^2{}_{|\bar{U}_2}. \quad (4.76)$$

Now putting exercise 15 together with Theorem 4.53 we see that if $\bar{B}_2(x)$ is a (local) 2-sector field on M, then

$$\tfrac{1}{2}CA^2[\bar{B}_2(x)]=d[W_{\bar{B}_2(x)}] \quad (4.77)$$

as 2-forms on M. This fact is easily checked directly using component formula (4.55).

Observation 4.60. If $f:(N, B^2)\to(M, A^2)$ is a smooth immersion with $f^*A^2=B^2$ and if V_2 is an S-sector at $x_0\in N$, then $T_2(f)V_2$ is an S-sector at $f(x_0)\in M$ and of course

$$CB^2\langle V_2\rangle=CA^2\langle T_2(f)V_2\rangle. \quad (4.78)$$

Further, for $\tau\in Kl(4)$, we have

$$CB^2\langle V_2\cdot\tau\rangle=(\operatorname{sgn}\tau)CB^2\langle V_2\rangle. \quad (4.79)$$

Proof. Formula (4.78) follows from the definition of f^* and (4.79) follows directly from (4.74). ■

We are now in a position to define the Riemann curvature tensor. Suppose that $(M, A^2(x))$ is a Riemannian manifold and that ϕ is a frame-germ at $x_0 \in M$. Let four tangent vectors at x_0 be given in pairs $[(X, Y); (Z, W)]$ with

$$X = [\phi_1, X^i], \qquad Y = [\phi_1, Y^j],$$
$$Z = [\phi_1, Z^k], \qquad W = [\phi_1, W^h].$$

Now we may use the affine connection Λ to construct an S-sector from these data.

Definition 4.61. For Riemannian manifold $(M, A^2(x))$ let $[(X, Y); (Z, W)]$ be an ordered pair of ordered pairs of tangent vectors at $x_0 \in M$. Define the *flat square* of $[(X, Y); (Z, W)]$ to be the S-sector with ϕ components

[ϕ_2,]

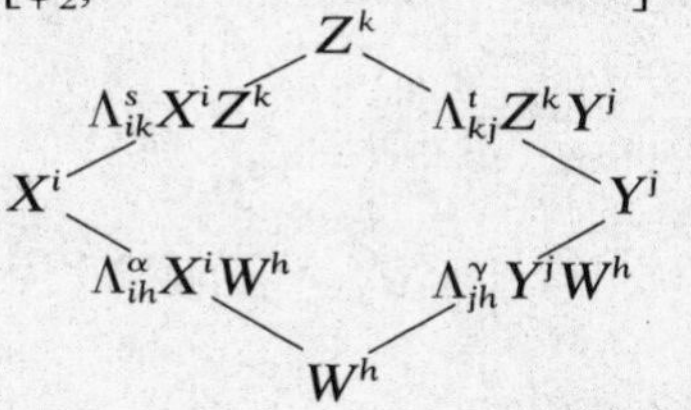

Here the coefficients Λ^r_{ij} depend, as always, on the frame ϕ_2. Denote *this* S-sector $Sq[(X, Y); (Z, W)]$. It obviously depends only on the metric and the 4-tuple of tangent vectors. If $\tau \in Kl(4)$, then it is clear that

$$Sq([(X, Y); (Z, W)]\tau) = (Sq[(X, Y); (Z, W)]\tau). \tag{4.80}$$

Now consider the association:

$$[(X, Y); (Z, W)] \xrightarrow{R} -\tfrac{1}{2}CA^2\langle Sq[(X, Y); (Z, W)]\rangle.$$

This gives a smooth real-valued mapping from $T^4_1 \to \mathbb{R}$. Denote $-\frac{1}{2}CA^2\langle Sq[(X, Y); (Z, W)]\rangle$ by $R[(X, Y); (Z, W)]$. We have the following relation with the action of $Kl(4)$ on $T^4_1(M)$:

1) $R[(X, Y); (Z, W)]S \;= R[(Y, X); (Z, W)] = -R[(X, Y); (Z, W)]$,
2) $R[(X, Y); (Z, W)]T \;= R[(X, Y); (W, Z)] = -R[(X, Y); (Z, W)]$,
3) $R[(X, Y); (Z, W)]ST = R[(Y, X); (W, Z)] = R[(X, Y); (Z, W)]$,

Further, R is 4-linear on each fiber. We see this via two observations. Firstly, the map $Y \to \Lambda(X, Y)$ is a *linear* map from $T_1 \to T_{2,2}$. Secondly, since the sectorform $-\frac{1}{2}CA^2$ is linear on the intrinsic spaces

$T_{4,1}, T_{4,2}, T_{4,3}, T_{4,4}$, we may conclude that its value on, say,

$$[\phi_2, \begin{array}{c} Z^k \\ \end{array}]$$

$$\begin{array}{ccc} & \Lambda^s_{ik}X^iZ^k & (\Lambda^t_{kj}Z^kY^j+\Lambda^t_{kj}Z^k\tilde{Y}^j) \\ X^i \diagdown & & \diagup (Y^j+\tilde{Y}^j) \\ & \Lambda^\alpha_{ih}X^iW^h & (\Lambda^\gamma_{jh}Y^jW^h+\Lambda^\gamma_{jh}\tilde{Y}^jW^h) \\ & W^h & \end{array}$$

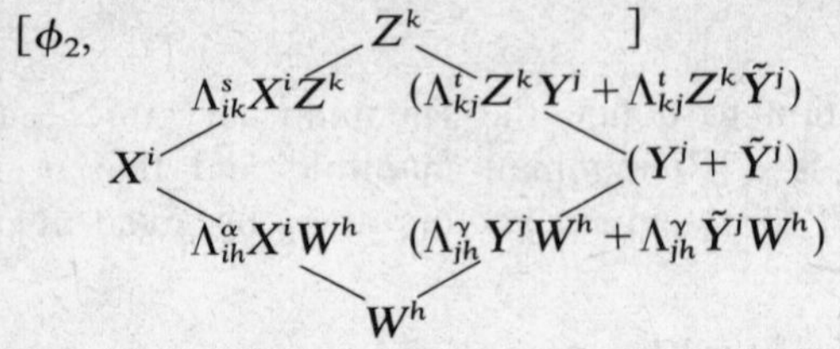

is $R[(X, Y); (Z, W)]+R[(X, \tilde{Y}); (Z, W)]$ using additivity on $T_{4,2}$.

Of course, in light of these facts we have defined a covariant (type $\binom{0}{4}$) tensor field on M. $R[(X, Y); (Z, W)]$ is called the *Riemann curvature tensor*.

We have the following naturality property. If $f:(M, A^2(x)) \to (N, B^2(x))$ is an *isometric immersion* $(f^*B^2=A^2)$, then for $[(X, Y); (Z, W)] \in T^4_1(M)$,

$$R[(X, Y); (Z, W)] = -\tfrac{1}{2}CB^2\langle T_2(f)(Sq[(X, Y); (Z, W)])\rangle. \tag{4.81}$$

Finally, calculating at a point in ϕ components, we have:

$$R[(X, Y); (Z, W)] = -\frac{1}{2}\left[\frac{\partial^2 \tilde{a}_{hj}}{\partial x^k \partial x^i}+\frac{\partial^2 \tilde{a}_{ik}}{\partial x^h \partial x^j}-\frac{\partial^2 \tilde{a}_{ih}}{\partial x^j \partial x^k}\right.$$

$$\left. -\frac{\partial^2 \tilde{a}_{jk}}{\partial x^i \partial x^h}\right]X^iY^jZ^kW^h - \tilde{a}_{s\gamma}(\Lambda^s_{ik}X^iZ^k)(\Lambda^\gamma_{jh}Y^jW^h)$$

$$+\tilde{a}_{\alpha t}(\Lambda^\alpha_{ih}X^iW^h)(\Lambda^t_{kj}Z^kY^j), \tag{4.82}$$

observing, for example, that

$$[ik; \gamma]X^iZ^k(\Lambda^\gamma_{jh}Y^jW^h)+a_{s\gamma}(\Lambda^s_{ik}X^iZ^k)(\Lambda^\gamma_{jh}Y^jW^h) = 0. \quad \blacksquare$$

It is not out of place to observe here that if $Oc(4)$ is the subgroup of $S(4)$ generated by $Kl(4)$ together with the (even) permutation

$$U: 1 \to 3$$
$$2 \to 4$$
$$3 \to 1$$
$$4 \to 2$$

then $U^2=e$ and we have $Oc(4)=\{e, S, T, U, ST, SU, TU, STU\}$. This can be thought of as the group of rigid motions of the square; it is also *the* subgroup of $S(4)$ which carries the subcomplex S to itself (as subcomplex of the tetrahedron), and it is called the *octic group*. A simple check with formula (4.82) shows that for τ in $Oc(4)$

$$R([(X, Y); (Z, W)]\tau) = (\text{sgn } \tau)R[(X, Y); (Z, W)].$$

We now take up the relationship between the curvature tensor R and the reduction operation

$$\tfrac{1}{2}CA^2[\bar{B}_2(x)] = d(W_{\bar{B}_2(x)})$$

(exercise 15) for $\bar{B}_2(x)$ a local 2-sector field on (M, A^2).

Suppose θ is a frame-germ at x_0 in (M, A^2) such that θ_2 is a *null frame* at x_0. In this case, formula (4.82) reduces to

$$R[(X, Y); (Z, W)] = -\frac{1}{2}\left[\frac{\partial^2 a_{hj}}{\partial x^k \partial x^i} + \frac{\partial^2 a_{ik}}{\partial x^h \partial x^j} - \frac{\partial^2 a_{ih}}{\partial x^j \partial x^k} - \frac{\partial^2 a_{jk}}{\partial x^i \delta x^h}\right] X^i Y^j Z^k W^h \tag{4.83}$$

with the partial derivatives computed with respect to θ.

Then for

$$\bar{B}_2(x) = [(\theta \circ t_z)_2, \tilde{Z}^k(z) \text{——} \tilde{K}^\beta(z) \text{——} \tilde{W}^h(z)]$$

a local 2-sector field with $\theta(0) = x_0$ above and with the tangent vectors X, Y, Z, W at x_0 with respective θ components $\tilde{X}^i(0)$, $\tilde{Y}^j(0)$, $\tilde{Z}^k(0)$, $\tilde{W}^h(0)$, we compute the contraction

$$d(W_{\bar{B}_2(x)})(x_0)\langle X, Y\rangle = -R[(X, Y); (Z, W)] + a_{\alpha t}(0)\left(\frac{\partial \tilde{W}^\alpha}{\partial x^i} X^i\right)\left(\frac{\partial \tilde{Z}^t}{\partial x^j} Y^j\right) - a_{\gamma s}(0)\left(\frac{\partial \tilde{W}^\gamma}{\partial x^j} Y^j\right)\left(\frac{\partial \tilde{Z}^s}{\partial x^i} X^i\right). \tag{4.84}$$

This follows using formula (4.75) together with the fact that the Christoffel symbols 'vanish' in a null frame. Observing that, in frame θ, $(\partial \tilde{Z}^t/\partial x^j) Y^j$ are just the components of $\nabla(Y \cup \tilde{Z}(x))$, we have the following theorem.

Theorem 4.62. *Let $(M, A^2(x))$ be a Riemannian manifold, $x_0 \in M$, $\bar{B}_2(x)$ a 2-sector field with $D(2)\bar{B}_2(x_0) = (\tilde{Z}, \tilde{W})$, then for (X, Y) a pair of tangent vectors at x_0:*

$$d(W_{\bar{B}_2(x)})(x_0)\langle X, Y\rangle + R[(X, Y); (\tilde{Z}, \tilde{W})] = -A^2(x_0)\langle \nabla(X \cup \tilde{Z}(x)), \nabla(Y \cup \tilde{W}(x))\rangle + A^2(x_0) \times \langle \nabla(X \cup \tilde{W}(x)), \nabla(Y \cup \tilde{Z}(x))\rangle. \tag{4.85}$$

■

Corollary 4.63. *If M is two-dimensional in Theorem* 4.62, *and if $\bar{B}_2(x)$ is* orthogonal, *then*

$$d(W_{\bar{B}_2(x)})(x_0)\langle X, Y\rangle = -R[(X, Y); (\tilde{Z}, \tilde{W})]. \tag{4.86}$$

Proof. $A^2\langle W, \nabla(X \cup \tilde{W}(x))\rangle = 0$ as follows easily from exercise 12. Since $W(x_0)$ and $Z(x_0)$ give an *orthogonal basis* for $T_1(x_0)$, it follows that $\nabla(X \cup \tilde{Z}(x))$ and $\nabla(Y \cup \tilde{W}(x))$ are also orthogonal. The result follows. ■

When the connection between Riemannian and Gaussian curvature is developed, this corollary will give an interpretation of the result of Theorem 4.56.

Now suppose for $(M, A^2(x))$ that X and Y are chosen in $T_1(M)$ at x_0. They determine a square (S-sector) $Sq[(X, Y); (X, Y)]$, and if they are linearly independent the corresponding $R[(X, Y); (X, Y)]$ are of interest. Notice that for all $\tau \in S(4)$ the only numbers $R([(X, Y); (X, Y)]\tau)$ which are not *a priori* zero are obtained from $\tau \in Oc(4)$.

Observation 4.64. Suppose that X and Y are as above (possibly linearly dependent) and that X', Y' is another pair of tangent vectors *in the space* spanned by X and Y. Let $X' = aX + bY$ and $Y' = cX + dY$ for scalars a, b, c, d. Then

$$R[(X, Y); (X, Y)](A^2\langle X', X'\rangle A^2\langle Y', Y'\rangle - A^2\langle X', Y'\rangle^2)$$
$$= R[(X', Y'); (X', Y')](A^2\langle X, X\rangle A^2\langle Y, Y\rangle - A^2\langle X, Y\rangle^2). \quad (4.87)$$

Therefore for any two-dimensional subspace $H \subset T_1(x_0)$, the number

$$K(H) = \frac{R[(X, Y); (X, Y)]}{A^2\langle X, X\rangle A^2\langle Y, Y\rangle - A^2\langle X, Y\rangle^2}$$

for (X, Y) *independent*, does not depend on the choice of independent set (X, Y). $K(H)$ is called the *sectional curvature* of the subspace H; it is entirely determined by the metric.

Proof. From our remarks concerning the action of the octic group $Oc(4)$, we have

$$R[(X', Y'); (X', Y')] = (ad - bc)^2 R[(X, Y); (X, Y)].$$

Then write the product

$$(ad - bc)^2 R[(X, Y); (X, Y)](A^2\langle X, X\rangle A^2\langle Y, Y\rangle - A^2\langle X, Y\rangle^2)$$

in two ways. ■

If the form of the sectional curvature is familiar, it is because for $M = E = \mathbb{R}^2$ and $\gamma : (E, A^2(x)) \to (\mathbb{R}^3, B^2(y))$, for $B^2(y)$ the Euclidean metric an isometric immersion, then the sectional curvature $K(T_1(x_0))$ is equal to the Gaussian curvature $K_\gamma(x_0)$. This is the content of the next theorem, the second version of Gauss' *Theorema Egregium*. Once this is proven, it will be clear *why* the Gaussian curvature should depend only on the metric. That dependence will be given from component formula (4.82).

Theorem 4.65 (Gauss' Theorema Egregium). (Second form.) *Let* $E=\mathbb{R}^2$ *and* $E'=\mathbb{R}^3$, *and let* E' *be equipped with the standard Euclidean metric* B^2. *Suppose that* $\gamma:(E, A^2(X, Y))\to(E', B^2)$ *is an isometric immersion with* (x, y) *the standard linear coordinates for* E. *Then for each* $(x_0, y_0)\in E$, *the* sectional curvature $K(T_1(x_0, y_0))$ *is equal to the* Gaussian curvature $K_\gamma(x_0, y_0)$.

Thus the Gaussian curvature is again seen to depend only on the induced metric. Now if the components of $A^2(x, y)$ *with respect to the global identity frame* ϕ *on* E *are* $\tilde{a}_{ij}(x, y)$ *and the components of the affine connection associated with each 2-frame* $(\phi\circ t_{(x,y)})_2$ *are* $\Lambda^{\gamma}_{\alpha\beta}(x, y)$, *and if* X *and* Y *are tangent vectors at* (x_0, y_0) *with respective components* X^i, Y^i, *and these are linearly independent, and finally, if we set*

$$\Delta = A^2\langle X, X\rangle A^2\langle Y, Y\rangle - A^2\langle X, Y\rangle^2$$

(the squared area of the parallelogram spanned by $T_1(\gamma)X$ *and* $T_1(\gamma)Y$*) then, evaluating everything at* (x_0, y_0), *we have*

$$\Delta K_\gamma(x_0, y_0) = R[(X, Y); (X, Y)]$$

and this is equal to

$$-\frac{1}{2}\left[\frac{\partial^2\tilde{a}_{ik}}{\partial x^j\partial x^h} - 2\frac{\partial^2\tilde{a}_{ih}}{\partial x^j\partial x^k} + \frac{\partial^2\tilde{a}_{hj}}{\partial x^i\partial x^k}\right]X^iY^jX^kY^h$$
$$+\tilde{a}_{\alpha t}(\Lambda^{\alpha}_{ih}X^iY^h)(\Lambda^{t}_{kj}X^kY^j) - \tilde{a}_{s\gamma}(\Lambda^{s}_{ik}X^iX^k)(\Lambda^{\gamma}_{jh}Y^jY^h). \quad (4.88)$$

Proof. Formula (4.88) will follow from (4.82) once we show that

$$R[(X, Y); (X, Y)] = \Delta K_\gamma(x_0, y_0).$$

Now it follows from Observation 4.60 that

$$R[(X, Y); (X, Y)] = -\tfrac{1}{2}CB^2\langle T_2(\gamma)(Sq[(X, Y); (X, Y)])\rangle.$$

Since B^2 is the Euclidean metric, the Riemann–Christoffel symbols and the second partial derivatives vanish in the calculation from formula (4.73). What remains of the right-hand side of the equation above is simply

$$B^2\langle\Omega(\gamma)(X, X), \Omega(\gamma)(Y, Y)\rangle - B^2\langle\Omega(\gamma)(X, Y), \Omega(\gamma)(X, Y)\rangle.$$

Formula (4.29) then gives the result. ■

Formula (4.88) is certainly more attractive than (4.34). Still it reduces to (4.34) when we take for X and Y the standard basis vectors with components $\binom{1}{0}$ and $\binom{0}{1}$, respectively, in the tangent fiber. The following exercises will make this clear.

Exercises

16. Show that

$$a_{s\gamma}(\Lambda^s_{ik}X^iZ^k)(\Lambda^\gamma_{jh}Y^jW^h) = a^{s\gamma}[ik;s][jh;\gamma]X^iY^jZ^kW^h.$$

17. Letting (M, A^2) be a two-dimensional Riemannian manifold, and $\phi:(\mathbb{R}^2, 0) \to (M, u_0)$ a frame-germ at u_0, suppose that (x, y) give standard linear coordinates for $\mathbb{R}^2$. Then according to a fairly obvious convention $[ij; k]$ is a $2\times2\times2$ matrix. Letting E, F, G be considered germs of functions of (x, y) as in Theorem 4.56, show that

$$[ij;k] = \; i\begin{bmatrix} \frac{1}{2}\frac{\partial E}{\partial x} & \frac{1}{2}\frac{\partial E}{\partial y} \\ \frac{1}{2}\frac{\partial E}{\partial y} & \left(\frac{\partial F}{\partial y} - \frac{1}{2}\frac{\partial G}{\partial x}\right) \end{bmatrix} k = 1$$

$$\longleftarrow j \longrightarrow$$

$$i\begin{bmatrix} \left(\frac{\partial F}{\partial x} - \frac{1}{2}\frac{\partial E}{\partial y}\right) & \frac{1}{2}\frac{\partial G}{\partial x} \\ \frac{1}{2}\frac{\partial G}{\partial x} & \frac{1}{2}\frac{\partial G}{\partial y} \end{bmatrix} k = 2.$$

$$\longleftarrow j \longrightarrow$$

18. Using exercises 16 and 17 deduce formula (4.34) from formula (4.88).

The last topic of this section will be an interpretation of curvature in terms of the Riemannian geometry *of the tangent bundle.* It will be convenient to introduce and study the 'contravariant' curvature tensor in this connection. That tensor will appear as a *bracket* of 3-sectors (in fact, as a covariant derivative in a sense that will be clear later). The starting point for these developments is the promoted metric δA^2 again; this time we construct certain analogs for the affine connection and the force associated with this metric for $T_1(M)$. Certain novelties arise in this context essentially for the reason that 2-sectors can be identified as elements of $T_1(T_1(M))$ in two different ways; 3-sectors have six realizations as elements of $T_1(T_2(M))$ and so forth.

Recall that the second promotion of the metric A^2, that is $\delta^2 A^2$, gives a metric on $T_2(M)$ by contracting $d_{1,2}A^2$ with sector complexes of the form:

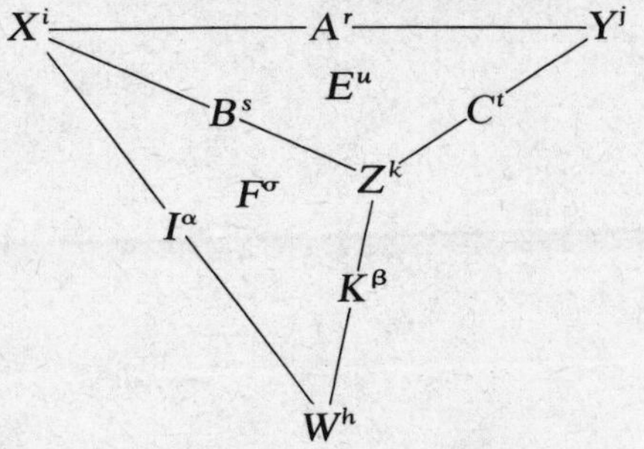

(4.89)

The symmetric, bilinear, non-degenerate pairing on intrinsic space $T_{3,2}$ is given in components in formula (4.70). The metric whose action is pictured above is thought of as the promotion of δA^2. Our aim here is to study some of the consequences of generalizing the constructions of §4.1 to the Riemannian manifold $(T_1(M), \delta A^2)$ for (M, A^2) a given Riemannian manifold. We have already defined an analog of the *cycle of* δA^2 which we called

$$d_1[DA^2] = d_{1,1}A^2 + d_{1,2}A^2 - d_{1,3}A^2,$$

and in fact the *cross* was defined in Definition 4.58 in terms of this 4 sectorform field

$$CA^2 = d_1DA^2 - S(d_1DA^2).$$

Now in order to maintain some consistency with previous constructions, we shall have to make systematic (but obvious) modifications of the pointing operators and these will be indicated largely by the diagrams, sometimes by special notation.

In Theorem 4.3 we defined the *cycle product pairing* as a D-sectorform (order 3) for $D \subset \Gamma(2)$ the subcomplex

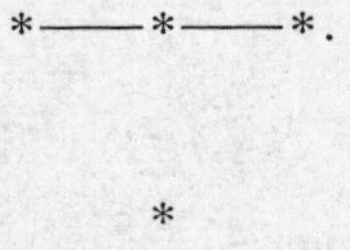

This gave a pairing of 1-sectors and 2-sectors. Now in order to generalize this to give a pairing of 1-sectors and 2-sectors *on the tangent bundle,* we define a *new* complex $D' \subset \Gamma(3)$, the one appropriate to the 4 sectorform d_1DA^2:

$D' =$ 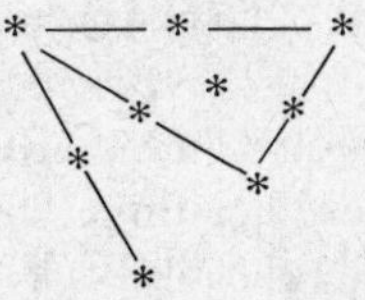

A D'-sector may be represented

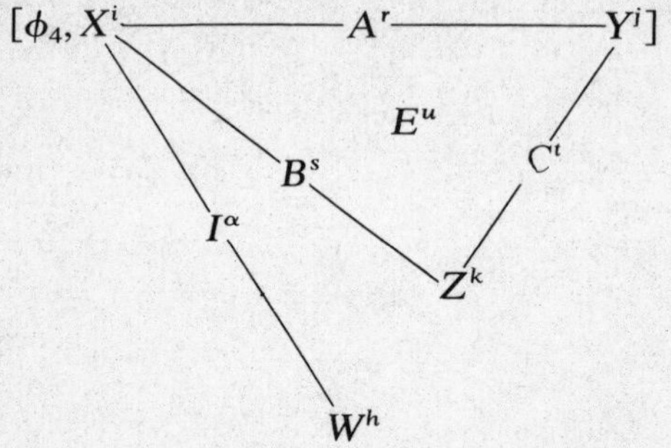

4.90)

If these components belong to a D'-sector in frame ϕ, then d_1DA^2 is a pairing of a 1-sector with a 2-sector *on* $T_1(M)$ *at* the tangent vector $[\phi_1, X^i]$ via the pointing at that tangent vector.

Definition 4.66. A D'-sector is entirely specified by the data of a triple $(\bar{A}_1, \bar{B}_2, \bar{C}_3)$ of a 1-sector, 2-sector, and 3-sector with $\bar{A}_1$ the *common* 1-*vertex* of $\bar{B}_2$ and $\bar{C}_3$. Now d_1DA^2 is a *pairing* of 2-*sectors* with 3-*sectors* which have *common* 1-*vertex*. We write $d_1DA^2\langle\bar{A}_1, \bar{B}_2, \bar{C}_3\rangle$ for the contraction of d_1DA^2 with (4.90) when

$$\bar{A}_1 = [\phi_1, X^i]$$

$$\bar{A}_2 = [\phi_2, X^i — I^\alpha — W^h]$$

$$\bar{A}_3 = [\phi_3, X^i — A^r — Y^i]$$

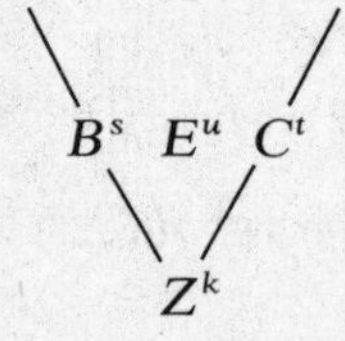

This pairing has these properties:

1) Holding $\bar{A}_3$ fixed, the mapping $\bar{A}_2 \to d_1DA^2\langle\bar{A}_1, \bar{A}_2, \bar{A}_3\rangle \in R$ is linear on $T_{2,2}$.
2) Holding $\bar{A}_2$ fixed, the mapping $\bar{A}_3 \to d_1DA^2\langle\bar{A}_1, \bar{A}_2, \bar{A}_3\rangle \in R$ is *linear* on intrinsic spaces $T_{3,2}$ and $T_{3,3}$.
3) For $f:(M, A^2(x)) \to (N, B^2(y))$ an *isometric immersion*, we have the naturality property

$$d_1DA^2\langle\bar{A}_1, \bar{B}_2, \bar{C}_3\rangle = d_1DB^2\langle T_1(f)\bar{A}_1, T_2(f)\bar{A}_2, T_3(f)\bar{A}_3\rangle. \qquad \blacksquare$$

This is the sense in which d_1DA^2 is an extension of DA^2. But there is an *important* difference. Given a 3-sector $\bar{C}_3$, there are three ways to pair it with 2-sectors, whereas a 2-sector is paired in only one way with tangent vectors. The pairings at the *other* vertices are given in an obvious way by 4 sectorforms $\sigma d_1 DA^2$ and by $\sigma^2 d_1 DA^2$, given by applying σ and σ^2 to $d_1 DA^2$ where

$$\begin{aligned}\sigma\colon 1 &\to 2\\ 2 &\to 3\\ 3 &\to 1\\ 4 &\to 4\end{aligned}$$

Next, we calculate the pairing $d_1DA^2\langle \bar{A}_1, \bar{A}_2, \bar{A}_3\rangle$ with ϕ components given in (4.90) and with $A^2(x) = \tilde{a}_{ij}(z)$ with respect to ϕ:

$$\begin{aligned} d_1DA^2\langle \bar{A}_1, \bar{A}_2, \bar{A}_3\rangle = {} & \left[\frac{\partial^2 \tilde{a}_{kh}}{\partial x^i \partial x^j} + \frac{\partial^2 \tilde{a}_{jh}}{\partial x^i \partial x^k} - \frac{\partial^2 \tilde{a}_{jk}}{\partial x^i \partial x^h}\right] X^i Y^j Z^k W^h \\ & + 2\frac{\partial \tilde{a}_{ht}}{\partial x^i} X^i W^h C^t + 2\tilde{a}_{uh} E^u W^h + 2\tilde{a}_{t\alpha} C^t I^\alpha \\ & + 2[js; h] Y^j W^h B^s \\ & + 2[rk; h] A^r Z^k W^h + 2[jk; \alpha] Y^j Z^k I^\alpha. \end{aligned} \tag{4.91}$$

We next extend the notion of 'force' (covariant derivative) to 3-sectors.

Definition 4.67. Let $(M, A^2(x))$ be a Riemannian manifold and let $\bar{C}_3(x_0)$ be a 3-sector at $x_0 \in M$. Suppose the 1-vertex of $\bar{C}_3(x_0)$ is denoted simply X, a tangent vector at x_0. Then there is a *unique* 2-sector $\bar{A}_2(x_0)$ with 1-vertex X with the property that for *any* 2-sector $\bar{B}_2(x_0)$ in the $T_{2,1}$ fiber of $\bar{A}_2(x_0)$

$$\tfrac{1}{2} d_1 DA^2\langle X, \bar{B}_2, \bar{C}_2\rangle = \delta A^2{}_{|X}\langle \bar{B}_2, \bar{A}_2\rangle. \tag{4.92}$$

Call the 2-sector $\bar{A}_2(x_0)$ the 1-force of $\bar{C}_3(x_0)$. The existence and uniqueness follow immediately from the non-degeneracy of $\delta A^2{}_{|X}$.

In general, for $i = 1, 2, 3$ define the *i-force* of $\bar{C}_3(x_0)$ to be the *unique* 2-sector with 1-vertex in common with the i-vertex of $\bar{C}_3(x_0)$ (call the vertex U and the 2-sector $\bar{A}_2(x_0)$) satisfying

$$\tfrac{1}{2}\sigma^{i-1} d_1 DA^2\langle U, \bar{B}_2, \bar{C}_3\rangle = \delta A^2{}_{|U}\langle \bar{A}_2, \bar{B}_2\rangle$$

for all $\bar{B}_2(x_0)$ with 1-vertex U. ■

We see that, in the presence of a Riemannian metric, every 3-sector determines *three* 2-sectors as its associated forces. Extending Definition 4.9, we say that a 3-sector is *i-flat* ($i = 1, 2, 3$) if its corresponding *i-force* is zero, as element of linear space $T_{2,2}$. This is the notion we intend to investigate here, both in its relation to the *cross* and for its geometric consequences. Roughly speaking, Riemannian curvature at a point gives the obstruction to finding 3-sectors which satisfy certain initial conditions and are simultaneously 1-flat and 2-flat.

In order to develop this idea, we need an extended idea of *affine connection.* This ought to be formulated in terms of the affine bundles introduced in Definition 2.21 and exercise 2.10 a)–d). Recall the names we gave certain star-complexes in $\Gamma(2)$:

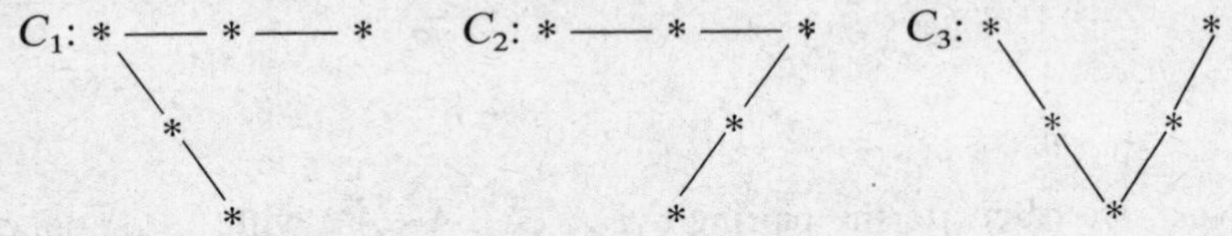

In the notation of Definition 2.21 let γ_1 be the 1-vertex of $\Gamma(2)$ and let $\bar{\gamma}_1$ be the 1-vertex of $\Gamma(1)$ and consider the fiber bundle projection

$$\begin{array}{c} T_3(M) \\ \downarrow F \\ T_3[M; C_1] \end{array}$$

with $T_3[M; C_1]$ the bundle of C_1-sectors. Next define the product of bundles

$$\begin{array}{ccc} & & T_3[M; C_1] \\ & & \downarrow \bar{q}_{3.1} \\ T_2(M) & \xrightarrow[D^2(2)]{} & T_1(M) \end{array}$$

denote the total space $T_3[M; C_1] \times_{T_1(M)} T_2(M)$ and consider the induced *vector* bundle projection

$$\begin{array}{c} T_3[M; C_1] \times_{T_1(M)} T_2(M) \\ \downarrow D^2(2)^* \\ T_3[M; C_1] \end{array}$$

Then we know from exercise 2.10 that the bundle with projection F is an affine bundle over the bundle with projection $D^2(2)^*$. We may therefore form a *generalized bracket* (in this case $[\quad ; \quad]_{\gamma_1, \bar{\gamma}_1}$) of 3-sectors in the same F fiber and the result will be a 2-sector with 1-vertex common to the 1-vertex of the 3-sectors. In components, if

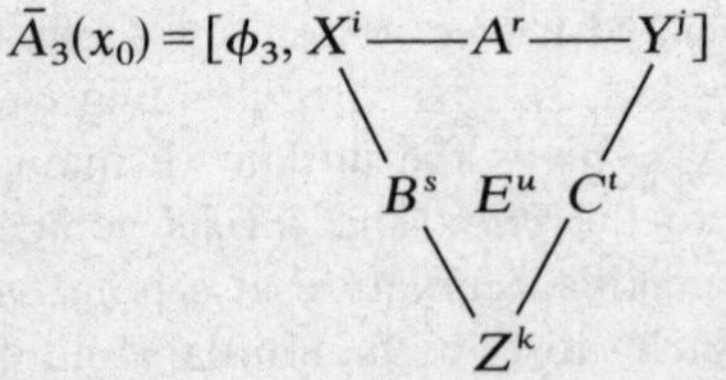

and

$$\bar{B}_3(x_0)=\begin{array}{l}[\phi_3,\ X^i \text{——} A^r \text{——} Y^j] \\ \qquad\quad \diagdown \qquad\quad \diagup \\ \qquad\quad B^s\ \ F^u\ \ G^t \\ \qquad\qquad \diagdown \ \diagup \\ \qquad\qquad\ \ Z^k \end{array}$$

then

$$[\bar{A}_3(x_0);\bar{B}_3(x_0)]_{\gamma_1,\bar{\gamma}_1}=[\phi_2,X^i \text{——}(F^u-E^u)\text{——}(G^t-C^t)]. \tag{4.93}$$

We shall want to choose for γ_1 the three vertices $(1,0,0)$, $(0,1,0)$, $(0,0,1)$ and $\bar{\gamma}_1$ will be fixed $(1,0)$ for these considerations. As a shorthand, then denote the generalized bracket over C_i $(i=1,2,3)$ as $[\ \ ,\ \ ]_{(i,1)}$, the case above being $[\ \ ,\ \ ]_{(1,1)}$.

Now we claim that for i fixed, the introduction of a Riemannian metric on M gives a *cross section* of the affine bundle

$$\begin{array}{c} T_3(M) \\ \downarrow^{F_i} \\ T_3[M;C_i] \end{array}$$

which precisely extends the 'affine connection' construction of §4.1. If the metric on M is A^2, then this section is *an* affine connection associated with the promoted metric δA^2. Since 3-sectors have six identifications as 2-sectors on $T_1(M)$, there are *three* distinct Levi-Civita connections associated with δA^2: a somewhat paradoxical state of affairs but one which leads directly to the contravariant curvature tensor.

Theorem 4.68. *Let (M,A^2) be a Riemannian manifold and let*

$$\begin{array}{c} T_3(M) \\ \downarrow^{F_i} \\ T_3[M;C_i] \end{array}$$

be one of the generalized boundary projections described above. There is, in each F_i fiber, a unique i-flat 3-sector. The association which takes each C_i-sector to the flat 3-sector in its fiber will be denoted $\Lambda[i]:T_3[M;C_i]\to T_3(M)$.

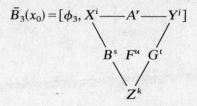

Proof. Here we argue the case $i=1$. The other two cases can be handled similarly. Let

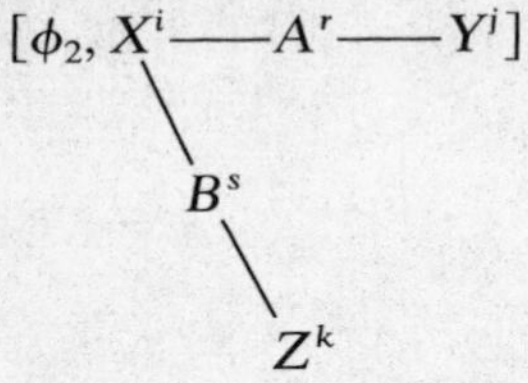

be given, and suppose that

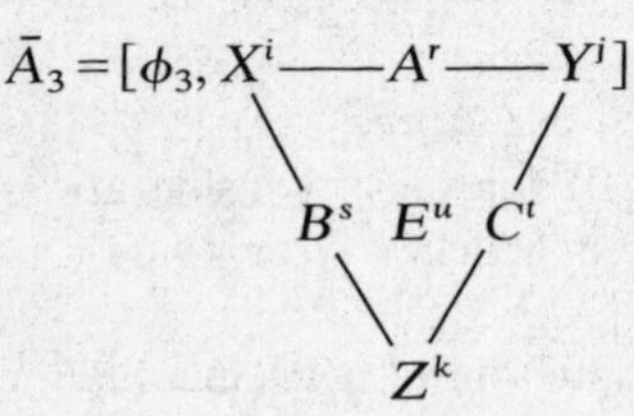

and

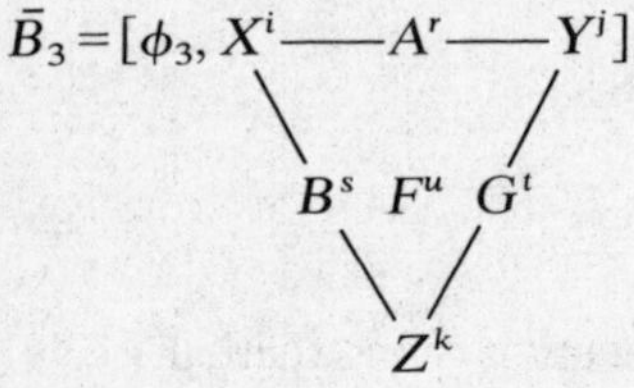

belong to its F_1 fiber. Let $\bar{A}_2 = [\phi_2, X^i \text{——} I^\alpha \text{——} W^h]$ be an *arbitrary* 2-sector with common 1-vertex with the original C_1-sector. Then it follows from formula (4.91) that (for X the common 1-vertex)

$$\begin{aligned}\tfrac{1}{2}d_1DA^2\langle X, \bar{A}_2, \bar{A}_3\rangle - \tfrac{1}{2}d_1DA^2\langle X, \bar{A}_2, \bar{B}_3\rangle \\ &= \frac{\partial \tilde{a}_{ht}}{\partial x^i} X^i W^h (C^t - G^t) + \tilde{a}_{uh} W^h (E^u - F^u) + \tilde{a}_{t\alpha} I^\alpha (C^t - G^t) \\ &= \delta A^2{}_{|X}\langle [\bar{B}_3; \bar{A}_3]_{(1,1)}, \bar{A}_2\rangle.\end{aligned}$$

This implies that

$$1\text{-force}\,(\bar{A}_3) - 1\text{-force}\,(\bar{B}_3) = [\bar{B}_3; \bar{A}_3]_{(1,1)}. \tag{4.94}$$

Appealing to the affine structure of the fiber, we see that a *pair* of 1-flat 3-sectors in the fiber would have zero bracket, and hence would be equal. This gives uniqueness. Then given $\bar{A}_3$, find $\bar{B}_3$ such that $[\bar{B}_3; \bar{A}_3]_{(1,1)}$ *is equal to* 1-force $(\bar{A}_3)$. Then from (4.94), 1-force $(\bar{B}_3)$ is zero. ■

Next we compute the *components* of the 1-flat 3-sector over

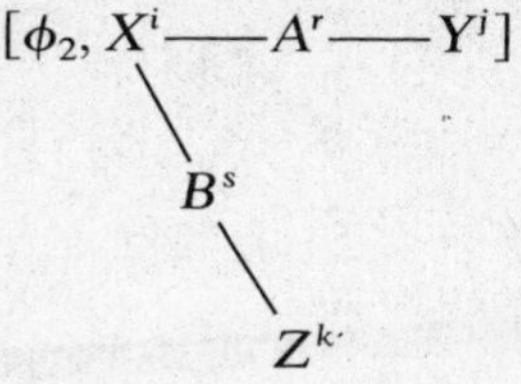

This amounts to computing the analog of 'connection coefficients' for $\Lambda[1]$, thinking of the C_1-sector above as a pair of tangent vectors at $X=[\phi_1, X^i]$. Call that C_1-sector U_2. Suppose that

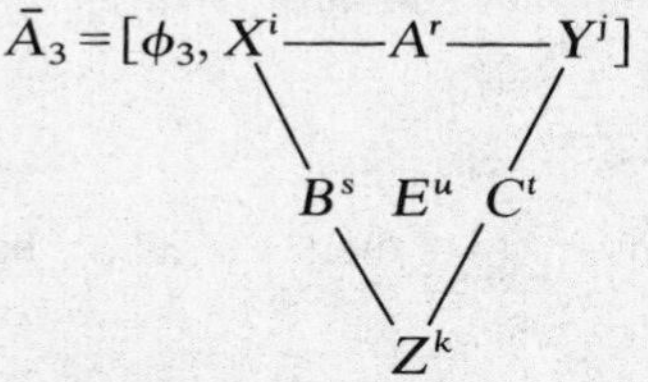

is $\Lambda[1](U_2)$. Then (E^u, C^t) are to be determined.

Let $\bar{A}_2=[\phi_2, X^i$——I^α——$W^h]$ (violating Convention 3.9) be an arbitrary 2-sector with 1-vertex X. Then $d_1DA^2\langle X, \bar{A}_2, \bar{A}_3\rangle=0$ since $\bar{A}_3$ is 1-flat. Letting $W^h=0$, formula (4.91) implies that (for frame-jet ϕ_2)

$$\tilde{a}_{t\alpha}C^tI^\alpha+[jk;\alpha]Y^jZ^kI^\alpha=0 \qquad \text{for all } I^\alpha. \tag{4.95}$$

This implies (formula (4.11) that $C^t=\Lambda^t_{jk}Y^jZ^k$.

Next, setting $I^\alpha=0$, (4.91) yields

$$\begin{aligned}\tilde{a}_{uh}E^uW^h=&-\frac{1}{2}\left[\frac{\partial^2\tilde{a}_{kh}}{\partial x^i\partial x^j}+\frac{\partial^2\tilde{a}_{jh}}{\partial x^i\partial x^k}-\frac{\partial^2\tilde{a}_{jk}}{\partial x^i\partial x^h}\right]X^iY^jZ^kW^h\\&-\frac{\partial\tilde{a}_{ht}}{\partial x^i}X^iW^h(\Lambda^t_{jk}Y^jZ^k)-[js;h]Y^jW^hB^s\\&-[rk;h]Z^kW^hA^r \qquad \text{for all } W^h.\end{aligned} \tag{4.96}$$

This determines E^u by non-degeneracy of the metric. In fact if we set

$$\bar{B}_2(x)=[(\phi\circ t_z)_2, \tilde{Y}^j(z)\text{——}\tilde{J}^\gamma(z)\text{——}\tilde{Z}^k(z)]$$

a local 2-sector field with $\phi(0)=x_0$ with $X\cup\bar{B}_2(x)$ in the F_1 fiber over U_2, then a straightforward calculation yields

$$\tilde{a}_{uh}E^uW^h=a_{uh}(0)\frac{\partial}{\partial x^i}[\tilde{\Lambda}^u_{jk}(z)\tilde{Y}^j(z)\tilde{Z}^k(z)]X^iW^h$$

from which

$$E^u=\frac{\partial}{\partial x^i}[\tilde{\Lambda}^u_{jk}(z)\tilde{Y}^j(z)\tilde{Z}^k(z)]X^i(0) \tag{4.97}$$

(in frame-germ $\phi:(E,0)\to(M,x_0)$). We may therefore write the following.

Observation 4.69. Letting

$$
\begin{array}{l}
U_2=[\phi_2, X^i \text{——} A^r \text{——} Y^j], \\
\qquad\quad \diagdown \\
\qquad\qquad B^s \\
\qquad\qquad\quad \diagdown \\
\qquad\qquad\qquad Z^k
\end{array}
$$

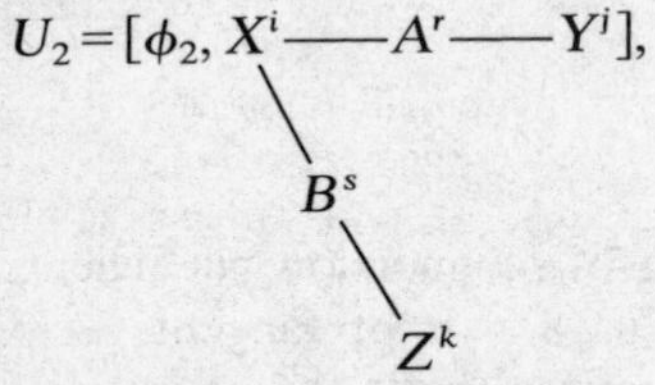

$\phi:(E,0)\to(M,x_0)$ at point x_0 in Riemannian manifold $(M, A^2(x))$, if $\bar{B}_2(x)$ is a local 2-sector field with

$$\bar{B}_2(x)=[(\phi\circ t_z)_2, \tilde{Y}^j(z)\text{——}\tilde{I}^\alpha(z)\text{——}\tilde{Z}^k(z)]$$

and with $X\cup\bar{B}_2(x)$ in the F_1 fiber over U_2, then $\Lambda[1](U_2)$ has ϕ components:

$$
\begin{array}{ccccc}
[\phi_3, X^i & \text{————————} & A^r & \text{————————} & Y^j] \\
 & \diagdown & & \diagup & \\
 & & \left(\dfrac{\partial}{\partial x^i}[\tilde{\Lambda}^u_{jk}(z)\tilde{Y}^j(z)\tilde{Z}^k(z)]X^i\right) & & \\
 & B^s & & (\Lambda^t_{jk}Y^jZ^k) & \\
 & & \diagdown \qquad \diagup & & \\
 & & Z^k & &
\end{array}
$$

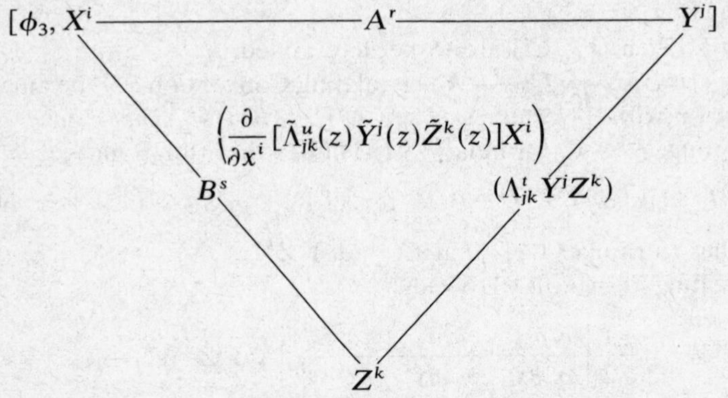

The sections $\Lambda[2]$ and $\Lambda[3]$ are handled similarly. ■

Notice that if ϕ_2 is a null frame, then $\Lambda^t_{jk}Y^jZ^k$ vanishes but the (111) component need not vanish.

We now develop the relation with *Riemannian curvature*. Let $\phi:(E,0)\to(M,x_0)$ be a frame-germ at x_0 in Riemannian manifold $(M, A^2(x))$. Let $[\bar{A}_2(x_0), Z(x_0)]$ be a given pair $\bar{A}_2(x_0)=[\phi_2, X^i\text{——}A^r\text{——}Y^j]$ and $Z(x_0)=[\phi_1, Z^k]$. Now construct C_1-sector U_2 and C_2-sector V_2 respectively:

$$
\begin{array}{l}
U_2=[\phi_2, X^i \text{——} A^r \text{——} Y^j] \\
\qquad\quad \diagdown \\
\qquad\quad \Lambda^s_{ik}X^iZ^k \\
\qquad\qquad\quad \diagdown \\
\qquad\qquad\qquad Z^k
\end{array}
$$

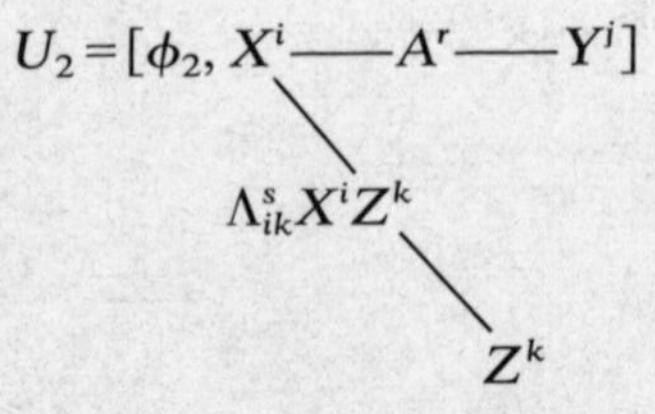

and

$$\begin{array}{l} V_2 = [\phi_2, X^i \text{———} A^r \text{———} Y^j] \\ \qquad\qquad\qquad\qquad\quad / \\ \qquad\qquad\qquad \Lambda^t_{jk} Y^j Z^k \\ \qquad\qquad\qquad\quad / \\ \qquad\qquad\quad Z^k \end{array}$$

U_2 and V_2 are of course entirely determined by the data $[\bar{A}_2, Z]$. From these, construct the 3-sectors $\Lambda[1](U_2)$ and $\Lambda[2](V_2)$. These also are completely determined by the initial data $[\bar{A}_2, Z]$ (together with the metric A^2).

Definition 4.70. For $(M, A^2(x))$ a Riemannian manifold and at $x_0 \in M$ a pair of sectors $[\bar{A}_2, Z]$ specified with ϕ components as above, the 1-flat 3-sector $\Lambda[1](U_2)$ and the 2-flat 3-sector $\Lambda[2](V_2)$ have an ordinary *bracket*, this being a *tangent vector* at x_0.

The bracket $[\Lambda[1](U_2): \Lambda[2](V_2)]$ depends only on the *ordered triple* (X, Y, Z) of tangent vectors at x_0 where (X, Y) is $D(2)\bar{A}_2$. Accordingly, denote that bracket $R(X, Y)Z$ to indicate the special role which Z plays in the construction. We shall show that the association $(X, Y, Z) \to R(X, Y)Z$ gives a tensor field of type $\binom{1}{3}$ with a certain antisymmetry property in the next observation which relates it with the (covariant) curvature tensor. In any case, we call $R(X, Y)Z$ the (contravariant) Riemann curvature tensor. ■

Exercises

19. Given a triple of tangent vectors (X, Y, Z) with components in frame ϕ as above, show that over the sector complex

$$\begin{array}{ccccc} [\phi_2, X^i & \text{———} & \Lambda^r_{ij} X^i Y^j & \text{———} & Y^j] \\ & \diagdown & & \diagup & \\ & \Lambda^s_{ik} X^i Z^k & & \Lambda^t_{jk} Y^j Z^k & \\ & & \diagdown \quad \diagup & & \\ & & Z^k & & \end{array}$$

there are three i-flat 3-sectors (one for each vertex) and that they have pairwise brackets. Deduce that

$$R(X, Y)Z + R(Y, Z)X + R(Z, X)Y = 0$$

and that

$$R(X, Y)Z + R(Y, X)Z = 0.$$

20. For $f : (M, A^2) \to (N, B^2)$ an isometric immersion with $T(f) : T_1(M) \to T_1(N)$, under what conditions is it true that

$$T_1(f)[R(X, Y)Z] = R(T_1(f)X, T_1(f)Y)T_1(f)Z.$$

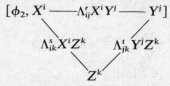

21. Show that $R(X, Y)Z$ is *identically zero* on a one-dimensional manifold.

22. For $(\tilde{X}^i(z), \tilde{Y}^j(z), \tilde{Z}^k(z))$ a triple of local vector fields with value (X, Y, Z) at $z=0$, show that

$$R(X, Y)Z = \frac{\partial}{\partial x^j}[\tilde{\Lambda}^u_{ik}(z)\tilde{X}^i(z)\tilde{Z}^k(z)]Y^j(0)$$

$$-\frac{\partial}{\partial x^i}[\tilde{\Lambda}^u_{jk}(z)\tilde{Y}^j(z)\tilde{Z}^k(z)]X^i(0)$$

in ϕ components.

Observation 4.71. For $(M, A^2(x))$ a Riemannian manifold and (X, Y, Z, W) a 4-tuple of tangent vectors at $x_0 \in M$,

$$A^2\langle R(X, Y)Z, W\rangle = -R[(X, Y); (Z, W)].$$

This shows that $R(X, Y)Z$ gives a tensor of type $\binom{1}{3}$ and also that

$$R(X, Y)Z = -R(Y, X)Z.$$

Proof. Consider the S-sector

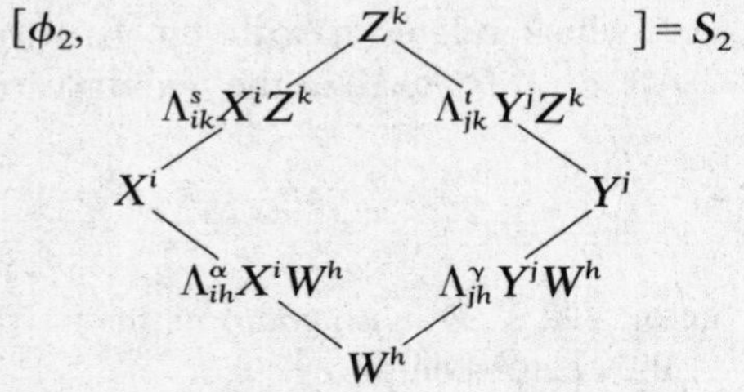

Then by definition

$$R[(X, Y); (Z, W)] = -\tfrac{1}{2}CA^2\langle S_2\rangle = \tfrac{1}{2}(Sd_1DA^2 - d_1DA^2)\langle S_2\rangle$$

(for S the transposition of 1 and 2 in $Kl(4)$).

Now consider $\Lambda[1](U_2)$ as it was computed in Definition 4.69, and let $\bar{A}_2 = [\phi_2, X^i \text{——} \Lambda^\alpha_{ih}X^iW^h \text{——} W^h]$ and $\bar{B}_2 = [Y^j \text{——} \Lambda^\gamma_{jh}Y^jW^h \text{——} W^h]$. Then

$$\tfrac{1}{2}(Sd_1DA^2 - d_1DA^2)(S_2)$$

$$= \tfrac{1}{2}Sd_1DA^2\langle Y, \bar{B}_2, \Lambda[1](U_2)\rangle - \tfrac{1}{2}d_1DA^2\langle X, \bar{A}_2, \Lambda[1](U_2)\rangle$$

and since

$$\tfrac{1}{2}d_1DA^2\langle X, \bar{A}_2, \Lambda[1](U_2)\rangle = 0,$$

we have

$$R[X, Y); (Z, W)] = \tfrac{1}{2}Sd_1DA^2\langle Y, \bar{B}_2, \Lambda[1](U_2)\rangle$$
$$= \tfrac{1}{2}\sigma d_1 DA^2\langle Y, \bar{B}_2, \Lambda[1](U_2)\rangle$$

(for σ introduced in Definition 4.67).

Now let V_2 be as in Definition 4.70. Then according to the calculation preceding formula (4.94) we have

$$\tfrac{1}{2}Sd_1DA^2\langle Y, \bar{B}_2, \Lambda[1](U_2)\rangle = \tfrac{1}{2}Sd_1DA^2\langle Y, \bar{B}_2, \Lambda[1](U_2)\rangle$$
$$-\tfrac{1}{2}Sd_1DA^2\langle Y, \bar{B}_2, \Lambda[2](V_2)\rangle$$
$$= \delta A^2{}_{|Y}\langle[\Lambda[2](V_2); \Lambda[1](U_2)]_{(2,1)}, \bar{B}_2\rangle.$$

Now observing that the 2-vertex of $[\Lambda[2](V_2): \Lambda[1](U_2)]_{(2,1)}$ is zero, we have the conclusion:

$$A^2\langle R(X, Y)Z, W\rangle = -R[(X, Y); (Z, W)]. \tag{4.98}$$

■

In order to see the geometric meaning of $R(X, Y)Z$, we formulate things in a slightly different way.

Suppose that $(M, A^2(x))$ is a Riemannian manifold, $\phi: O \to U \subset M$ a frame on M with $\phi(z_0) = x_0$. And let $\tilde{Z}(x) = \tilde{Z}^k(z)$ be a local smooth vector field on U. Suppose that $X^i(z_0)$ is a tangent vector at $x_0 \in M$. Then as we observed, the suspension $X \cup \tilde{Z}$ is the 2-sector generated by 'pushing' X along the local flow for $\tilde{Z}$ (discussion following Definition 2.14). Now it is easy to arrange that, for a fixed X, a $\tilde{Z}$ be found with $\nabla(X \cup \tilde{Z}) = 0$, that is, with $X \cup \tilde{Z}$ flat.

Interesting things happen when this procedure is extended to the two-dimensional case. Thus, suppose that instead of tangent vector $X(x_0)$ we are given 2-sector $\bar{A}_2(x_0)$. Then for local vector field $\tilde{Z}$ on U we have defined the suspension of the vector field $\tilde{Z}$ by the 2-sector $\bar{A}_2(x_0)$; it has the interpretation that it is generated by 'pushing' $\bar{A}_2(x_0)$ along the flow of $\tilde{Z}$ (Definition 3.20 and Observation 3.21). In local components, we may write for

$$\bar{A}_2 = [(\phi \circ t_{z_0})_2, X^i \text{——} A^r \text{——} Y^j]$$

that

$$A_2(x_0) \cup \tilde{Z}(x) = \left[(\phi \circ t_{z_0})_2,\ \begin{array}{ccccc} X^i & \text{————} & A^r & \text{————} & Y^j \\ & \diagdown & & \diagup & \\ \dfrac{\partial Z^s}{\partial x^i}\Big|_{z_0} X^i & & (*) & & \dfrac{\partial Z^t}{\partial x^j}\Big|_{z_0} Y^j \\ & \diagdown & & \diagup & \\ & & Z^k(z_0) & & \end{array}\right] \tag{4.99}$$

where

$$(*)=\left[\frac{\partial Z^k}{\partial x^r}A^u+\frac{\partial^2 Z^u}{\partial x^i \partial x^j}X^iY^j\right]_{|z_0}.$$

Notice that this is *not* the suspension of a 2-sector field by a tangent vector; we consider $\bar{A}_2(x_0)$ to be fixed. It will be useful to keep in mind one way in which $\bar{A}_2(x_0)$ might be given. A pair of vector fields $\tilde{X}$ and $\tilde{Y}$ with zero Lie bracket on U (having locally commuting flows) will define a 2-sector field, thus determining *at each point* a candidate for $\bar{A}_2(x_0)$. We may thus think of the above suspension as an infinitesimal deformation of the two-dimensional coordinate system at a point.

Now given the data of the *pair* $(\bar{A}_2(x_0), Z(x_0))$ then, if a vector field $\tilde{Z}$ can be found extending Z such that $\bar{A}_2 \cup \tilde{Z}$ is 1-*flat* and 2-*flat*, it is clear that the curvature $R(X, Y)Z$ is zero at x_0. A moderately weakened converse will also be shown to be true. This will require an extension of the ∇ operator in an interesting new direction; in fact, the result that we are after is usually interpreted as a commutativity condition on ∇ iterations being equivalent to zero curvature.

Definition 4.72. Let $(M, A^2(x))$ be a Riemannian manifold and suppose that $\bar{A}_2(x_0)$ is a 2-sector at $x_0 \in M$ and that with respect to a local frame $\phi: O \to U \subset M$, $\phi(z_0)=x_0$, $\tilde{Z}=\tilde{Z}^k(z)$ is a local vector field on U.

Define the *tangent vector* at x_0 denoted $\nabla\nabla_1[\bar{A}_2(x_0) \cup \tilde{Z}(x)]$ in the following way. Suppose that $\bar{A}_2 \cup \tilde{Z}$ has the components (4.99) above. Let

$$\begin{array}{l} U_2=[(\phi\circ t_{z_0})_2, X^i \text{———} A^r \text{———} Y^i] \\ \qquad\qquad\qquad\quad \diagdown \\ \qquad\qquad\qquad\qquad \dfrac{\partial Z^s}{\partial x^i}_{|z_0} X^i \\ \qquad\qquad\qquad\qquad\qquad \diagdown \\ \qquad\qquad\qquad\qquad\qquad Z^k(z_0) \end{array}$$

be a C_1-sector at x_0.

Form $\Lambda[1](U_2)$ in the usual way (Theorem 4.68) to get the 1-flat 3-sector over U_2. Then the *generalized bracket* (Definition 2.21) of the elements of the F_1 fiber (4.99) and $\Lambda[1](U_2)$ is the 2-sector with components:

$$\Bigg[(\phi\circ t_{z_0})_2, X^i \text{——}\left[\frac{\partial Z^r A^u}{\partial x^u}_{|z_0}+\frac{\partial^2 Z^r}{\partial x^i \partial x^j}_{|z_0}X^iY^j-\frac{\partial}{\partial x^i}_{|z_0}[\tilde{\Lambda}^r_{jk}\tilde{Y}^j\tilde{Z}^k]X^i\right]$$

$$\text{——}\frac{\partial Z^j}{\partial x^v}_{|z_0}Y^v-\Lambda^j_{ik}Y^iZ^k\Bigg]. \quad (4.100)$$

(Here we let $\tilde{X}$ and $\tilde{Y}$ be the choice of any pair of local fields with $X(x_0) \cup \tilde{Y}(x) = \tilde{X}(x) \cup Y(x_0) = \bar{A}_2(x_0)$.) This gives the 1-force of 3-sector (4.99).

Since the generalized bracket is over $T_3[M; C_1]$, we denote this 2-sector as $\nabla_1[\bar{A}_2(x_0) \cup \tilde{Z}(x)]$. Notice that the 2-vertex (1-side) is the (ordinary) force of the 1-side of (4.99).

Now the force of *this* 2-sector is the desired tangent vector at x_0. We call that tangent vector $\nabla\nabla_1[\bar{A}_2(x_0) \cup \tilde{Z}(x)]$ and its components are

$$\nabla\nabla_1[\bar{A}_2(x_0) \cup \tilde{Z}(x)] = \Bigg[(\phi \circ t_{z_0})_1, \frac{\partial Z^r A^u}{\partial x^u}\bigg|_{z_0} + \frac{\partial^2 Z^r}{\partial x^i \partial x^j}\bigg|_{z_0} X^i Y^j - \frac{\partial}{\partial x^i}\bigg|_{z_0} [\tilde{\Lambda}^r_{jk} \tilde{Y}^j \tilde{Z}^k] X^i - \Lambda^r_{ij} X^i \bigg[\frac{\partial Z^j}{\partial x^v}\bigg|_{z_0} Y^v - \Lambda^j_{ik} Y^i Z^k\bigg]\Bigg]. \qquad (4.101)$$

■

In a perfectly general way, we may define the function $\nabla\nabla_1$ with $\nabla\nabla_1 : T_3(M) \to T_1(M)$ a fiber bundle map over M. Then $\nabla\nabla_1[\bar{A}_3(x_0)]$ is simply the *iteration* $\nabla(\nabla_1[\bar{A}_3(x_0)])$. Thus given

$$\bar{A}_3(x_0) = \left[\phi_3, \begin{array}{ccccc} X^i & \text{——} & A^r & \text{——} & Y^j \\ \backslash & & & & / \\ & B^s & E^u & C^t & \\ & \backslash & & / & \\ & & Z^k & & \end{array}\right],$$

letting $\tilde{X}, \tilde{Y}, \tilde{Z}$ be local vector fields with vanishing pairwise brackets at $x_0 = \phi(0)$, and with $X = X^i(0)$, $Y = Y^j(0)$, $Z = Z^k(0)$, $D^1(3)\bar{A}_3 = Y \cup \tilde{Z}$ $(= \tilde{Y} \cup Z)$, $D^2(3)\bar{A}_3 = Z \cup \tilde{X}$, and $D^3(3)\bar{A}_3 = X \cup \tilde{Y}$ at x_0, the components of $\nabla\nabla_1[\bar{A}_3(x_0)]$ are

$$\left[\phi_1, E^u - \frac{\partial}{\partial x^i}[\tilde{\Lambda}^u_{jk} \tilde{Y}^j \tilde{Z}^k] X^i - \Lambda^u_{it} X^i [C^t - \Lambda^t_{jk} Y^j Z^k]\right]. \qquad (4.102)$$

This map is linear on the subspaces of the fiber $T_{3,1}$, $T_{3,2}$, and $T_{3,3}$.

Now if $\bar{A}_3(x_0)$ is as in the discussion above, then $\nabla\nabla_1 \bar{A}_3(x_0)$ has the following interesting property. If $W = [\phi_1, W^h]$ is an *arbitrary* tangent vector at x_0 and we construct the 2-sector $[\phi_2, X^i$——$\Lambda^\alpha_{ih} X^i W^h$——$W^h] = \bar{B}_2(x_0)$ then

$$\tfrac{1}{2} d_1 D A^2 \langle X, \bar{B}_2, \bar{A}_3 \rangle = A^2 \langle W, \nabla\nabla_1 \bar{A}_3 \rangle. \qquad (4.103)$$

This follows simply on calculation of $\delta A^2{}_{|X} \langle \nabla_1 \bar{A}_3(x_0), \bar{B}_2(x_0) \rangle$ in a *null frame.*

Now we interpret the operation $\nabla\nabla_1$ for the case studied in Definition 4.72. $\nabla\nabla_1[\bar{A}_2 \cup \tilde{Z}]$ can then be interpreted in the following manner. Let $\tilde{X}, \tilde{Y}$ be a locally commuting pair of vector fields giving (by suspension)

$\bar{A}_2$. They define for chart $\phi : O \to U \subset M$ the germ of a map $f : I \times I \to O$ with $I = (-\varepsilon, \varepsilon)$ for some small positive ε, and with the property that $\phi \circ f(0, 0) = x_0$. Letting (s, t)be standard linear coordinate functions for $I \times I$, we have an f such that

$$\frac{\partial f}{\partial s}_{|(s,t)} = X^i[f(s, t)] \quad \text{and} \quad \frac{\partial f}{\partial t}_{|(s,t)} = Y^j[f(s, t)],$$

$X = X^i[f(0, 0)]$ and $Y = Y^j[f(0, 0)]$.

The mapping $s \mapsto \nabla[Y(\phi \circ f(s, 0)) \cup \tilde{Z}]$ then defines a 1-variation of the path $s \to \phi \circ f(s, 0)$ taking $I \to M$. That 1-variation defines a 2-sector at $s = 0$ $(t = 0)$, and $\nabla\nabla_1[\bar{A}_2 \cup \tilde{Z}]$ is then interpreted as the force of that 2-sector. The remarkable thing is that it is *independent of all choices*; in the end, it depends only on the 3-sector $\bar{A}_2 \cup \tilde{Z}$ at x_0. In this light, it is a sequence of 'covariant derivatives' first in the Y direction and then in the X direction. $R(X, Y)Z$ will give the *precise* measure of the extent to which these differentiations give different results when the *order* in which they are applied is reversed.

Thus, with the same data as in Definition 4.72 except that U_2 is replaced with

$$\begin{array}{llll} V_2 = [\phi \circ t_{z_0})_2, X^i & \text{———} A^r \text{———} & Y^j] \\ & & / \\ & \dfrac{\partial Z^t Y^j}{\partial x^j}_{|z_0} & \\ & / & \\ Z^k(z_0) & & \end{array}$$

and with C_1 replaced by C_2, we may define the generalized bracket of (4.99) and $\Lambda[2](V_2)$ (the 2-flat 3-sector over V_2). This bracket will be called $\nabla\nabla_2[\bar{A}_2(x_0) \cup \tilde{Z}]$. The mapping $\nabla\nabla_2$ is also a mapping of fiber bundles $T_3(M) \to T_1(M)$ and the properties of this mapping are parallel to those already described for $\nabla\nabla_1$. In particular, if $\bar{A}_3(x_0)$ is as in formula (4.103) and

$$\bar{C}_2(x_0) = [\phi_2, Y^j \text{——} \Lambda^{\gamma}_{jh} Y^j W^h \text{——} W^h],$$

then we have

$$A^2\langle W, \nabla\nabla_2 \bar{A}_3(x_0)\rangle = \tfrac{1}{2} S d_1 D A^2\langle Y, \bar{C}_2, \bar{A}_3\rangle. \tag{4.104}$$

Theorem 4.73. *For* (M, A^2) *a Riemannian manifold,* $\bar{A}_3(x_0)$ *an* arbitrary 3-*sector at* x_0

$$\nabla\nabla_1 \bar{A}_3(x_0) - \nabla\nabla_2 \bar{A}_3(x_0) = R(X, Y)Z. \tag{4.105}$$

Proof. We compute $A^2\langle W, \nabla\nabla_1\bar{A}_3(x_0)-\nabla\nabla_2\bar{A}_3(x_0)\rangle$ according to (4.103) and (4.104). Letting $\bar{B}_2$ and $\bar{C}_2$ be as in the previous discussion, use formula (4.91) to evaluate

$$\tfrac{1}{2}d_1DA^2\langle X, \bar{B}_2, \bar{A}_3\rangle-\tfrac{1}{2}Sd_1DA^2\langle Y, \bar{C}_2, \bar{A}_3\rangle$$

and compare with (4.82) (after a few reductions with Riemann–Christoffel symbols) to obtain

$$-R[(X, Y); (Z, W)]=A^2\langle W, R(X, Y)Z\rangle.$$

Since W was arbitrary, the result follows. ■

We see then that $R(X, Y)Z$ gives the *precise* measure of the (non-) commutativity of iterated covariant differentiations. Observing that $R(X, Y)Z$ depends only on the tangent vectors (X, Y, Z), we have a rather surprising consequence of the theorem above which gives yet another geometric interpretation of the cross.

Corollary 4.74. *Let (M, A^2) be a Riemannian manifold and suppose $\phi:(E, 0)\to(M, x_0)$ is a frame-germ at x_0. Let $X=[\phi_1, X^i]$, $Y=[\phi_1, Y^j]$, $Z=[\phi_1, Z^k]$ and $W=[\phi_1, W^h]$ be given tangent vectors. And let S_2 be* any *S-sector of the form:*

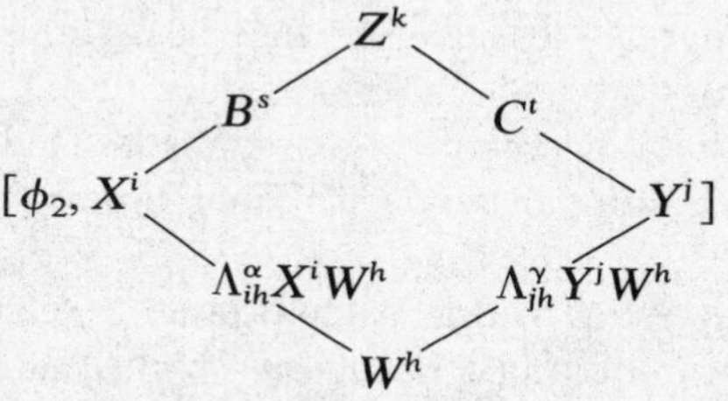

the B^s and C^t components being arbitrary. Then we have

$$\tfrac{1}{2}CA^2\langle S_2\rangle=-R[(X, Y); (Z, W)]. \tag{4.106}$$

In particular if $(X, Y)=D(2)\bar{A}_2(x_0)$ for arbitrary *$\bar{A}_2(x_0)$ and if Z has extension to local vector field $Z(x)$, then*

$$\tfrac{1}{2}CA^2\langle S_2\rangle=A^2\langle(\nabla\nabla_1-\nabla\nabla_2)(\bar{A}_2(x_0)\cup Z(x)), W\rangle \tag{4.107}$$

and this is independent *of the extensions $\bar{A}_2(x_0)$ and $Z(x)$.* ■

The last topic of this section is concerned with the (im)possibility of locally defining a version of *arc length* for general Riemannian manifolds $(2\leq \dim M)$. Recall that for a curve $g:\mathbb{R}\to\mathbb{R}^n$ with non-vanishing derivative, if $\mathbb{R}$ is given the pull-back of the standard Euclidean metric in $\mathbb{R}^n$ then there is a diffeomorphism $\phi:\mathbb{R}\to\mathbb{R}$ with the property that (if ϕ is thought of as a global frame) each 2-frame $(\phi\circ t_z)_2$ is a null frame with respect to the metric (see the discussion following formula (4.20).) With reference to the *fundamental section* (Definition 4.8)

$$\begin{array}{c} F_2(\mathbb{R}) \\ s\uparrow \quad \downarrow \pi_{2,1} \\ F_1(\mathbb{R}) \end{array}$$

this can be interpreted as saying that

$$s[(\phi \circ t_z)_1] = (\phi \circ t_z)_2 \qquad \text{for all } z. \tag{4.108}$$

Condition (4.108) holds for global frame ϕ if and only if the pull-back metric under ϕ has constant components, that is ϕ is a 'multiple' of arc length parametrization. In the case of a general smooth manifold M, we cannot expect a global parametrization by a Euclidean space, but we can ask the following question.

If (M, A^2) is a Riemannian manifold and $x_0 \in M$, under what circumstances is there a neighborhood U of x_0 and a frame $\phi: V \to U \subset M$ with the property that

$$s[(\phi \circ t_z)_1] = (\phi \circ t_z)_2 \qquad \text{for all } z \in V?$$

When it is possible to find such a local frame ϕ, we shall say that the *fundamental section is realized by ϕ.* We shall see that in a sense this condition is the most natural extension of the notion of (local) arc length parametrization to the case of general Riemannian manifolds. A Riemannian manifold is said to be *flat* on an open set U if the curvature tensor vanishes *at every point in U.*

Now the answer to the above question is this: such a frame ϕ realizing near x_0 the fundamental section can be found *if and only if* the manifold is flat in some neighborhood of x_0. This is the sense in which the *Riemannian curvature gives the obstruction to defining a local generalization of arc length for Riemannian manifolds.* We prove this assertion in a series of steps (see also Spivak [22]).

Theorem 4.75. *For a Riemannian manifold $(M, A^2(x))$ with $x_0 \in M$, the following are equivalent:*

1) *There is a frame $\phi: V \to U \subset M$ with $x_0 \in U$ such that the fundamental section is realized by ϕ.*
2) *There is a frame $\phi: V \to U \subset M$ with $x_0 \in U$ such that for* each *pair of tangent vectors $[(\phi \circ t_z)_1, X^i(z)] = X(x)$ and $[(\phi \circ t_z)_1, Y^i(z)] = Y(x)$ with $z \in V$,*

$$\Lambda[X(x), Y(x)] = [(\phi \circ t_z)_2, X^i(z) \text{——} 0 \text{——} Y^j(z)].$$

3) *There is a frame* $\phi: V \to U \subset M$ *with* $x_0 \in U$ *such that the* ϕ *components of* $A^2(x)$ *are constant.*
4) *The Riemann curvature tensor* $R(X, Y)Z$ vanishes *on* U *for some* U *a neighborhood of* x_0: (M, A^2) *is* flat *on a neighborhood of* x_0.
5) *There is a frame* $\phi: V \to U \subset M$ *with* $x_0 \in U$ *such that for* any *tangent vector* $X(x)$ *at a point in* U *there is a local vector field* $\tilde{X}(x)$ *extending it with the property that* $\nabla(\tilde{X} \cup Y) = 0$ *for all tangent vectors* Y *at points in* U. *Such a local vector field is said to be* parallel *on* U *(any solution to the 'equation of variations' for* $\tilde{X}(x)$ *is a parallel translation).*
6) *There is a frame* $\phi: V \to U \subset M$ *with* $x_0 \in U$ *and a set* $\tilde{X}_1(x), \tilde{X}_2(x), \ldots, \tilde{X}_m(x)$ *of* parallel *local vector fields on* U *which gives at each point of* U *a basis for the tangent space.*

Proof. 1) $\leftrightarrow$ 2) If $s[(\phi \circ t_z)_1] = (\phi \circ t_z)_2$ for the fundamental section s and frame $\phi: V \to U$ and for all $z \in V$, then $(\phi \circ t_z)_2$ is a null frame for each z and so $[(\phi \circ t_z)_2, X^i(z)$——$0$——$Y^i(z)]$ is *flat* for any pair $X(x)$ and $Y(x)$.

On the other hand, given a local frame with the property that all 2-sectors of the form $[(\phi \circ t_z)_2, X^i(z)$——$0$——$Y^i(z)]$ are flat, then it follows that the connection coefficients for $(\phi \circ t_z)_2$, that is $\Lambda^r_{ij}(z)$, are identically zero. Hence for $f: F_2(U) \to \gamma_2(m)$ the $\gamma_2(m)$-equivariant map introduced in Definition 4.8, we have $f[(\phi \circ t_z)_2] = \text{id}$. Thus $(\phi \circ t_z)_2$ is a null frame.

2) $\to$ 3) Given frame $\phi: V \to U$ satisfying 2), the $\Lambda^r_{ij}(z)$ vanish on V. Therefore since $[ij; k] + a_{rk}\Lambda^r_{ij} = 0$ the $[ij; k](z)$ vanish on V. But

$$[ij; k] + [ik; j] = \frac{\partial a_{jk}}{\partial x^i}\bigg|_z$$

at each point $z \in V$. These being 0, the components of A^2 on V are constant.

3) $\to$ 4) This is immediate from Corollary 4.74 and formula 4.73.

4) $\to$ 5) Let Z be a tangent vector at a point $x_0 = \phi(z_0)$ of U for frame $\phi: V \to U$ with V *connected.* Let $(s^1, s^2, \ldots, s^m)$ be the standard linear coordinates in E, and say that $z_0 = (0, 0, \ldots, 0)$ for simplicity. Now for $(s^1_0, s^2_0, \ldots, s^m_0)$ an arbitrary point w_0 of V, let $\tilde{Z}^k(w_0)$ be the components of the tangent vector obtained by parallel translation of Z from $\phi(z_0)$ to $\phi(w_0)$ along a finite sequence of arcs $\phi \circ g$ with g a path in Z keeping all coordinates s^k constant with the exception of one of them.

We must show that this defines at $\phi(w_0)$ a unique tangent vector. Smoothness will follow from remarks made earlier on smoothness of parallel translation. Thus suppose we are given a 'rectangle' on V

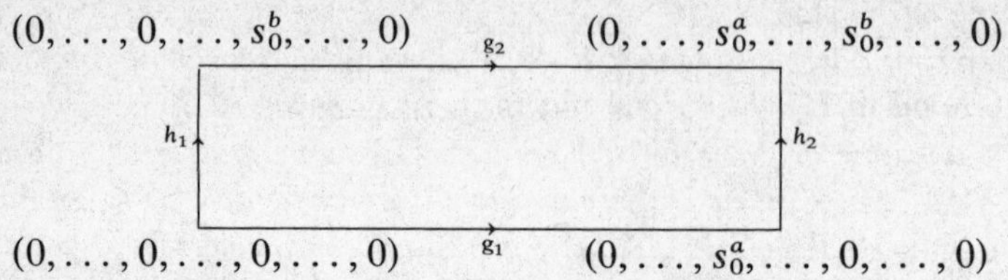

The lower left vertex is $(0, \ldots, 0)$ for simplicity, and the argument is independent of the choice of component functions s^a and s^b. Let the variables σ vary from 0 to s_0^a and τ vary from 0 to s_0^b. The paths $h(\sigma)$ are defined in the obvious way ($h(0) = h_1$ and $h(s_0^a) = h_2$), and the paths $g(\tau)$ are defined similarly with $g(0) = g_1$ and $g(s_0^b) = g_2$. Finally, let $\tilde{X}$ and $\tilde{Y}$ be local vector fields on U with the property that $\tilde{X}^i(0, \ldots, \sigma, \ldots, \tau, \ldots, 0)$ (henceforth denoted simply $\tilde{X}^i(\sigma, \tau)$ satisfies

$$\tilde{X}^i(\sigma, \tau)) = \frac{dg(\tau)}{d\sigma}_{|\sigma}$$

and

$$\tilde{Y}^i(\sigma, \tau) = \frac{dh(\sigma)}{d\tau}_{|\tau} .$$

These vector fields have zero Lie bracket on the image of the rectangle and can obviously be extended to a 2-sector field which is adapted to the pair: call it $\bar{A}_2(x) = X \cup \tilde{Y} = \tilde{X} \cup Y$. Then we have from the vanishing of the curvature on U.

$$\nabla\nabla_1[\bar{A}_2(x) \cup \tilde{Z}] = \nabla\nabla_2[\bar{A}_2(x) \cup \tilde{Z}] \tag{4.109}$$

(see formula (4.107)).

Assume that $\tilde{Z}$ is parallel along g_1 and $h(\sigma)$ for all $0 \leqslant \sigma \leqslant s_0^a$, then from (4.109) and the fact that $\nabla\nabla_1[\bar{A}_2(x) \cup \tilde{Z}] = 0$, we have that $\nabla\nabla_2[\bar{A}_2(x) \cup \tilde{Z}]$ *vanishes identically* on the rectangle. This means $\nabla[Y \cup (\tilde{\nabla} X \cup \tilde{Z})] = 0$ (or $\nabla_Y(\nabla_X \tilde{Z}) = 0$ in terms of Koszul connections).

Now suppose that the local vector field $\tilde{\nabla}(X \cup \tilde{Z})$ we call $\tilde{V}$. We have $\nabla(Y \cup \tilde{V}) = 0$ on the rectangle and $\tilde{V}$ vanishes on the image of g_1. Along any path $h(\sigma)$ the fact that $\nabla(Y \cup \tilde{V}) = 0$ is equivalent to the statement that V *and* the parallel translates of $V[\phi \circ g_1(\sigma)]$ along $h(\sigma)$ are solutions to the *same* linear variation equation satisfying the same initial condition at $\phi \circ g_1(\sigma)$. Therefore $\tilde{V}$ vanishes also on the image of g_2, and this implies that $\tilde{Z}$ is parallel along g_2. We conclude that parallel translation along g_1 and h_2 yields the same result as parallel translation along h_1 then g_2 for the vector Z.

5) $\to$ 6) This is obvious as parallel translation preserves the inner product (Theorem 4.45).

6) → 2) Given the set of parallel vector fields $\tilde{X}_1(x), \ldots, \tilde{X}_m(x)$ on U, to construct a frame realizing the fundamental section we show that these fields have zero Lie bracket, then applying Froebenius' theorem (which generalizes Theorem 2.19) together with the inverse mapping theorem we get a local frame $\phi: V' \to U' \subset M$ such that at each $x_0 = \phi(z_0)$,

$$\tilde{X}_a(x_0) = [(\phi \circ t_{z_0})_1, e^i_a]$$

with

$$e^i_a = \begin{cases} 0 & \text{if } i \neq a, \\ 1 & \text{if } i = a. \end{cases}$$

Then $X_a \cup \tilde{X}_b = [(\phi \circ t_{z_0})_2, e^i_a \text{——} 0 \text{——} e^j_b]$ for each pair (a, b). Since these are parallel we also have $\Lambda(X_a, X_b) = X_a \cup \tilde{X}_b$. Linearity properties of the suspension operator then will give the result.

Now the bracket $[\tilde{X}_a, \tilde{X}_b] = 0$ simply because the 2-sectors $X_a \cup \tilde{X}_b$ and $\tilde{X}_a \cup X_b$ *are* equal to $\Lambda(X_a, X_b)$ being flat. ■

Bibliography

1. Abraham, R. and Marsden, J. E., *Foundations of mechanics*, revised ed, Benjamin/Cummings, Menlo Park, Calif., 1978.
2. Arnold, V. I., *Ordinary differential equations*, MIT Press, Cambridge, Mass., 1973.
3. Arnold, V. I., *Mathematical methods of classical mechanics*, Springer-Verlag, Berlin, 1978.
4. Bruter, C. P., *Topologie et perception*, Maloine-Doin, Paris, 1974.
5. Cartan, E., *La géométrie des espaces de Riemann*, Gauthier-Villars, Paris 1925.
6. Cartan, H., *Differential forms*, Hermann, Paris, 1970.
7. Do Carmo, M., *Differential geometry of curves and surfaces*, Prentice-Hall, Englewood Cliffs, N.J., 1976.
8. Dodson, C. and Poston, T., *Tensor geometry*, Pitman, London, 1977.
9. Golibutsky, M. and Guillemin, V., *Stable mappings and their singularities*, Springer, Berlin, 1974.
10. Hermann, R., *Ricci and Levi-Civita's tensor analysis paper* (Méthodes de calcul differential absolu et leurs applications), Math. Sci. Press, 1975.
11. Hirsch, M. and Smale, S., *Differential equations, dynamical systems, and linear algebra*, Academic Press, New York, 1974.
12. Lang, S., *Introduction to differentiable manifolds*, Wiley, New York, 1962.
13. Lang, S., *Algebra*, Addison-Wesley, Reading, Mass., 1965.
14. Lefschetz, S., *Differential equations: geometric theory*, Wiley, New York, 1963.
15. Levi-Civita, T., *The absolute differential calculus*, Blackie, Glasgow, 1926.
16. Maclane, S., Hamiltonian mechanics and geometry, *Am. Math. Mon.*, **77,** 570–586, 1970.
17. May, J. P., *Simplicial objects in algebraic topology*, Van Nostrand, Princeton, N.J., 1967.
18. Poston, T. and Stewart, I., *Catastrophe theory and its applications*, Pitman, London, 1977.
19. Smale, S., Global analysis and economics I, in *Dynamical systems*, ed. M. M. Peixoto, Academic Press, New York, 1973.

20. Spivak, M., *Calculus on manifolds*, Benjamin/Cummings, Menlo Park, Calif., 1965.
21. Spivak, M., *A comprehensive introduction to differential geometry*, Vol. I, Publish or Perish, 1970.
22. Spivak, M., *A comprehensive introduction to differential geometry*, Vol. II, Publish or Perish, 1970.
23. Steenrod, M., *The topology of fibre bundles*, Princeton University Press, Princeton, N.J., 1951.
24. Sternberg, S., *Lectures on differential geometry*, Prentice-Hall, Englewood Cliffs, N.J., 1963.
25. Thom, R., *Structural stability and morphogenesis*, Benjamin/Cummings, Menlo Park, Calif. 1975.
26. Thom, R. and Levine, H., Singularities of differentiable mappings, in *Proceedings of the Liverpool Singularities Symposium*, ed. C. T. C. Wall, Springer, Berlin, 1971.
27. Yano, K. and Ishihara, S., *Tangent and cotangent bundles*, Marcel Dekker, New York, 1973.

Index